20 - 99 ✓

Applied Digital El...

Applied Digital Electronics

Fourth Edition

D. C. Green
Mtech, CEng, MIEE

LONGMAN

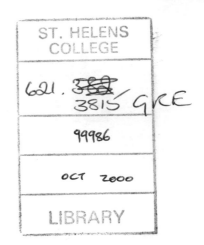
Pearson Education Limited
Edinburgh Gate
Harlow, Essex CM20 2JE, England

Published in the United States of America
by Pearson Education Inc., New York

Many of the designations used by manufacturers and sellers to
distinguish their products are claimed as trademarks. Pearson
Education Limited has made every attempt to supply trademark
information about manufacturers and their products mentioned
in this book. A list of the trademark designations and their
owners appears on page x.

First published in Great Britain 1982 by Pitman Books Ltd
Second impression published by Longman Scientific & Technical 1987
Second edition 1988
Second impression 1990
Third impression 1991
Third edition 1993
Fourth edition 1999

ISBN 0 582 35632 6

British Library Cataloguing-in-Publication Data
A catalogue record for this book is
available from the British Library.

Set by 35 in 10/12pt Times
Produced by Addison Wesley Longman Singapore (Pte) Ltd.,
Printed in Singapore

Contents

Preface to the Fourth Edition

This book provides a comprehensive coverage of the basic techniques and circuits employed in modern digital electronics.

Chapter 1 discusses the various logic technologies used in modern digital electronics and the terms employed in data sheets. This chapter also introduces the reader to the ways in which devices from different logic families may be interfaced together. Chapters 2 through to 4 consider the circuits that are employed in combinational logic designs, such as gates, code converters, adders, multiplexers and decoders. Throughout these chapters the emphasis has been placed on the 74HC logic family because devices in this family are readily available, are cheap, and devices in the 74LS logic family are gradually being phased out by the manufacturers. Chapters 5, 6 and 7 consider MSI devices such as multiplexers, comparators and sequential logic circuits (flip-flops, counters and shift registers), while Chapters 8 and 9 cover the design of synchronous counters and sequential systems. Programmable logic devices are the concern of Chapter 10, and both analogue-to-digital and digital-to-analogue converters are the subject of Chapter 11. Chapter 12 deals with the transmission of digital signals between one digital device, or circuit, and another. An introduction to the design of digital systems is given in Chapter 13 and the final chapter presents a number of exercises that can be carried out by a reader with access to the electronic design software package *Electronics Workbench* and a PC. The Electronics Workbench files can be downloaded from Addison Wesley Longman's website ftp://ftp.awl.co.uk/pub/awl-he/engineering/green

Throughout the book digital electronic devices are shown using either one, or both, of their IEC symbols and pinouts. The IEC symbols are the standard method of representing devices and they are increasingly employed. Any student of digital electronics needs to become well acquainted with their use.

A large number of worked examples are given throughout the text to illustrate various points. Most chapters include several practical exercises; these exercises require only the use of components and equipment that should be available in any laboratory where such practical work is carried out. Each chapter concludes with a number of exercises

Publisher's note: Texas Instruments data sheets have been reproduced exactly and differ from the use of subscripts and italics in this book.

and all but purely descriptive questions are provided with a worked solution that will be found at the back of the book.

I wish to express my thanks to Lattice Semiconductor Corporation and to Texas Instruments, Inc. for their permission to use, and reproduce, in this book, their copyright material.

D.C.G.

Acknowledgements

Reproduction of Figs 10.19, 10.25 and 10.26, and use of registered trademarks GAL, ispLSI and pLSI, granted courtesy of Lattice Semiconductor UK Limited.

Table 1.1, Figs 1.4, 1.5, 1.6, 2.2, 4.17, 5.2, 5.10, 5.12, 6.3, 6.4, 6.11, 6.18, 6.20, 7.4, 12.15 and 12.19 reproduced courtesy of Texas Instruments, Inc.

Trademark notice

GAL, ispLSI and pLSI are registered trademarks of Lattice Semiconductor Corporation. Electronics Workbench is a registered trademark of Interactive Image Technologies Ltd.

1 Logic technologies, data sheets and interfacing

After reading this chapter you should be able to:

- Read and understand the terminology used in the data sheets of digital ICs. Determine the voltage and current ratings of a device from its data sheet.
- Determine the parameters of a digital IC from its data sheet.
- Understand the meanings of the terms SSI, MSI and LSI.
- Compare the characteristics of the various 74TTL and 74CMOS devices.
- Calculate the noise margin of a device using its data sheet parameters.
- Calculate the power dissipated in a digital device.
- Explain why 3.3 V devices are increasingly employed in modern digital systems.
- Understand the considerations for interfacing devices from different logic families.

Most modern electronic circuits consist of one, or more, integrated circuits (ICs) that are connected together to provide a required circuit function. A *digital* IC is one whose output signal(s) can only be at either the logical 1, or the logical 0, voltage level and must bear a logical relationship to the input signal(s). Digital ICs come in various levels of complexity and they are categorized by the number of gates, or equivalent circuits, that they contain. Those ICs with fewer than 10 gates are known as *small-scale integration* (SSI) devices; those with between 10 and 100 gates are called *medium-scale integration* (MSI) devices; ICs with between 100 and 5000 gates are known as *large-scale integration* (LSI) circuits; lastly, *very large-scale integration* (VLSI) devices contain the equivalent of more than 5000 gates. Such a wide variety of MSI/LSI/VLSI digital ICs are available that SSI devices are rarely used for other than simple interconnection purposes.

While the set of MSI/SSI ICs is both increasing and diverging, the role of random logic is becoming more important. Complex systems with a number of large functional blocks need to use more and more SSI devices to interface between the blocks.

System design using MSI or LSI ICs consists of the correct interconnection of such chips and consequently a good knowledge of their functions is necessary. Usually when LSI devices are employed there are some interfacing and/or decoding tasks that can best be carried out using standard MSI ICs. Digital system design using VLSI devices can also use any available standard ICs but also often involves the use of ICs known as application specific circuits (ASICs). These devices include full- and semi-custom circuits as well as programmable circuits such as programmable logic arrays (PLAs), programmable array logic (PAL), etc. Nowadays, ASICs are also available to perform both analogue circuit and mixed analogue/digital functions.

Digital circuits are available in a number of integrated circuit logic technologies. There are two main technologies in common use: namely, transistor–transistor logic (TTL) and complementary metal oxide semiconductor (CMOS) logic. A third technology, emitter coupled logic (ECL), is used for a few very high-speed systems but it is not considered in this book. Both the TTL and CMOS technologies include a number of logic families. Also, another technology, known as BICMOS, that combines the best features of TTL and CMOS is employed for bus interface functions.

CMOS devices dissipate negligible power when they are not switching from one state to the other. When a device is switched its power dissipation increases in direct proportion with increase in the clock frequency, and at some frequency it will become greater than the power dissipation of the corresponding TTL device. For example, the 25×10^{-7} mW static power dissipation of a 74HC device rises to about 0.17 mW at 100 kHz. However, system level power savings are still obtained at most clock frequencies.

Low power dissipation is a very important consideration in a digital system since it saves power, reduces costs, simplifies the power supplies – perhaps allowing a battery to be employed – and increases reliability because less heat is generated. Also, a cooler PCB (printed circuit board) allows the component packing density to be greater with consequent reductions in both equipment size and weight.

It is an advantage if the power supply voltage to a digital IC is of low voltage since it then allows a lithium battery to be used to provide a back-up supply. Also, many memory chips use voltages which are below 5 V.

3.3 V power supplies

There is an ever-increasing demand for digital circuits that can operate from a 3.3 V power supply instead of the traditional 5 V. The main benefit of working from 3.3 V is the reduced power dissipation without a fall-off in performance that results. Most of the power dissipated in a digital circuit is a result of the charging and discharging of internal capacitances and the external load capacitance. The power dissipation is proportional to the square of the supply voltage V_{CC} and hence a small reduction in V_{CC} brings a significant reduction in power dissipation. Reducing the power dissipation provides two main benefits:

(a) Less heat is generated, which increases the reliability of components (including ICs), and either eliminates the need for a cooling fan or allows a smaller and cheaper fan to be employed.
(b) The power taken from the supply is reduced; this is (obviously) very important for battery-operated equipment and for other equipment allows a smaller power supply to be used.

The manufacturers of LSI devices, such as DRAMs and SRAMs, are now producing 3.3 V devices in order to get the high packing densities necessary for 16 Mb and higher capacity memories.

To address the increasing demand for low-voltage digital devices some new logic families have been developed, e.g. low-voltage CMOS technology (LV and LVC) and advanced low-voltage CMOS technology (ALVC). Devices in some of the 5 V logic technologies, such as 74HC and 74AHC, can be operated at 3.3 V, but at the expense of increased propagation delay.

Logic families

The logic families that employ TTL technology are:

(a)   Standard TTL; the 74-- series.
(b)   Schottky TTL; the 74S-- series.
(c)   Low-power Schottky; the 74LS-- series.
(d)   Advanced Schottky; the 74AS-- series.
(e)   Advanced low-power Schottky; the 74ALS-- series.
(f)   Fast; the 74F-- series.

The use of TTL technology is in decline and such devices probably should not be used for new designs. With increased packing densities, higher operating frequencies, and a need for reduced heat generation (partly to reduce the need for cooling fans), there is a requirement for reduced power dissipation. This points the way to using CMOS devices, particularly those that can operate from a 3.3 V power supply. However, the 74LS series, in particular, contains a greater variety of different devices than any other technology and many existing systems employ such circuits.

The logic families that employ some form of CMOS technology are:

(a)   High-speed CMOS; the 74HC--/74HCT-- series.
(b)   Advanced CMOS; the 74AC--/74ACT-- series.
(c)   Advanced high-speed CMOS; the 74AHC--/74AHCT-- series.
(d)   Low-voltage CMOS; the 74LV-- series.
(e)   Advanced low-voltage CMOS; the 74ALVC-- series.

Digital circuits designed for bus interface applications often use BICMOS technology which is a combination of bipolar and CMOS techniques. BICMOS combines the output drive capability and low noise characteristics of bipolar technology with the low static power dissipation of CMOS technology. BICMOS logic families are:

(a)   BICMOS logic; the 74BCT-- series.
(b)   64-series BICMOS; the 64BCT-- series.
(c)   Advanced BICMOS; the 74ABT-- series.
(d)   Advanced low-voltage BICMOS; the 74ALB-- series.

The choice of the logic family for a particular application is usually made after a careful consideration of the following criteria:

(a)   Switching speed.
(b)   Power dissipation.
(c)   Cost.
(d)   Availability.
(e)   Noise immunity.

Transistor–transistor logic

The 74-- series was the original member of the TTL technology and it is no longer recommended by the manufacturers for new designs. 74-- series devices may still be in use in some existing circuitry but the number of devices now available from manufacturers is much reduced. Currently, Texas Instruments, for example, offer only 33 devices. Typically a 74-- series device has a propagation delay of 9 ns and a power dissipation of 10 mW.

Schottky logic

Schottky logic (74S) devices are faster than standard TTL devices but, like standard TTL, are not employed for new designs.

Low-power Schottky logic

Although few, if any, new designs employ 74LS devices nowadays, and hence the overall demand for them is declining, this is still a significant logic family. It has the advantages of being reasonably fast to operate with a propagation delay of typically 7 ns, fairly low power dissipation (≈ 2 mW), and low noise. This logic family still offers the largest number of different devices which are cheap, easy to source, and easy to use. Texas Instruments say they will continue to supply LS devices as long as there is a demand for them.

Advanced Schottky logic

There are more than 90 different circuits in the 74AS logic family including the more common circuits such as gates, flip-flops, counters and transceivers. The propagation delay is 1.5 ns and the power dissipation is 2 mW.

Advanced low-power Schottky logic

The 74ALS logic family provides more than 130 different circuits, such as gates, flip-flops, counters and transceivers. It is currently one of the most commonly employed logic families. It has a propagation delay of 4 ns and a power dissipation of 1 mW. ALS devices are often combined with AS devices in the same system; AS circuits are used in speed-critical paths and ALS wherever the maximum speed is not essential. This practice allows a system to be built having both its speed and power performances optimized.

High-speed CMOS logic

The 74HC/74HCT logic families offer more than 100 circuit functions with a very low power dissipation and low switching noise. The output drive capability is 8 mA, the typical propagation delay is 18 ns, and the static current is 80 μA. This is the logic family that superseded the original CMOS family that was known as the 4000 series. Devices in the 4000 series are not used for new designs today. The 74HCT series devices have TTL compatible inputs and the same pinouts as the corresponding devices in the 74LS series and they can be used as direct replacements for them.

Advanced CMOS logic

The 74AC logic family of devices includes digital circuits of all kinds including gates, flip-flops and transceivers. The output driving capability is up to 24 mA when $V_{CC} = 5$ V and up to 12 mA when $V_{CC} = 3.3$ V; the propagation delay is 3.5 ns for 5 V operation and 5 ns for 3.3 V operation. The quiescent current taken from the power supply is 40 μA. The 74AC family is split into two sub-groups, (a) 74AC devices that have CMOS compatible inputs and outputs and (b) 74ACT devices that have TTL compatible inputs.

Advanced high-speed CMOS logic

74AHC/74AHCT devices are a further development of the 74CMOS logic family that provide extra speed and are able to operate from a 3.3 V power supply. Devices in this logic family are normally used to replace devices in the slower 74HC/74HCT families. Available devices include gates, flip-flops and transceivers. The propagation delay is typically 5.4 ns at $V_{CC} = 5$ V and 8.3 ns at $V_{CC} = 3.3$ V. The driving capability is 8 mA at 5 V and 4 mA at 3.3 V. The quiescent current taken from the power supply is 40 μA. 74AHC circuits are compatible with CMOS 4000 devices, and 74AHCT devices are TTL compatible.

Low-voltage CMOS technology

The low-voltage CMOS technology (74LV series) supplies devices that are designed to work from either a 5 or 3.3 V power supply. Up to 8 mA output drive current can be supplied with a maximum propagation delay of 18 ns. The static current taken from the power supply is only 20 μA. The 74LV family is the 3.3 V equivalent of the 5 V 74HC series.

A higher performance version of the 74LV series is the 74LVC series. This has a drive capability of 24 mA and 6.5 ns maximum propagation delay. The 74LVC series corresponds with the 5 V 74AC series.

All CMOS devices require any unused inputs to be tied to either V_{CC} or earth via a 1 kΩ resistor and not left floating. If the unused inputs are not held either HIGH or LOW, but are allowed to float, output glitches may appear. Some LVC devices incorporate a feature known as *bus hold*. Bus hold allows unused inputs to be left floating since internal circuitry holds the last known state of the input until such time as the next input arrives. This is an advantage since it means that the extra cost and space occupancy of resistors is eliminated. The inclusion of the bus hold facility in a device is indicated by the letter H added to the device label, i.e. 74LVCH--.

Advanced low-voltage CMOS logic

74ALVC is a 3.3 V bus interface logic family that employs CMOS technology. The typical propagation delay is less than 3 ns, the driving capability is 24 mA, and the quiescent current is 40 μA. Devices available in the 74ALVC family include buffer/drivers, transceivers and D latches and flip-flops. The corresponding 5 V logic family is the 74AC-- series.

Microgate logic

Equipment such as laptop and notebook computers, which must be designed to be as small and light in weight as possible, use only the odd gate here and there. The usual quad 2-input gates occupy unnecessary room when only one of the gates is required. To overcome this problem a range of microgates have been produced which consist of a single gate in a package. Such a device is designated as 74AC1G--, where 1G indicates one gate. Currently, there are 14 different devices in the microgate logic family.

BICMOS technology

A BICMOS device uses a bipolar transistor output stage while the rest of the circuitry uses CMOS technology. The use of a bipolar transistor output stage gives the following advantages:

(a) The voltage swing is smaller than for the comparable CMOS output stage. This means that the power dissipated as internal and external capacitances are discharged and charged is reduced.

(b) Bipolar transistors are able to turn OFF more efficiently than CMOS transistors. This reduces the current taken from the power supply when the output stage is non-active.

Although a bipolar output stage has a larger static power dissipation, its dynamic power dissipation is more or less constant whereas the power dissipation of a CMOS stage increases with increase in frequency. Hence, a bipolar stage has a smaller power dissipation at high frequencies. This results in BICMOS devices having better overall power performance than either TTL or CMOS devices since dynamic power dissipation provides most of a device's power consumption.

The original family, known as 74BCT, operates from a 5 V ± 10 per cent power supply voltage and has a propagation delay of 4.8 ns typically and 7 ns maximum. The family includes over 50 bus interface circuits such as drivers, transceivers and latches. The output drive capability is up to I_{OL} = 64 mA and I_{OH} = 15 mA and hence lines of characteristic impedance as low as 35 Ω can be driven.

Advanced BICMOS technology

Advanced BICMOS technology (74ABT) devices are designed to operate from a 5 V ± 10 per cent power supply voltage and provide bus interface functions. Devices presently available in the 74ABT logic family include buffer gates, buffer/drivers, bus transceivers, and both D flip-flops and latches. Both the inputs and outputs of 74ABT devices are fully TTL compatible. The 74ABT family features a low power dissipation, and propagation delays of 2.9 ns typical and 4.6 ns maximum, and a maximum output current driving capability of $I_{OH} = 32$ mA and $I_{OL} = 64$ mA.

Advanced low-power BICMOS

Advanced low-power BICMOS (74ALB) is a 3.3 V logic family designed for bus interface functions. It provides 25 mA drive at 3.3 V with a maximum propagation delay of 2.2 ns. A buffer/driver and a bus transceiver are presently available.

Low-voltage BICMOS technology

The low-voltage BICMOS (LVT) logic family has been designed for 3.3 V interface circuit applications and corresponds to the 5 V ABT logic family having the same drive characteristics and propagation delay. Features of LVT are supply voltage 2.7–3.6 V, standard TTL output drives of $V_{OH} = 2$ V at $I_{OH} = -32$ mA, $V_{OL} = 0.55$ V at $I_{OL} = 64$ mA, and a propagation delay of less than 4.6 ns.

Noise immunity

Because of the presence of unavoidable noise voltages a gap must always be left between the threshold voltage V_T and the defined limits of the input and output logic levels. The *noise immunity* (or *noise margin*) of a digital IC is either equal to the difference between maximum input voltage $V_{IL(max)}$ that is recognized as logic 1 and the maximum output voltage $V_{OL(max)}$ that will be recognized as logic 0, or equal to the difference between the minimum input voltage $V_{OH(min)}$ that will be recognized as logic 0 and the maximum input voltage $V_{IH(min)}$ that is recognized as logic 1. These voltages are shown in Fig. 1.1.

$$\text{Noise margin} = V_{IL(max)} - V_{OL(max)} \qquad (1.1)$$

$$\text{Noise margin} = V_{OH(min)} - V_{IH(min)} \qquad (1.2)$$

Numerically, the noise margin is equal to a percentage of its power supply voltage. For LS devices this is:

(a) low level, 8 per cent of $V_{CC} = 0.4$ V
(b) high level, 14 per cent of $V_{CC} = 0.7$ V

giving a noise margin of $0.7 - 0.4 = 0.3$ V.
    For HC devices the corresponding figures (for $V_{DD} = 5$ V) are:

(a) low level, 18 per cent of $V_{CC} = 0.9$ V
(b) high level, 28 per cent of $V_{CC} = 1.4$ V

giving a noise margin of $1.4 - 0.9 = 0.5$ V.

Fig. 1.1 *Noise immunity (margin)*

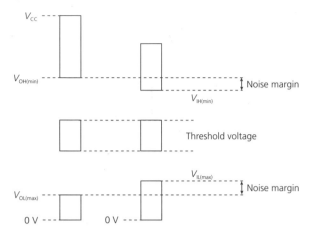

Choice of a logic family

The choice of a logic family depends upon the intended application and its main requirements. Any one, or more, of the following characteristics may be of great importance:

(a) High speed of operation.
(b) High drive capability.
(c) Low power dissipation.
(d) Ease of use.

Some of the logic families that are available have been designed solely for use as bus interface circuits and they do not include any circuits such as gates, flip-flops or counters.

High speed of operation

(a) For a bus interface circuit the chosen logic family, in order of speed, is ALB, ABT, LVT and AHC/AHCT.
(b) For general logic functions the chosen logic family should be 74F, AS, AC or ACT and AHC or AHCT.

High drive capability

(a) Bus interface circuits. In order of merit: ABT, LVT, ALB, ALVC and LVC.
(b) General logic circuits. Again in order of merit: LVC, AHC, ALVC, LV and LS.

All the general logic families include a quad 2-input NAND gate IC. This IC is variously labelled as: 74AC00, 74AHC00, 74AHCT00, 74ALS00, 74AS00, 74F00, 74HC00, 74HCT00, 74LS00, 74LV00, 74LVC00, 74S00 and 7400.

Data sheets

The manufacturers of digital ICs publish information in the form of *data sheets* to assist the circuit designer to choose the most suitable device for a particular application. The data sheet for an IC provides all the necessary technical information about the device in the form of tables, graphs and diagrams. Before the information given can be understood and made use of, it is necessary for the user to understand the operation of the circuit and the meanings of the various terms that are employed. The technical specification of an IC, obtained from its data sheet, ought not to be the only factor which is considered in the selection of a particular device. Also of importance are such factors as its cost, its ready availability, and whether or not it is *second-sourced*. Many ICs, although originally introduced by one manufacturer, are now produced by more than one firm and they are then said to be second-sourced. Second-sourcing implies pin compatibility although the parameters of the device may not be identical. Second sourcing is advantageous to the user since it means that future supplies of the IC are better assured.

The first section of a data sheet gives a macro description of the device. The information given may include:

(a) The number and title of the IC.
(b) An outline of its main features.
(c) Package options and pinouts.
(d) A description of the circuit and an outline description of its operation.
(e) A function table.
(f) The logic symbol for the device.

The next part usually presents the absolute maximum ratings of the IC; these include the current, temperature and voltage figures that must not be exceeded during the operation of the circuit if the IC is not to suffer damage. The recommended operating conditions are then given which specify recommended figures for input, output, and supply currents and voltages, and for the ambient temperature. If these figures are not exceeded the manufacturer guarantees the operation of the device. All currents are defined with respect to current flow into the terminal in question. This means that a current that flows out of a terminal is quoted as a negative value.

The maximum allowable rate of change of the input voltage waveform, both LOW to HIGH and HIGH to LOW, is often also given. The third part of the data sheet gives the guaranteed electrical characteristic limits of the device when it is operated under the recommended conditions. These electrical characteristics include the minimum high-level and the maximum low-level, input/output voltages and the input/output currents and capacitances.

The data sheets of sequential logic devices will then have sections that give the timing requirements and the switching characteristics. The timing requirements specify:

(a) The maximum clock frequency.
(b) The minimum pulse duration.
(c) The minimum hold and set-up times.

The switching characteristics give the maximum propagation delays for various conditions.

Most of the terms quoted in data sheets are reasonably obvious and require only a brief specification. Two terms, however, are somewhat more complex and so need further explanation.

Absolute maximum ratings

Supply voltage V_{CC}
The supply voltage V_{CC} is the maximum voltage that can be safely applied to the V_{CC} terminal.

Input voltage V_i
This is the maximum voltage that may be safely applied to an input terminal. The quoted figure may be exceeded if the quoted input clamp current rating is observed (see below).

Output voltage V_o
The output voltage V_o is the maximum voltage that can be safely applied to an output terminal. The quoted figure can be exceeded if the output clamp current rating is observed (see below).

Input clamp current I_{iK}
I_{iK} is the maximum current that can safely flow into, or out of, an input terminal when the input voltage is either less than 0 V (i.e. negative) or greater than V_{CC} volts.

Output clamp current I_{oK}
I_{oK} is the maximum current that may safely flow into, or out of, an output terminal when the voltage at that terminal is either less than 0 V or greater than V_{CC} volts.

Continuous output current I_o
This is the maximum output source or sink current that can safely flow into, or out of, an output terminal when the voltage at that terminal is between 0 and V_{CC} volts.

Continuous current through V_{CC} or earth
This is the maximum current that may flow safely into, or out of, the V_{CC} or GND (earth) terminals.

Storage temperature range
This figure gives the range of temperatures over which the device may be safely stored.

Recommended operating conditions

Supply voltage V_{CC}
The supply voltage V_{CC} is the range of power supply voltages over which the correct operation of the device within its specified limits is guaranteed.

High-level input voltage V_{IH}

V_{IH} is the least positive value of input voltage that will be recognized by the circuit as representing logic 1 in positive logic.

Low-level input voltage V_{IL}

V_{IL} is the most positive value of input voltage that is guaranteed to be recognized as representing logic 0 in a positive logic system.

Input voltage V_i

This is the range of input voltages over which the circuit is designed to operate.

Output voltage V_o

V_o is the range of output voltages over which the gate has been designed to operate.

High-level output current I_{OH}

This is the current flowing into, or out of, an output terminal that will give a high level at that output.

Low-level output current I_{OL}

I_{OL} is the current that flows into, or out of, an output terminal that will give a low level at that terminal.

Operating free-air temperature T_A

The range of temperatures over which the device has been designed to operate.

Electrical characteristics

High-level output voltage V_{OH}

V_{OH} is the voltage at an output terminal with inputs applied to specification that will establish a high level at the output. Several different test conditions are normally specified.

Low-level output voltage V_{OL}

V_{OL} is the voltage at an output terminal with inputs applied according to specification that will establish a low level at the output. Several test conditions are normally specified.

Logic switching levels

Figure 1.2 shows the logic switching levels for (a) the 5 V TTL logic families, 74, 74F, 74S, 74LS, 74AS and 74ALS, (b) the 5 V CMOS logic families, 74HC, 74HCT, 74AHC, 74AHCT, and (c) the 3.3 V CMOS logic families, 74LV, 74AC, and 74LVC. The switching levels for 5 V TTL and 3.3 V CMOS are the same but the 5 V CMOS levels are different. This may cause difficulties when interfacing devices from differing logic families and this is considered later in this chapter.

Fig. 1.2 *Logic switching levels*

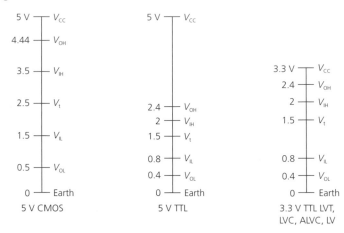

Input current I_i
This is the current flowing into, or out of, an input terminal with first V_{CC} and then 0 V applied to that terminal. It guarantees the maximum input current for any input voltage in the normal range of operation.

Supply current I_{CC}
I_{CC} is the current flowing into, or out of, the V_{CC} terminal under static no-load conditions.

Supply current change ΔI_{CC}
This parameter is used only for CMOS devices that have TTL compatible inputs. When a HCT/AHCT/ACT device is driven by a TTL circuit, both p-channel and n-channel transistors are turned ON and hence a path is created between the V_{CC} and earth lines. This current is specified as ΔI_{CC}. It is the increase in the supply current for each input that is at one of the specified TTL voltage levels instead of either 0 V or V_{CC}. If n inputs are at voltages other than 0 V or V_{CC} the increase in the supply current is $n\Delta I_{CC}$.

Input capacitance C_i
The input capacitance is the internal capacitance of an input terminal.

Output capacitance C_o
This is the output capacitance of an output terminal.

Switching characteristics

Propagation delay, high-to-low output
Propagation delay is the time it takes for an input pulse to travel through the circuit to the output. The propagation delay t_{PHL} is the measured time between a specified point on the output voltage transition from HIGH to LOW (usually half-voltage), and the same

Fig. 1.3 *(a) Pulse width, (b) propagation delay and (c) set-up time and hold time for the flip-flop*

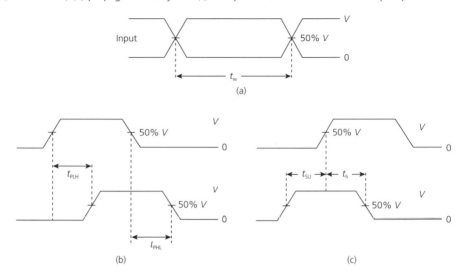

measured point on the input transition that causes the output voltage to change. The propagation delay is largely determined by the load capacitance C_L. Hence the total propagation delay is

$$t_D + 0.5R_oC_L \tag{1.3}$$

where t_D is the inherent delay of the gate, R_o is the output resistance of the gate and is approximately equal to V_{CC}/I_{OS} (I_{OS} is the output current when the output terminals are short-circuited).

Propagation delay, low-to-high output

The propagation delay t_{PLH} is the time measured between the half-voltage point on the output voltage transition from LOW to HIGH and the half-voltage point on the input transition causing the change. Both t_{PHL} and t_{PLH} are defined graphically in Fig. 1.3.

EXAMPLE 1.1

A gate has $t_{PHL} = t_{PLH} = 18$ ns and $I_{OS} = 18$ mA when $V_{CC} = 5$ V. Calculate the total propagation delay when the load capacitance is 42 pF.

Solution

$$t_D = 18 \times 10^{-9} + [5/(18 \times 10^{-3}) \times 0.5 \times 42 \times 10^{-12}] = 23.8 \text{ ns } (Ans.)$$

Fig. 1.4 *(a) Amplitude, risetime, falltime and pulse width and (b) transition time. (Courtesy of Texas Instruments)*

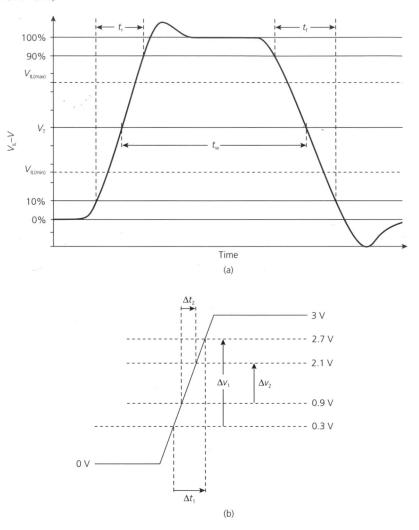

Input transition time and power dissipation capacitance

As mentioned earlier, two parameters mentioned in data sheets are somewhat more complex.

Input transition time ΔV/Δt

The normal definitions of the amplitude, risetime, falltime, and width of a pulse are shown in Fig. 1.4(a). Overshoot and undershoot are ignored. If, however, the switching threshold voltage is not 50 per cent of the amplitude of the pulse waveform the rise- and falltimes need to be modified. The important voltages for the correct operation of a

Table 1.1 *Transition rise/fall rates of logic circuits*

Series	V_{CC} (V)	$V_{IL(max)}$ (V)	$V_{IH(min)}$ (V)	V_T (V)	dt/dv (ns/V)
SN74	4.75–5.25	0.8	2	1.4	100
SN74LS	4.75–5.25	0.8	2	1.4	50
SN74S	4.75–5.25	0.8	2	1.4	50
SN74ALS	4.5–5.5	0.8	2	1.4	15
SN74AS	4.5–5.5	0.8	2	1.4	8
SN74F	4.5–5.5	0.8	2	1.4	8
SN74HC	2	0.3	1.5	1.4	625
	4.6	0.9	3.15	2.25	110
	6	1.2	4.2	3	80
SN74HCT	4.5–5.5	0.8	2	1.4	125
74AC	3	0.9	2.1	1.5	10
	4.5	1.35	3.15	2.25	10
	5.5	1.65	3.85	2.75	10
74ACT	4.5–5.5	0.8	2	1.4	10
SN74BCT	4.5–5.5	0.8	2	1.4	10
SN74ABT	4.5–5.5	0.8	2	1.4	5/10
SN74LV	2.7–3.6	0.8	2	≈1.5	100
SN74LVC	2.7–3.6	0.8	2	≈1.5	5/10
SN74LVT	3.0–3.6	0.8	2	1.4	10

Courtesy of Texas Instruments.

digital circuit are the maximum input voltage $V_{IL(max)}$ that is recognized as logic 0, and the minimum input voltage $V_{IH(min)}$ that is recognized as logic 1. Hence, it is better to define the risetimes and falltimes over the voltage range between $V_{IL(max)}$ and $V_{IH(min)}$. The transition time is the time required for the output voltage to change from LOW to HIGH (see Fig. 1.4(b)).

For a 74AC device at $V_{CC} = 3$ V, $V_{IL(max)} = 0.9$ V and $V_{IH(min)} = 2.1$ V. The pulse amplitude is 3 V and the risetime is 19 ns (typical). Then

transition time $= (19 \times 10^{-9})/[(0.9 \times 3) - (0.1 \times 3)] = 7.92$ ns/V

Table 1.1 gives typical values for V_{CC}, $V_{IL(max)}$, $V_{IH(min)}$, V_T and dt/dv for all the logic families.

Power dissipation capacitance C_{PD}

C_{PD} is the equivalent capacitance that is used in the calculation of the transient, or dynamic, power dissipated in a CMOS device. This is the power that is dissipated during the time that the transistors are switching from one logic state to the other. During this short time both transistors are conducting current and so some power is dissipated. Further power dissipation occurs because of the charging and discharging of the external load capacitance(s). The power dissipation occurs in the resistive part of the device as currents flow into, and out of, it to charge and discharge the total capacitance.

For a device with CMOS 4000 compatible inputs, such as a 74AC-- device, the total power dissipation P_T per gate is equal to the sum of the static and the dynamic power dissipations, thus:

$$P_T = V_{CC}I_{CC} + C_{PD}V_{CC}^2 f_i + C_L V_{CC}^2 f_o \qquad (1.4)$$

where f_i = input frequency, f_o = output frequency, C_L = load capacitance, and I_{CC} = quiescent supply current.

The higher the frequency at which a gate is switched ON and OFF, the greater the power dissipation but the static power is very small. The total power dissipated in an IC is the sum of the power dissipations of each of the individual circuits within the IC. The power dissipation capacitance C_{PD} is usually quoted in the data sheet for a device; if it is not it can be calculated using the expression

$$C_{PD} = [I_{CC(dyn)}/(V_{CC}f)] - C_L \qquad (1.5)$$

where $I_{CC(dyn)}$ is the current into the device when it is being switched at frequency f.

The expression for the total power dissipated in a CMOS device with TTL compatible inputs is more complex because the TTL voltage levels cause both input transistors to be partially turned ON.

$$P_T = V_{CC}(I_{CC} + N_i \Delta I_{CC}) + N_o(C_{PD}V_{CC}^2 f_i + C_L V_{CC}^2 f_o) \qquad (1.6)$$

where N_i = number of driven inputs, N_o = number of switching outputs, and ΔI_{CC} has the same meaning as before, i.e. increase in supply current.

EXAMPLE 1.2

Calculate the total power dissipated in a 74AC00 quad 2-input NAND gate when all four gates are switched on and off at a frequency of 1 MHz. V_{CC} = 5 V, C_{PD} = 40 pF, C_L = 50 pF and I_{CC} = 20 μA.

Solution

$$P_T = 5 \times 20 \times 10^{-6} + 90 \times 10^{-12} \times 25 \times 1 \times 10^6 = 2.35 \text{ mW } (Ans.)$$

Table 1.1 lists typical values for power supply voltage V_{CC}, $V_{IL(max)}$, $V_{IH(min)}$, and switching threshold voltage V_T for each general-purpose logic family. Values for the transition time $\Delta v/\Delta t$ are also listed.

NAND gate data sheets

Figure 1.5 gives the data sheet for the Texas Instruments SN74AC00 quad 2-input NAND gate IC (also the 54 series military version). The meanings of all the terms/ symbols used in the sheet have been explained earlier.

Fig. 1.5 *Data sheet for the 74AC00 quad 2-input NAND gate. (Courtesy of Texas Instruments)*

SN54AC00, SN74AC00
QUADRUPLE 2-INPUT POSITIVE-NAND GATES

SCAS524C – AUGUST 1995 – REVISED SEPTEMBER 1996

- *EPIC*™ (Enhanced-Performance Implanted CMOS) 1-μm Process
- Package Options Include Plastic Small-Outline (D), Shrink Small-Outline (DB), Thin Shrink Small-Outline (PW), DIP (N) Packages, Ceramic Chip Carriers (FK), Flat (W), and DIP (J) Packages

description

The 'AC00 contain four independent 2-input NAND gates. Each gate performs the Boolean function of $Y = \overline{A \cdot B}$ or $Y = \overline{A} + \overline{B}$ in positive logic.

The SN54AC00 is characterized for operation over the full military temperature range of −55°C to 125°C. The SN74AC00 is characterized for operation from −40°C to 85°C.

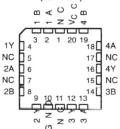

SN54AC00 . . . J OR W PACKAGE
SN74AC00 . . . D, DB, N, OR PW PACKAGE
(TOP VIEW)

1A	1	14	V$_{CC}$
1B	2	13	4B
1Y	3	12	4A
2A	4	11	4Y
2B	5	10	3B
2Y	6	9	3A
GND	7	8	3Y

SN54AC00 . . . FK PACKAGE
(TOP VIEW)

NC – No internal connection

FUNCTION TABLE
(each gate)

INPUTS		OUTPUT
A	B	Y
H	H	L
L	X	H
X	L	H

logic symbol†

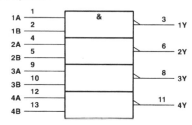

1A — 1	
1B — 2	&
	3 — 1Y
2A — 4	
2B — 5	6 — 2Y
3A — 9	
3B — 10	8 — 3Y
4A — 12	
4B — 13	11 — 4Y

† This symbol is in accordance with ANSI/IEEE Std 91-1984 and IEC Publication 617-12.
Pin numbers shown are for the D, DB, J, N, PW, and W packages.

logic diagram (positive logic)

A	
B	— Y

absolute maximum ratings over operating free-air temperature range (unless otherwise noted)†

Supply voltage range, V$_{CC}$	−0.5 V to 7 V
Input voltage range, V$_I$ (see Note 1)	−0.5 V to V$_{CC}$ + 0.5 V
Output voltage range, V$_O$ (see Note 1)	−0.5 V to V$_{CC}$ + 0.5 V
Input clamp current, I$_{IK}$ (V$_I$ < 0 or V$_I$ > V$_{CC}$)	±20 mA
Output clamp current, I$_{OK}$ (V$_O$ < 0 or V$_O$ > V$_{CC}$)	±20 mA
Continuous output current, I$_O$ (V$_O$ = 0 to V$_{CC}$)	±50 mA
Continuous current through V$_{CC}$ or GND	±200 mA
Maximum power dissipation at T$_A$ = 55°C (in still air) (see Note 2): D package	1.25 W
DB package	0.5 W
N package	1.1 W
PW package	0.5 W
Storage temperature range, T$_{stg}$	−65°C to 150°C

† Stresses beyond those listed under "absolute maximum ratings" may cause permanent damage to the device. These are stress ratings only, and functional operation of the device at these or any other conditions beyond those indicated under "recommended operating conditions" is not implied. Exposure to absolute-maximum-rated conditions for extended periods may affect device reliability.

Fig. 1.5 *(Cont'd)*

recommended operating conditions (see Note 3)

			SN54AC00 MIN	SN54AC00 MAX	SN74AC00 MIN	SN74AC00 MAX	UNIT
V_{CC}	Supply voltage		2	6	2	6	V
V_{IH}	High-level input voltage	V_{CC} = 3 V	2.1		2.1		V
		V_{CC} = 4.5 V	3.15		3.15		
		V_{CC} = 5.5 V	3.85		3.85		
V_{IL}	Low-level input voltage	V_{CC} = 3 V		0.9		0.9	V
		V_{CC} = 4.5 V		1.35		1.35	
		V_{CC} = 5.5 V		1.65		1.65	
V_I	Input voltage		0	V_{CC}	0	V_{CC}	V
V_O	Output voltage		0	V_{CC}	0	V_{CC}	V
I_{OH}	High-level output current	V_{CC} = 3 V		−12		−12	mA
		V_{CC} = 4.5 V		−24		−24	
		V_{CC} = 5.5 V		−24		−24	
I_{OL}	Low-level output current	V_{CC} = 3 V		12		12	mA
		V_{CC} = 4.5 V		24		24	
		V_{CC} = 5.5 V		24		24	
$\Delta t/\Delta v$	Input transition rise or fall rate		0	8	0	8	ns/V
T_A	Operating free-air temperature		−55	125	−40	85	°C

NOTE 3: Unused inputs must be held high or low to prevent them from floating.

electrical characteristics over recommended operating free-air temperature range (unless otherwise noted)

PARAMETER	TEST CONDITIONS	V_{CC}	T_A = 25°C MIN	T_A = 25°C TYP	T_A = 25°C MAX	SN54AC00 MIN	SN54AC00 MAX	SN74AC00 MIN	SN74AC00 MAX	UNIT
V_{OH}	I_{OH} = −50 µA	3 V	2.9			2.9		2.9		V
		4.5 V	4.4			4.4		4.4		
		5.5 V	5.4			5.4		5.4		
	I_{OH} = −12 mA	3 V	2.56			2.4		2.46		
	I_{OH} = −24 mA	4.5 V	3.86			3.7		3.76		
		5.5 V	4.86			4.7		4.76		
	I_{OH} = −50 mA†	5.5 V				3.85				
	I_{OH} = −75 mA†	5.5 V						3.85		
V_{OL}	I_{OH} = 50 µA	3 V		0.002	0.1		0.1		0.1	V
		4.5 V		0.001	0.1		0.1		0.1	
		5.5 V		0.001	0.1		0.1		0.1	
	I_{OL} = 12 mA	3 V			0.36		0.5		0.44	
	I_{OL} = 24 mA	4.5 V			0.36		0.5		0.44	
		5.5 V			0.36		0.5		0.44	
	I_{OL} = 50 mA†	5.5 V					1.65			
	I_{OL} = 75 mA†	5.5 V							1.65	
I_I	V_I = V_{CC} or GND	5.5 V			±0.1		±1		±1	µA
I_{CC}	V_I = V_{CC} or GND, I_O = 0	5.5 V			2		40		20	µA
C_i	V_I = V_{CC} or GND	5 V		2.6						pF

† Not more than one output should be tested at a time, and the duration of the test should not exceed 2 ms.

switching characteristics over recommended operating free-air temperature range, V_{CC} = 3.3 V ± 0.3 V (unless otherwise noted) (see Figure 1)

PARAMETER	FROM (INPUT)	TO (OUTPUT)	T_A = 25°C MIN	T_A = 25°C TYP	T_A = 25°C MAX	SN54AC00 MIN	SN54AC00 MAX	SN74AC00 MIN	SN74AC00 MAX	UNIT
t_{PLH}	A or B	Y	2	7	9.5	1	11	2	10	ns
t_{PHL}			1.5	5.5	8	1	9	1	8.5	

Fig. 1.5 *(Cont'd)*

switching characteristics over recommended operating free-air temperature range, $V_{CC} = 5\ V \pm 0.5\ V$ (unless otherwise noted) (see Figure 1)

PARAMETER	FROM (INPUT)	TO (OUTPUT)	$T_A = 25°C$			SN54AC00		SN74AC00		UNIT
			MIN	TYP	MAX	MIN	MAX	MIN	MAX	
t_{PLH}	A or B	Y	1.5	6	8	1	8.5	1.5	8.5	ns
t_{PHL}			1.5	4.5	6.5	1	7	1	7	

operating characteristics, $V_{CC} = 5\ V$, $T_A = 25°C$

PARAMETER		TEST CONDITIONS	TYP	UNIT
C_{pd}	Power dissipation capacitance	$C_L = 50\ pF$, $f = 1\ MHz$	40	pF

EXAMPLE 1.3

For the SN74AC00 quad 2-input NAND gate IC answer the following questions assuming that $V_{CC} = 5.5$ V:

(a) What is the value of $V_{IL(max)}$?
(b) What current flows at an output terminal when the output is:
   (i) LOW
   (ii) HIGH? Does the current flow into or out of the terminal?
(c) When $I_{OH} = -24$ mA what is
   (i) the minimum value of V_{OH}
   (ii) the maximum value of V_{OL}? Using these two figures calculate the noise margin of a gate.
(d) Calculate the static power dissipation.

Solution

(a) $V_{IL(max)} = 1.65$ V *(Ans.)*
(b) (i) $I_{OL(max)} = 24$ mA out of the gate *(Ans.)*
   (ii) $I_{OH(max)} = 24$ mA into the gate. [The actual current in each case depends upon the number and types of loads connected to the output.] *(Ans.)*
(c) (i) $V_{OH(min)} = 4.76$ V, (ii) $V_{OL(max)} = 0.44$ V *(Ans.)*
   Noise margin = (i) $4.76 - 3.85 = 0.91$ V; (ii) $1.65 - 0.44 = 1.21$ V *(Ans.)*
(d) Static power dissipation $= 5.5 \times 20 \times 10^{-6} = 110$ µW *(Ans.)*

PRACTICAL EXERCISE 1.1

Aim: to determine the differences between quad 2-input NAND gates in different logic families.
Procedure:
Refer to the manufacturer's data sheet for each of the following devices: 7400, 74AC00, 74ACT00, 74AHC00, 74AHCT00, 74LV00, 74LVC00, and 74LS00.

D-type flip-flop data sheets

Figure 1.6 shows the data sheet for the SN74AHCT74 dual leading-edge-triggered D-type flip-flop IC. The device is provided with active-LOW PRESET and CLEAR terminals. Most of the terms used in the data sheet have been met previously, but there are a few new ones. In the logic diagram each of the boxes marked as TG is a transmission gate. This is just a digitally controlled CMOS switch. When the transmission gate is open the impedance between its input and output terminals is very high, but when the transmission gate is closed the impedance is very low. In the function table ↑ indicates operation at the leading edge of a clock pulse. Additional terms, relative to the NAND gate data sheet in Fig. 1.5, are:

(a) The pulse duration t_W.
(b) The set-up time t_{SU}.
(c) The hold time t_H.
(d) The maximum frequency f_{max}.

Also, there are more values of propagation delay quoted, since there are three different inputs, $\overline{PRE}$, $\overline{CLEAR}$ and CLK.

Pulse duration
The pulse duration t_W is the time interval between specified points on the leading and trailing edges of a pulse waveform. The quoted value is the shortest time for which correct operation is guaranteed.

Set-up time
This is the time interval between the application of a signal at a data input terminal and a consecutive transition at the active clock edge. This means that the flip-flop will look back one set-up time to determine the logical state of its D input (or the J and K inputs of a J-K flip-flop).

Hold time
The hold time t_H is the time for which the logical state of the D (or J-K) input must be held constant after the leading edge (or trailing edge for some devices) of the clock has occurred. t_H is often quoted as being zero; this means that the data input(s) does

not have to be held after the clock leading edge has passed. The set-up and hold times are shown on p. 13.

Maximum clock frequency

The maximum clock frequency f_{max} is the highest clock frequency at which correct operation of the circuit is guaranteed.

Fig. 1.6 *Data sheet for the 74AHCT74 dual D flip-flop. (Courtesy of Texas Instruments)*

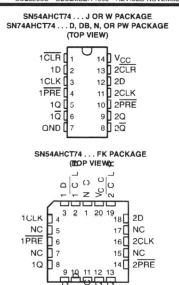

SN54AHCT74, SN74AHCT74
DUAL POSITIVE-EDGE-TRIGGERED D-TYPE FLIP-FLOPS
WITH CLEAR AND PRESET
SCLS263C – DECEMBER 1995 – REVISED NOVEMBER 1996

- Inputs Are TTL-Voltage Compatible
- *EPIC*™ (Enhanced-Performance Implanted CMOS) Process
- High Latch-Up Immunity Exceeds 250 mA Per JEDEC Standard JESD-17
- ESD Protection Exceeds 2000 V Per MIL-STD-883, Method 3015; Exceeds 200 V Using Machine Model (C = 200 pF, R = 0)
- Package Options Include Plastic Small-Outline (D), Shrink Small-Outline (DB), Thin Shrink Small-Outline (PW), and Ceramic Flat (W) Packages, Ceramic Chip Carriers (FK), and Standard Plastic (N) and Ceramic (J) 300-mil DIPs

description

The 'AHCT74 are dual positive-edge-triggered D-type flip-flops.

A low level at the preset ($\overline{PRE}$) or clear ($\overline{CLR}$) inputs sets or resets the outputs regardless of the levels of the other inputs. When $\overline{PRE}$ and $\overline{CLR}$ are inactive (high), data at the data (D) input meeting the setup time requirements is transferred to the outputs on the positive-going edge of the clock pulse. Clock triggering occurs at a voltage level and is not directly related to the rise time of the clock pulse. Following the hold-time interval, data at the D input can be changed without affecting the levels at the outputs.

The SN54AHCT74 is characterized for operation over the full military temperature range of –55°C to 125°C. The SN74AHCT74 is characterized for operation from –40°C to 85°C.

FUNCTION TABLE

INPUTS				OUTPUTS	
PRE	CLR	CLK	D	Q	$\overline{Q}$
L	H	X	X	H	L
H	L	X	X	L	H
L	L	X	X	H†	H†
H	H	↑	H	H	L
H	H	↑	L	L	H
H	H	L	X	Q_0	$\overline{Q}_0$

† This configuration is nonstable; that is, it does not persist when $\overline{PRE}$ or $\overline{CLR}$ returns to its inactive (high) level.

Fig. 1.6 *(Cont'd)*

logic symbol†

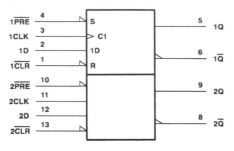

† This symbol is in accordance with ANSI/IEEE Std 91-1984 and IEC Publication 617-12.
Pin numbers shown are for the D, DB, J, N, PW, and W packages.

logic diagram, each flip-flop (positive logic)

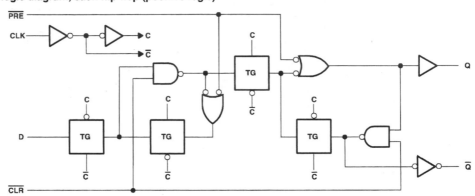

absolute maximum ratings over operating free-air temperature range (unless otherwise noted)†

Supply voltage range, V_{CC} .... −0.5 V to 7 V
Input voltage range, V_I (see Note 1) .... −0.5 V to 7 V
Output voltage range, V_O (see Note 1) .... −0.5 V to V_{CC} + 0.5 V
Input clamp current, I_{IK} ($V_I < 0$ or $V_I > V_{CC}$) .... −20 mA
Output clamp current, I_{OK} ($V_O < 0$ or $V_O > V_{CC}$) .... −20 mA
Continuous output current, I_O (V_O = 0 to V_{CC}) .... ±25 mA
Continuous current through V_{CC} or GND .... ±50 mA
Maximum power dissipation at T_A = 55°C (in still air) (see Note 2): D package .... 1.25 W
DB or PW package .... 0.5 W
N package .... 1.1 W
Storage temperature range, T_{stg} .... −65°C to 150°C

† Stresses beyond those listed under "absolute maximum ratings" may cause permanent damage to the device. These are stress ratings only, and
functional operation of the device at these or any other conditions beyond those indicated under "recommended operating conditions" is not
implied. Exposure to absolute-maximum-rated conditions for extended periods may affect device reliability.

recommended operating conditions (see Note 3)

		SN54AHCT74		SN74AHCT74		UNIT
		MIN	MAX	MIN	MAX	
V_{CC}	Supply voltage	4.5	5.5	4.5	5.5	V
V_{IH}	High-level input voltage	2		2		V
V_{IL}	Low-level input voltage		0.8		0.8	V
V_I	Input voltage	0	5.5	0	5.5	V
V_O	Output voltage	0	V_{CC}	0	V_{CC}	V
I_{OH}	High-level output current		−8		−8	mA
I_{OL}	Low-level output current		8		8	mA
$\Delta t/\Delta v$	Input transition rise or fall rate		20		20	ns/V
T_A	Operating free-air temperature	−55	125	−40	85	°C

NOTE 3: Unused inputs must be held high or low to prevent them from floating.

Fig. 1.6 *(Cont'd)*

electrical characteristics over recommended operating free-air temperature range (unless otherwise noted)

PARAMETER	TEST CONDITIONS		V_{CC}	$T_A = 25°C$			SN54AHCT74		SN74AHCT74		UNIT
				MIN	TYP	MAX	MIN	MAX	MIN	MAX	
V_{OH}	$I_{OH} = -50\ \mu A$		4.5 V	4.4	4.5		4.4		4.4		V
	$I_{OH} = -8\ mA$			3.94			3.8		3.8		
V_{OL}	$I_{OL} = 50\ \mu A$		4.5 V			0.1		0.1		0.1	V
	$I_{OL} = 8\ mA$					0.36		0.5		0.44	
I_I	$V_I = V_{CC}$ or GND		5.5 V			±0.1		±1		±1	μA
I_{CC}	$V_I = V_{CC}$ or GND,	$I_O = 0$	5.5 V			2		20		20	μA
$\Delta I_{CC}‡$	One Input at 3.4 V, Other inputs at GND or V_{CC}		5.5 V			1.35		1.5		1.5	mA
C_i	$V_I = V_{CC}$ or GND		5 V		2	10				10	pF

‡ This is the increase in supply current for each input at one of the specified TTL voltage levels rather than 0 V or V_{CC}.

timing requirements over recommended ranges of supply voltage and operating free-air temperature (unless otherwise noted) (see Figure 1)

PARAMETER			$T_A = 25°C$		SN54AHCT74		SN74AHCT74		UNIT
			MIN	MAX	MIN	MAX	MIN	MAX	
t_w	Pulse duration	PRE or CLR low	5		5		5		ns
		CLK	5		5		5		
t_{su}	Setup time before CLK↑	Data	5		5		5		ns
		PRE or CLR inactive	3.5		3.5		3.5		
t_h	Hold time, data after CLK↑		0		0		0		ns

switching characteristics over recommended ranges of supply voltage and operating free-air temperature (unless otherwise noted) (see Figure 1)

PARAMETER	FROM (INPUT)	TO (OUTPUT)	LOAD CAPACITANCE	SN54AHCT74					UNIT
				$T_A = 25°C$			MIN	MAX	
				MIN	TYP	MAX			
f_{max}			$C_L = 15\ pF$	100	160		80		MHz
			$C_L = 50\ pF$	80	140		65		
t_{PLH}^*	PRE or CLR	Q or Q̄	$C_L = 15\ pF$		7.6	10.4	1	12	ns
t_{PHL}^*					7.6	10.4	1	12	
t_{PLH}^*	CLK	Q or Q̄	$C_L = 15\ pF$		5.8	7.8	1	9	ns
t_{PHL}^*					5.8	7.8	1	9	
t_{PLH}	PRE or CLR	Q or Q̄	$C_L = 50\ pF$		8.1	11.4	1	13	ns
t_{PHL}					8.1	11.4	1	13	
t_{PLH}	CLK	Q or Q̄	$C_L = 50\ pF$		6.3	8.8	1	10	ns
t_{PHL}					6.3	8.8	1	10	

* On products compliant to MIL-PRF-38535, this parameter is ensured but not production tested.

switching characteristics over recommended ranges of supply voltage and operating free-air temperature (unless otherwise noted) (see Figure 1)

PARAMETER	FROM (INPUT)	TO (OUTPUT)	LOAD CAPACITANCE	SN74AHCT74					UNIT
				$T_A = 25°C$			MIN	MAX	
				MIN	TYP	MAX			
f_{max}			$C_L = 15\ pF$	100	160		80		MHz
			$C_L = 50\ pF$	80	140		65		
t_{PLH}	PRE or CLR	Q or Q̄	$C_L = 15\ pF$		7.6	10.4	1	12	ns
t_{PHL}					7.6	10.4	1	12	
t_{PLH}	CLK	Q or Q̄	$C_L = 15\ pF$		5.8	7.8	1	9	ns
t_{PHL}					5.8	7.8	1	9	
t_{PLH}	PRE or CLR	Q or Q̄	$C_L = 50\ pF$		8.1	11.4	1	13	ns
t_{PHL}					8.1	11.4	1	13	
t_{PLH}	CLK	Q or Q̄	$C_L = 50\ pF$		6.3	8.8	1	10	ns
t_{PHL}					6.3	8.8	1	10	

Fig. 1.6 *(Cont'd)*

noise characteristics, $V_{CC} = 5$ V, $C_L = 50$ pF, $T_A = 25°C$ (see Note 4)

PARAMETER		SN74AHCT74 MIN	SN74AHCT74 MAX	UNIT
$V_{OL(P)}$	Quiet output, maximum dynamic V_{OL}		0.8	V
$V_{OL(V)}$	Quiet output, minimum dynamic V_{OL}		−0.8	V
$V_{OH(V)}$	Quiet output, minimum dynamic V_{OH}	4		V
$V_{IH(D)}$	High-level dynamic input voltage	2		V
$V_{IL(D)}$	Low-level dynamic input voltage		0.8	V

NOTE 4:   Characteristics are determined during product characterization and ensured by design for surface-mount packages only.

operating characteristics, $V_{CC} = 5$ V, $T_A = 25°C$

PARAMETER		TEST CONDITIONS		TYP	UNIT
C_{pd}	Power dissipation capacitance	No load,	f = 1 MHz	32	pF

EXAMPLE 1.4

A 74ACT374 octal leading-edge-triggered D flip-flop has the following parameters: $t_{W(min)} = 5$ ns, $t_{SU(min)} = 5.5$ ns, $t_H = 1.5$ ns, $t_{PLH} = 8.5$ ns and $t_{PHL} = 8$ ns. Draw the timing diagram.

Solution

The timing diagram is shown in Fig. 1.7.

Fig. 1.7

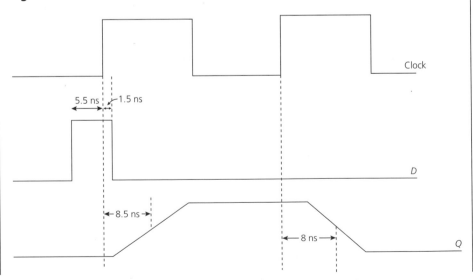

EXAMPLE 1.5

How fast can a 74AHC74 be clocked when the load capacitance is:

(a)   30 pF.
(b)   100 pF?

Solution

From the data sheet, when C_L = 15 pF, f_{max} = 80 MHz, when C_L = 50 pF, f_{max} = 65 MHz.
Hence, $80 \times 10^6 = k/(C_{in} + 15 \times 10^{-12})$ and $65 \times 10^6 = k/(C_{in} + 50 \times 10^{-12})$
$C_{in}(80 - 65) \times 10^6 = 2.05 \times 10^{-3}$, or C_{in} = 136.7 pF. Then, k = 12.136 $\times$ 10^{-3}

(a) $f_{max} = (12.136 \times 10^{-3})/[(136.7 + 30) \times 10^{-12}]$ = 72.8 MHz (*Ans.*)
(b) $f_{max} = (12.136 \times 10^{-3})/[(136.7 + 100) \times 10^{-12}]$ = 51.27 MHz (*Ans.*)

Interfacing between logic families

The term 'interfacing' means connecting the output(s) of one circuit to the input of another circuit which has different input current/voltage requirements. Often a direct connection between the circuits cannot be made because of differences between the current/voltage supplied by the driving circuit and the input current/voltage requirements of the driven circuit. An interface circuit, such as a pull-up resistor or a *level-shifter*, may be necessary to ensure compatibility between the two circuits. Two ICs from the same logic family can always be connected together without difficulty provided the fanout of the driving circuit is not exceeded.

Digital systems often include devices from more than one logic family because the designer has taken full advanatage of the best features of each family. Some examples of this practice are:

(a)   74 ALS devices are used in part of a system where high-speed operation is required, and 74HCT devices are used in the rest of the system where speed is of less importance than the minimum power dissipation.
(b)   A system may be designed using both devices that operate from a 3.3 V power supply and devices that require a 5 V power supply. With the introduction of the 3 V logic families many systems are designed to take advantage of the new technology. In some cases, all the devices required for a design may not have been available (and may still not be) and hence it may have been necessary to employ some 5 V devices in the system.
(c)   A designer may be using 74AHC devices (for example) for a circuit but a particular device that is required may have been available only in the 74ALS family.

The interface between devices from two different logic families must ensure that:

(a)   The driving circuit is able to source, or sink, enough current to meet the total requirements of all the driven circuits.

Table 1.2

Logic family	74LS	74ALS	74AS	74HC/HCT	74AC/ACT	74AHC/AHCT
$I_{IH(max)}$	20 µA	20 µA	200 µA	1 µA	1 µA	1 µA
$I_{IL(max)}$	0.4 mA	0.2 mA	2 mA	1 µA	1 µA	1 µA
$I_{OH(max)}$	0.4 mA	0.4 mA	2 mA	4 mA	24 mA	8 mA
$I_{OL(max)}$	8 mA	4 mA	20 mA	4 mA	24 mA	8 mA

Table 1.3

Logic family	74LS	74ALS	74AS	74HC/HCT	74AC/ACT	74AHC/AHCT
$V_{IH(max)}$	2	2	2	3.5/2	3.5/2	3.85/2
$V_{IL(max)}$	0.8	0.8	0.8	1/0.8	1.5/0.8	1.65/2
$V_{OH(max)}$	2.7	2.7	2.7	4.9/4.9	4.9/4.9	4.4/3.15
$V_{OL(max)}$	0.4	0.4	0.5	0.1/0.1	0.1/0.1	0.44/0.1

(b) The HIGH and LOW output voltages of the driving circuit are within the specified range of HIGH and LOW input voltage for the driven circuit.

When devices from two different logic families are to be interfaced the data sheet of each device should be consulted to determine the input/output current/voltage requirements. A summary of the input and output current/voltages of the main logic families is given in Tables 1.2 and 1.3.

The input voltage levels for the 74 CMOS logic families are higher than for the TTL families, except for the TTL compatible versions HCT/ACT/AHCT.

74TTL driving 74CMOS

An examination of Table 1.2 shows that the input current $I_{IH(max)}$ and $I_{IL(max)}$ requirements of all the CMOS devices are much smaller than the output current $I_{OH(max)}$ and $I_{OL(max)}$ capabilities of the TTL devices. This means that the TTL devices are easily able to satisfy the CMOS input current requirements.

When the output voltage of a TTL device is LOW Table 1.3 shows that no difficulties arise. However, the maximum TTL ouput voltage $V_{OH(max)}$ of 2.7 V is less than the maximum input voltage $V_{IH(max)}$ of 3.5 V (or 3.85 V) of 74HC/74AC/74AHC. The TTL output voltage can be increased by the use of a pull-up resistor connected to V_{CC} as shown in Fig. 1.8. The use of a pull-up resistor has the disadvantages that (a) the resistor takes up valuable space on the printed circuit board, and (b) it dissipates power and so generates unwanted heat. The problem can be overcome by the use of a device with TTL compatible inputs, i.e. a device from the 74HCT/74ACT/74AHCT logic family when a direct interconnection may be made. Unfortunately, the use of TTL compatible inputs results in a device that is slower and has a higher power consumption.

Fig. 1.8 *Use of a pull-up resistor when interfacing 74LS to 74HC*

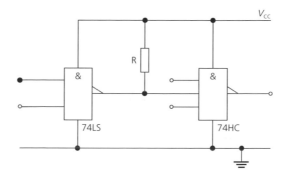

Pull-up resistor

When the output of the TTL driving IC is in the LOW state it must be able to sink current via the pull-up resistor as well as from the n CMOS inputs to which it is connected. Therefore, the required resistance value is given by

$$R = V_{CC(min)}/(I_{OL(TTL(max))} + nI_{IL(CMOS)}) \qquad (1.7)$$

EXAMPLE 1.6

A 74LS gate is to be connected to three 74HC gates with a supply voltage of 5 V. Calculate the minimum value for the pull-up resistor.

Solution

$I_{OL(max)} = 8$ mA, $I_{IL(max)} = -1$ µA (a sink current). Therefore,

$$R = 5/(8 \times 10^{-3} - 3 \times 1 \times 10^{-6}) = 625 \ \Omega \ (Ans.)$$

74CMOS driving 74TTL

Reference to Table 1.2 shows that when an output is HIGH all CMOS devices are able to source sufficient current to meet the TTL input requirements. When an output is HIGH the smallest sink current capability is that of 74HC/HCT, i.e. 4 mA. This is enough current to drive two 74AS devices and up to ten 74LS devices. Also, the CMOS $V_{OL(max)} = 0.1$ V value is less than the TTL $V_{IL(max)} = 0.8$ value and so there is no interfacing problem.

EXAMPLE 1.7

(a) Calculate how many 74LS inputs can be driven by
     (i) a 74HC output
     (ii) a 74AHC output.
(b) Repeat (a) for 74ALS inputs.

Solution

(a) For 74LS $I_{IL(max)} = 0.4$ mA.
  (i) For 74HC $I_{OL(max)} = 4$ mA. Therefore, fanout = 4/0.4 = 10 (*Ans.*)
  (ii) For 74AHC $I_{OL(max)} = 8$ mA. Therefore, fanout = 8/0.4 = 20 (*Ans.*)
(b) For 74ALS $I_{IL(max)} = 0.2$ mA.
  (i) Fanout = 4/0.2 = 20 (*Ans.*)
  (ii) Fanout = 8/0.2 = 40 (*Ans.*)

EXAMPLE 1.8

A 74HC00 quad 2-input NAND gate is to drive four 74LS devices. Can the connection be made directly?

Solution

Current: $I_{OL(max)} = 4$ mA, $I_{OH(max)} = 4$ mA, $I_{IL(max)} = 0.4$ mA, and $I_{IH(max)} = 20$ µA. The total I_{IL} current = 1.6 mA and the total I_{IH} current = 80 µA, so there is no problem with a direct connection (*Ans.*)
Voltage: $V_{OL(max)} = 0.1$ V, $V_{OH(max)} = 4.9$ V, $V_{IL(max)} = 0.8$ V, and $V_{IH(max)} = 2$ V. Again, a direct connection is all right (*Ans.*)

Low-voltage families interfacing with 5 V TTL and CMOS families

74LVC and LVT logic

A device in either the 74LVC or the 74LVT logic families is able to accept 5 V logic levels at its input terminals because the switching levels are the same. Hence, it may be connected directly to 74LS/74AS/74ALS devices. At the output, LVC/LV devices are able to drive TTL, and CMOS with TTL compatible inputs, devices directly. The basic circuit is shown in Fig. 1.9(a). The output voltages of a 74LVC IC are not acceptable to the V_{IH} requirements of a device in one of the 5 V CMOS logic families, such as 74HC.

Fig. 1.9 *Interfacing: (a) 74TTL to 74LVC/LVT and (b) 74TTL to 74LV/ALVC to 74HC/AC*

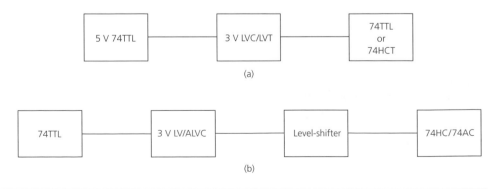

74ALVC/LV logic

Devices in these two logic families cannot accept input voltages that are higher than $V_{CC} + 0.5$ V, i.e. 4.1 V if V_{CC} is 3.6 V. If a 74ALVC/LV device is driven by a 5 V 74LS/AS/ALS device the TTL V_{OH} voltage will be well within the V_{IH} ALVC/LV figure. At the output, both ALVC and LV devices are able to drive TTL and TTL compatible inputs. This means that the 3.3 V ICs can be connected directly between TTL ICs, as shown in Fig. 1.9(a). If an ALVC/LV IC is to drive a 74CMOS IC without TTL compatible inputs it will be necessary to employ a *level-shifter* as shown in Fig. 1.9(b). The 74CMOS voltage levels are $V_{IL} = 0.3V_{CC}$ and $V_{IH} = 0.7V_{CC}$. The driving IC must be able to provide V_{OL} less than 1.35 V for $V_{CC} = 4.5$ V and V_{OH} greater than 3.85 V for $V_{CC} = 5.5$ V. These requirements cannot be satisfied reliably and so it is necessary to use a level-shifter.

Level-shifter

Sometimes it is necessary to be able to convert the logic levels of a 3 V LV/ALVC/LVC/LVT device to the logic levels of a 5 V 74CMOS device. This is the function of a level-shifter; the IC is connected to both the 3 V and the 5 V power supply voltages and it is able to interface the two sets of logic levels. Two level-shifters are available: the LVC425 and the ALVC16425.

EXERCISES

1.1 Explain the significant features of each of the following logic families, 74--, 74LS--, 74AC--, 74ACT--, 74AHC-- and 74AHCT--. Draw a graph to show propagation delay against power consumption for each of these families.

From the data sheet of 74AC02 IC determine:
(a) The function, logic symbol, and function table of the device.
(b) The logic levels.
(c) The noise margin.
(d) The power dissipation.
(e) The propagation delay.

1.2 The waveform shown in Fig. 1.10(a) is applied to the input of a 74AHC00 gate. The output waveform is shown in Fig. 1.10(b). Calculate:
(a) The frequency of the input and output pulse waveforms.
(b) The propagation delay times t_{PHL} and t_{PLH}.
(c) The risetime and falltime and hence determine the transition time.

1.3 Answer the following questions about the 74AHCT74 dual D flip-flop IC
(a) For how long must the $\overline{CLR}$ input be held LOW to guarantee that the flip-flop will clear?
(b) Determine the minimum set-up time for
  (i) input data
  (ii) $\overline{PRE}$.

Fig. 1.10

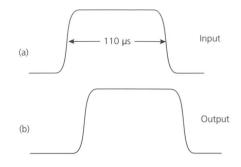

(a) ← 110 μs → Input

(b) Output

Fig. 1.11

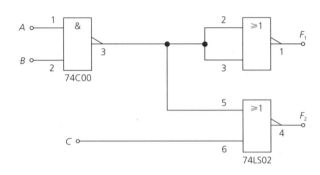

(c) Draw waveforms and label them with values from the data sheet for
   (i) t_{SU}
   (ii) t_H
   (iii) t_{PLH} and t_{PHL}.

1.4 Match up the following logic families and their characteristics

1	74--	A Bus applications
2	74ALS	B Very fast, low power dissipation
3	74LS	C Speed like LS. Low voltage/low power
4	74AC	D High speed, low power
5	74AHC	E Low power, fast as AS
6	74BCT	F Low power, faster than 74--
7	74HCT	G Basic circuits
8	74F	H Very high speed

1.5 (a) Determine Boolean expressions for the outputs F_1 and F_2 of the circuit shown in Fig. 1.11.
   (b) Refer to Tables 1.2 and 1.3 and determine if the circuit will work correctly.
   (c) If it doesn't say why, and then suggest a modification that will correct the fault.

1.6 A 74LS02 quad 2-input NOR gate is driving eight 74HC inputs. Calculate the pull-up resistance that is required.

1.7 (a) List the advantages of using a 3.3 V power supply instead of 5 V.
(b) Give three examples of systems in which high performance is more important than low power.
(c) Give three examples of systems in which low power is more important than high performance.

1.8 A 74HC00 quad 2-input NAND gate IC is operated from a 5 V power supply. The total d.c. current taken from the supply is 16 μA. Each gate drives a load of 8 pF capacitance and each is operated at a frequency of 1 MHz. $I_{OS} = 17$ mA. Calculate:
(a) The total power dissipation.
(b) The internal propagation delay of a gate.
(c) The propagation delay of each gate.

1.9 Compare the relative merits of the various logic technologies. Explain why the 74HC devices are replacing 74LS devices in digital systems. List the various 74CMOS logic families and compare their merits. What is meant by saying that a CMOS device is TTL compatible?

1.10 The data sheet of the 74HC138 3-to-8 line decoder includes: propagation time $V_{CC} = 2$ V, 200 ns; $V_{CC} = 4.5$ V, 40 ns; $V_{CC} = 6$ V, 34 ns for an input transition time of no more than 6 ns and a load capacitance of 50 pF. Determine the propagation delay if the device is to supply 10 HC loads at 4.5 V. Each HC load has an input capacitance of 10 pF.

1.11 Refer to the data sheet of the 64HC573
(a) What kind of device is it?
(b) What is its maximum operating frequency?
(c) The IC is used with $V_{CC} = 5$ V. Calculate its no-load dynamic power dissipation.
(d) If $V_{CC} = 44.5$ V what are the values of $V_{IH(min)}$ and $V_{IL(max)}$?

1.12 (a) A gate has $V_{OH(min)} = 3.5$ V and $V_{IH(min)} = 4.9$ V. Calculate its noise margin.
(b) The gate also has a propagation delay of 8 ns and an average power dissipation of 4.8 mW. Calculate the speed–power product.
(c) Discuss why the maximum voltage ratings of a digital IC must not be exceeded.
(d) A gate has $t_{PLH} = 10$ ns and $t_{PHL} = 8$ ns. Calculate the average propagation delay.

1.13 The I_{OL} value of a gate is 8 mA. The gate is to be connected to a number of similar gates whose I_L value is 0.2 mA. How many gates may be connected (i.e. what is the fanout)?

Table 1.4

	A	B
$V_{IH(min)}$	2.1 V	2.2 V
$V_{IL(max)}$	0.9 V	0.9 V
$V_{OH(min)}$	2.9 V	2.46 V
$V_{OL(max)}$	0.1 V	0.44 V
t_{PLH}	9.5 ns	10 ns
t_{PHL}	13 ns	8.5 ns
I_{CC}	20 µA	20 µA

1.14 Table 1.4 lists the characteristics of NAND gate ICs from two different logic families.
  (a) Which can operate at the higher frequency?
  (b) Which dissipates the most static power?
  (c) Which has the best LOW noise margin?

2 Combinational logic equations

After reading this chapter you should be able to:

- Describe the operation of the various kinds of gate.
- Derive both SOP and POS Boolean equations from a truth table.
- Use the data sheet of a gate to ascertain its main features.
- Draw the logic symbol for each gate.
- Use logic rules to simplify Boolean equations.
- State and use both of De Morgan's rules.
- Understand the use of duality to simplify equations and to obtain the complement of a function.
- Employ the decimal representation of both SOP and POS equations.
- Recognize and use some of the more important digital codes.
- Design a code converter.

A gate is a circuit that is able to perform a logical function on two, or more, input variables. The International Electrotechnical Commission (IEC) symbols for each of the different types of gate are given in Fig. 2.1. Gates are available in each of the different logic families other than BICMOS. The low-power Schottky TTL (74LS series) offers the greatest variety of devices but it is rarely used for new designs. The various 74 CMOS series of logic families offer a more restricted choice of gates.

Fig. 2.1 *British and International Standard gate symbols*

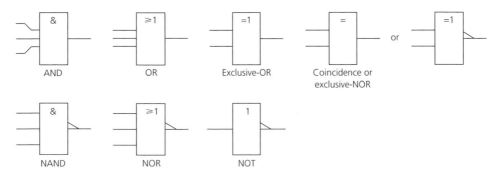

Table 2.1 *Four-input AND gate truth table*

A	0	1	0	1	0	1	0	1	0	1	0	1	0	1	0	1
B	0	0	1	1	0	0	1	1	0	0	1	1	0	0	1	1
C	0	0	0	0	1	1	1	1	0	0	0	0	1	1	1	1
D	0	0	0	0	0	0	0	0	1	1	1	1	1	1	1	1
F	0	0	0	0	0	0	0	0	0	0	0	0	0	0	0	1

Review of gates

AND gate

The AND gate is a logic element having two or more input terminals and one output terminal. Its output is at logical 1 *only* when *all* of its inputs are at logical 1. If one or more inputs are at logic 0, the output of the gate will also be at 0. The output F of a 4-input AND gate can be expressed using Boolean algebra as

$$F = A \cdot B \cdot C \cdot D$$

The symbol for the AND logical function is the dot, as shown, but the dot is often omitted, and the AND function is expressed as

$$F = ABCD$$

The operation of a logic circuit can be described by means of its *truth table*. This is a table which shows the output state of the circuit for all the possible combinations of the input variable states. The truth table of a 4-input AND gate is given in Table 2.1. The number of terms required is 2^n, where n is the number of input variables.

The AND gates presently available in the various logic families are: quad 2-input: 74ALS08, 74AS08, 74LS08, 74AC08, 74ACT08, 74AHC08, 74AHCT08, 74HC08, 74HCT08, 74LV08 and the 74LVC08; triple 3-input: 74ALS11, 74AS11, 74LS11, 74AC11, 74ACT11 and 74HC11. Dual 4-input AND gate ICs (74 '21) are available in the ALS, AS, LS, and HC logic families. Before finalizing the design of a circuit that uses a gate with multiple inputs it is desirable to check if the required gate exists in the logic family that is to be employed. A 4-input AND gate is readily obtained using three 2-input AND gates, which requires the use of one 74HC08 quad 2-input AND gate IC.

OR gate

The output of an OR gate is at the logic 1 level whenever any one, or more, of its inputs is/are at logic 1. The Boolean equation for a 4-input OR gate is

$$F = A + B + C + D$$

The truth table of a 4-input OR gate is given in Table 2.2.

Table 2.2 Four-input OR gate truth table

A	0	1	0	1	0	1	0	1	0	1	0	1	0	1	0	1
B	0	0	1	1	0	0	1	1	0	0	1	1	0	0	1	1
C	0	0	0	0	1	1	1	1	0	0	0	0	1	1	1	1
D	0	0	0	0	0	0	0	0	1	1	1	1	1	1	1	1
F	0	1	1	1	1	1	1	1	1	1	1	1	1	1	1	1

Table 2.3 Four-input NAND gate truth table

A	0	1	0	1	0	1	0	1	0	1	0	1	0	1	0	1
B	0	0	1	1	0	0	1	1	0	0	1	1	0	0	1	1
C	0	0	0	0	1	1	1	1	0	0	0	0	1	1	1	1
D	0	0	0	0	0	0	0	0	1	1	1	1	1	1	1	1
F	1	1	1	1	1	1	1	1	1	1	1	1	1	1	1	0

The quad 2-input 74--32 OR gate is to be found in all of the logic families. Other OR gate ICs are not so common and their availability needs to be checked. A 3-input, or 4-input, OR gate is easily obtained using three 2-input OR gates, i.e. one 74HC32 quad 2-input OR gate IC.

NAND gate

The output of a NAND gate is at logic 0 only when all its inputs are at logic 1. If any one, or more, of its inputs are at the logic 0 level the output F will be at 1. The Boolean equation for a NAND gate is

$$\bar{F} = ABCD$$

or

$$F = \overline{ABCD}$$

The truth table of a 4-input NAND gate is given in Table 2.3.

A single 8-input NAND gate, the 74--30, is available in the TTL logic families only. The 74--00 quad 2-input NAND gate is available in all the logic families and the 74--10 triple 3-input NAND gate is listed in most families.

The data sheet of the 74HC00 quad 2-input NAND gate is shown in Fig. 2.2. The information given in the data sheet follows the same order as for the 74AC00 given in Fig. 1.5. The description part of the data sheet gives the pinouts of two packages, the function table and the logic symbol of the device. The remainder of the sheet gives the absolute maximum ratings, the recommended operating conditions, and the electrical and switching characteristics.

Fig. 2.2 *Data sheet for the 74HC00 quad 2-input NAND gate. (Courtesy of Texas Instruments)*

SN54HC00, SN74HC00
QUADRUPLE 2-INPUT POSITIVE-NAND GATES

SCLS181B – DECEMBER 1982 – REVISED MAY 1997

- **Package Options Include Plastic Small-Outline (D), Thin Shrink Small-Outline (PW), and Ceramic Flat (W) Packages, Ceramic Chip Carriers (FK), and Standard Plastic (N) and Ceramic (J) 300-mil DIPs**

SN54HC00 . . . J OR W PACKAGE
SN74HC00 . . . D, N, OR PW PACKAGE
(TOP VIEW)

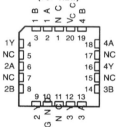

```
1A  [ 1      14 ]  VCC
1B  [ 2      13 ]  4B
1Y  [ 3      12 ]  4A
2A  [ 4      11 ]  4Y
2B  [ 5      10 ]  3B
2Y  [ 6       9 ]  3A
GND [ 7       8 ]  3Y
```

description

These devices contain four independent 2-input NAND gates. They perform the Boolean function $Y = \overline{A \cdot B}$ or $Y = \overline{A} + \overline{B}$ in positive logic.

The SN54HC00 is characterized for operation over the full military temperature range of –55°C to 125°C. The SN74HC00 is characterized for operation from –40°C to 85°C.

SN54HC00 . . . FK PACKAGE
(TOP VIEW)

FUNCTION TABLE
(each gate)

INPUTS		OUTPUT
A	**B**	**Y**
H	H	L
L	X	H
X	L	H

NC – No internal connection

logic symbol†

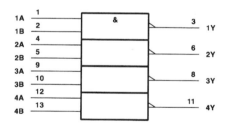

```
1A  1
1B  2      &        3   1Y
2A  4
2B  5               6   2Y
3A  9
3B  10              8   3Y
4A  12
4B  13              11  4Y
```

absolute maximum ratings over operating free-air temperature range†

Supply voltage range, V_{CC} .... –0.5 V to 7 V
Input clamp current, I_{IK} ($V_I < 0$ or $V_I > V_{CC}$) (see Note 1) .... ±20 mA
Output clamp current, I_{OK} ($V_O < 0$ or $V_O > V_{CC}$) (see Note 1) .... ±20 mA
Continuous output current, I_O ($V_O = 0$ to V_{CC}) .... ±25 mA
Continuous current through V_{CC} or GND .... ±50 mA
Package thermal impedance, θ_{JA} (see Note 2): D package .... 127°C/W
                                     N package .... 78°C/W
                                     PW package .... 170°C/W
Storage temperature range, T_{stg} .... –65°C to 150°C

† Stresses beyond those listed under "absolute maximum ratings" may cause permanent damage to the device. These are stress ratings only, and functional operation of the device at these or any other conditions beyond those indicated under "recommended operating conditions" is not implied. Exposure to absolute-maximum-rated conditions for extended periods may affect device reliability.

Fig. 2.2 *(Cont'd)*

recommended operating conditions

			SN54HC00			SN74HC00			UNIT
			MIN	NOM	MAX	MIN	NOM	MAX	
V_{CC}	Supply voltage		2	5	6	2	5	6	V
V_{IH}	High-level input voltage	V_{CC} = 2 V	1.5			1.5			V
		V_{CC} = 4.5 V	3.15			3.15			
		V_{CC} = 6 V	4.2			4.2			
V_{IL}	Low-level input voltage	V_{CC} = 2 V	0		0.5	0		0.5	V
		V_{CC} = 4.5 V	0		1.35	0		1.35	
		V_{CC} = 6 V	0		1.8	0		1.8	
V_I	Input voltage		0		V_{CC}	0		V_{CC}	V
V_O	Output voltage		0		V_{CC}	0		V_{CC}	V
t_t	Input transition (rise and fall) time	V_{CC} = 2 V	0		1000	0		1000	ns
		V_{CC} = 4.5 V	0		500	0		500	
		V_{CC} = 6 V	0		400	0		400	
T_A	Operating free-air temperature		−55		125	−40		85	°C

electrical characteristics over recommended operating free-air temperature range (unless otherwise noted)

PARAMETER	TEST CONDITIONS		V_{CC}	T_A = 25°C			SN54HC00		SN74HC00		UNIT
				MIN	TYP	MAX	MIN	MAX	MIN	MAX	
V_{OH}	V_I = V_{IH} or V_{IL}	I_{OH} = −20 μA	2 V	1.9	1.998		1.9		1.9		V
			4.5 V	4.4	4.499		4.4		4.4		
			6 V	5.9	5.999		5.9		5.9		
		I_{OH} = −4 mA	4.5 V	3.98	4.3		3.7		3.84		
		I_{OH} = −5.2 mA	6 V	5.48	5.8		5.2		5.34		
V_{OL}	V_I = V_{IH} or V_{IL}	I_{OL} = 20 μA	2 V		0.002	0.1		0.1		0.1	V
			4.5 V		0.001	0.1		0.1		0.1	
			6 V		0.001	0.1		0.1		0.1	
		I_{OL} = 4 mA	4.5 V		0.17	0.26		0.4		0.33	
		I_{OL} = 5.2 mA	6 V		0.15	0.26		0.4		0.33	
I_I	V_I = V_{CC} or 0		6 V		±0.1	±100		±1000		±1000	nA
I_{CC}	V_I = V_{CC} or 0, I_O = 0		6 V			2		40		20	μA
C_i			2 V to 6 V		3	10		10		10	pF

switching characteristics over recommended operating free-air temperature range, C_L = 50 pF (unless otherwise noted) (see Figure 1)

PARAMETER	FROM (INPUT)	TO (OUTPUT)	V_{CC}	T_A = 25°C			SN54HC00		SN74HC00		UNIT
				MIN	TYP	MAX	MIN	MAX	MIN	MAX	
t_{pd}	A or B	Y	2 V		45	90		135		115	ns
			4.5 V		9	18		27		23	
			6 V		8	15		23		20	
t_t		Y	2 V		38	75		110		95	ns
			4.5 V		8	15		22		19	
			6 V		6	13		19		16	

operating characteristics, T_A = 25°C

PARAMETER		TEST CONDITIONS	TYP	UNIT
C_{pd}	Power dissipation capacitance per gate	No load	20	pF

EXAMPLE 2.1

Show how (a) a 3-input NAND gate and (b) a 4-input NAND gate can be obtained using the 74HC00 quad 2-input NAND gate IC. [This might be necessary if 74AC/ACT or 74AHC/AHCT devices are used.]

Solution

(a) Connect inputs A and B to the pins *1A* and *1B* of the 74HC00. Connect pin *1Y* to both inputs *2A* and *2B* and connect *2Y* to *3A* and input C to *3B*. Then output *3Y* will give the function $F = \overline{ABC}$. The required circuit is shown in Fig. 2.3(a) (*Ans.*)

(b) Connect inputs A and B to pins *1A* and *1B* and connect inputs C and D to pins *2A* and *2B*. Connect *1Y* to *3A* and *3B* and connect *2Y* to *4A* and *4B*. Then connect *3Y* and *4Y* to pins *1A* and *1B* of another 74HC00 and then the output on pin *1Y* gives $F = \overline{ABCD}$. The circuit is shown in Fig. 2.3(b) (*Ans.*)

Fig. 2.3

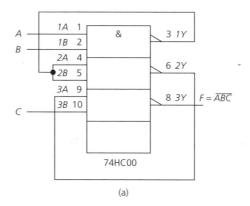

(a)

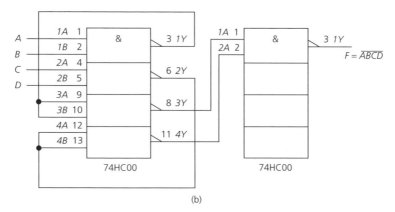

(b)

EXAMPLE 2.2

The data sheet for the 74HC00 quad 2-input NAND gate is given in Fig. 2.2. Answer each of the following questions:

(a) What is the difference between the 54 and the 74 versions of the device?
(b) What is the noise margin when $V_{CC} =$
    (i) 2 V?
    (ii) 4.5 V?
    (iii) 6 V?
(c) What is the delay between a HIGH appearing at the A input and the output going LOW when
    (i) $V_{CC} = 2$ V?
    (ii) 4.5 V?
    (iii) 6 V?
(d) What is the static power dissipation per gate when $V_{CC} = 6$ V?
(e) How is the power dissipation capacitance used in the calculation of dynamic power? Why does the load capacitance have an effect on power dissipation when a capacitance dissipates zero power?
(f) When might a 74HCT, a 74AC, or a 74AHC device be used instead of the 74HC00?

Solution

(a) The 54 can operate over a wider temperature range (*Ans.*)
(b) Noise margin $= V_{OH(min)} - V_{IH(min)}$. The three V_{CC} values are usable only when $I_{OH} = -20$ μA
    (i) 2 V: $N = 1.9 - 1.5 = 0.4$ V (*Ans.*)
    (ii) 4.5 V: $N = 4.4 - 3.15 = 1.25$ V (*Ans.*)
    (iii) 6 V: $N = 5.9 - 4.2 = 1.7$ V (*Ans.*)
(c) (i) 115 ns (*Ans.*)
    (ii) 23 ns (*Ans.*)
    (iii) 20 ns (*Ans.*)
(d) $P_S = 6 \times 20 \times 10^{-6} = 120$ μW (*Ans.*)
(e) The dynamic power dissipated is proportional to the value of the power dissipation capacitance, and also to the load capacitance. This is because the total capacitance determines how much current flows into and out of the gate (*Ans.*)
(f) 74HCT when TTL compatible inputs are an advantage; 74AHC when very good performance is required, including reduction in noise by placing the V_{CC} and earth pins in the middle of the package to reduce package inductance. Hence, 74AHC devices are not pin-for-pin compatible with TTL or 74HC devices. The 74AC is used when higher speed and lower power dissipation are required without a large price premium (*Ans.*)

Table 2.4 *Four-input NOR gate truth table*

A	0	1	0	1	0	1	0	1	0	1	0	1	0	1	0	1
B	0	0	1	1	0	0	1	1	0	0	1	1	0	0	1	1
C	0	0	0	0	1	1	1	1	0	0	0	0	1	1	1	1
D	0	0	0	0	0	0	0	0	1	1	1	1	1	1	1	1
F	1	0	0	0	0	0	0	0	0	0	0	0	0	0	0	0

Table 2.5 *Two-input exclusive-OR truth table*

A	0	1	0	1
B	0	0	1	1
F	0	1	1	0

NOR gate

The output F of a NOR gate is at logical 1 only when all its inputs are at logical 0. The Boolean expression for a 4-input NOR gate is

$$F = A + B + C + D$$

The truth table for the device is given in Table 2.4.

Quad 2-input NOR gates are available in most of the logic families, e.g. the 74HC02 and the 74LS02. A triple 3-input NOR gate IC is available only in the TTL logic families plus the 74HC27. No 4-input NOR gates exist. A 3-input, or 4-input, NOR gate can be derived from 2-input 74HC02 gates using the same method as shown in Fig. 2.3(a) and (b), but note the different pinouts shown in Fig. 2.4.

Exclusive-OR gate

An exclusive-OR gate has two inputs and one output. The output is at logical 1 only when one input is at 1 and the other input is at 0. If both inputs are at 0, or are at 1, the output of the gate will be 0. The Boolean expression describing the operation of the gate is

$$F = A\bar{B} + \bar{A}B$$

The truth table for an exclusive-OR gate is given in Table 2.5.

The 74--86 exclusive-OR gate is available in most of the logic families.

Exclusive-NOR gate

The exclusive-NOR gate performs the inverse logic function to the exclusive-OR gate. Its output will be at 1 only when both inputs are either at 1 or at 0, otherwise the output will be at 0. The Boolean expression for an exclusive-NOR gate is

$$F = AB + \bar{A}\bar{B}$$

Table 2.6 *Two-input exclusive-NOR truth table*			
A	0	1	0	1
B	0	0	1	1
F	1	0	0	1

The truth table is given in Table 2.6. Two exclusive-NOR gate ICs exist – the 74LS266 and the 74HC266.

Figure 2.4 shows the logic symbols for each of the more commonly employed gates.

Fig. 2.4 *Logic symbols for gates: (a) 7400 quad 2-input NAND, (b) 7410 triple 3-input NAND, (c) 7420 dual 4-input NAND, (d) 7404 hex inverter, (e) 7402 quad 2-input NOR, (f) 7427 triple 3-input NOR, (g) 7408 quad 2-input AND, (h) 7411 triple 3-input AND, (i) 7432 quad 2-input OR and (j) 7486 quad 2-input exclusive-OR*

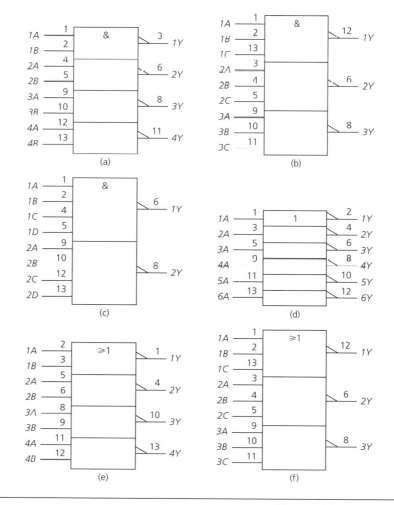

Fig. 2.4 *(Cont'd)*

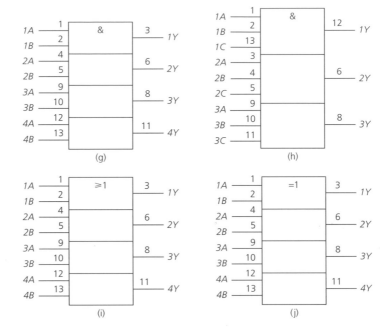

(g)

(h)

(i)

(j)

PRACTICAL EXERCISE 2.1

Aim: to investigate the use of exclusive-OR gates in parity checking circuits.
Components and equipment: two 74HC86 (or 74LS86) quad 2-input exclusive-OR gate
ICs, one LED and one 270 Ω resistor. Breadboard. Power supply. Perhaps some switches
for connecting either the logical 1 or logical 0 voltage levels to various input terminals.

In a 4-bit digital system a fifth bit can be added to each word and used as a
parity bit for the detection of 1-bit errors. In an *odd-parity* system the parity bit is
added before the digital word is transmitted to make the number of 1 bits in the
word an odd number. In an *even-parity* system the parity bit is added to make the
sum of the 1 bits in a word an even number. At the sending end of the system a
parity generator circuit generates the required bit, odd or even, and at the receiving
end a *parity checking* circuit determines whether the received digital word has the
correct parity. One bit in error will be detected.

Procedure:

(a) Connect up the parity generator circuit shown in Fig. 2.5(a). The logical state of
the output of the circuit is indicated by an LED that will glow visibly when the
output is at logic 1. The 270 Ω resistor is required to limit the current flowing
in the LED to a safe value; the resistance value is not critical. The inputs to the
circuit will need to be changed from 1 to 0 and back again. This can be done
using wander leads connected to the 5 V power supply or to earth as required,
or switched connections can be used.

(b) Make $A = C = 1$, and $B = D = 0$. Note the logical state of the LED. Is the parity
bit 1 or 0? Is this circuit an even-, or an odd-parity bit generator?

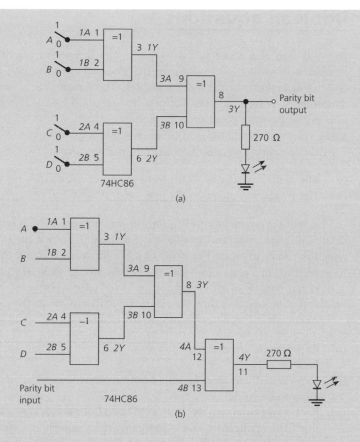

Fig. 2.5

(a)

(b)

(c)　Make $A = 1$, $B = C = D = 0$ and repeat procedure (b). Is the parity bit 1 or 0 now?

(d)　Try other combinations of the input variables and each time check whether or not the circuit always ensures that the sum of the 5 bits is even.

(e)　Do not disconnect circuit (a). Build the parity checking circuit shown in Fig. 2.5(b).

(f)　Connect inputs A, B, C and D to the corresponding inputs of Fig. 2.5(a) and then connect the output of that circuit to the parity bit input terminal of the circuit shown in Fig. 2.5(b).

(g)　Make $A = C = 1$, and $B = D = 0$ and note the logical states of the two LEDs. An error is indicated by the parity checking circuit's LED glowing visibly. Has an error occurred? Repeat for several other combinations of A, B, C and D.

(h)　Introduce an error into the system by disconnecting the link between the A inputs of the two circuits. Make $C = 1$, $B = D = 0$ for both circuits and then make $A = 1$ for the parity generator circuit and $A = 0$ for the parity checking circuit. Repeat procedure (g). Is an error indicated?

(i)　Repeat procedure (h). Now restore the connection between the two A inputs and introduce an error into one of the other data lines. Check that the circuit detects the error.

　A simulation of the circuit is to be found in Electronics Workbench exercise EWB 2.1.

Boolean equations

In the design of a combinational logic circuit the truth table of the required operation is written down and used to derive the Boolean equation that expresses the output of the circuit in terms of the input variables. The equation thus obtained can generally be simplified to reduce the complexity of the required circuit. Very often the implementation of the Boolean equation is carried out using either NAND or NOR gates only, or using either a multiplexer or a programmable logic device.

Boolean equations are often written in either one of two standard forms:

(a) The *sum-of-products* (SOP) form, e.g. $F = A\bar{B} + \bar{A}B$.
(b) The *product-of-sums* (POS) form, e.g. $F = (A + \bar{B})(\bar{A} + B)$.

Each of the product terms in a sum-of-products equation is known as a *minterm*, while each sum term in a product-of-sums equation is known as a *maxterm*. Any Boolean equation can be expressed in either of these forms and one, or the other, may be simpler to implement in a particular case. If any minterm in an SOP equation is equal to logical 1 then the sum F of that equation is also 1. Conversely, if any maxterm in a POS equation is at logical 0 then the product F will also be at 0.

The standard forms are employed because:

(a) Expressions are derived from truth tables in standard form, most often the SOP form.
(b) Expressions can then be implemented with just two levels of gates (not counting any input inverters).
(c) Each expression requires only AND and OR gates and inverters for its implementation. The application of De Morgan's rules allows an expression in the SOP form to be built using NAND gates only, and an expression in the POS form to be built using NOR gates only (see Chapter 4).
(d) The SOP form of equations is used to specify equations to be implemented by MSI/LSI devices such as programmable logic devices, multiplexers and ROMs.

To obtain an SOP equation from a truth table:

(a) Every 1 in the output F column (or row) must be represented by a product term.
(b) Each variable that is at 1 must be written in its true form.
(c) Each variable that is at 0 must be written in its inverted form.
(d) The SOP equation that represents the circuit is then given by the sum of the product terms thus obtained.

To obtain a POS equation from a truth table:

(a) Every 0 in the output column (or row) must be represented by a sum term.
(b) Each variable that is at 1 must be written in its inverted form.
(c) Each variable that is at 0 must be written in its true form.
(d) The POS equation that represents the circuit is then given by the product of the sum terms thus obtained.

EXAMPLE 2.3

Obtain:

(a) The SOP equation.
(b) The POS equation that represents the circuit whose truth table is given in Table 2.7.

Table 2.7

A	0	1	0	1	0	1	0	1
B	0	0	1	1	0	0	1	1
C	0	0	0	0	1	1	1	1
F	1	0	0	0	1	1	0	1

Solution

(a) The SOP equation is $F = \bar{A}\bar{B}\bar{C} + \bar{A}\bar{B}C + A\bar{B}C + ABC$ (Ans.)
(b) The POS equation is $F = (\bar{A} + B + C)(A + \bar{B} + C)(\bar{A} + \bar{B} + C)(A + \bar{B} + \bar{C})$ (Ans.)

PRACTICAL EXERCISE 2.2

Aim: to confirm that the SOP and POS equations for a circuit give the same result.
Components and equipment: one triple 3-input 74HC11 AND gate IC, two dual 2-input 74HC08 AND gate ICs, two quad 2-input 74HC32 OR gate ICs and one 74HC04 hex inverter, two LEDs and two 270 Ω resistors. Breadboard. Power supply.
Procedure:

(a) Determine the SOP and POS equations for the circuit whose truth table is given in Table 2.8.
(b) Build the two circuits using a 270 Ω resistor in series with an LED to indicate the logical state of the output of each circuit.
(c) Apply the input variable combinations given in the table to each of the circuits in turn and confirm that they both give the same results.

See Electronics Workbench exercise EWB 2.2.

Table 2.8

A	0	1	0	1	0	1	0	1
B	0	0	1	1	0	0	1	1
C	0	0	0	0	1	1	1	1
F	0	0	1	1	1	0	0	1

Boolean algebra

When the Boolean equation describing the logic operation of a circuit has been obtained, an attempt is often made to simplify, or minimize, the equation.

An equation is said to be minimized when it (a) contains the lowest possible number of input variables and (b) the lowest possible number of terms. The minimized equation will usually require the fewest number of gates possible for its implementation, although this may not necessarily also give the least number of IC packages. Further, the minimal solution may well be subject to *race hazards*.

There are three main methods by which a Boolean equation can be simplified; these are the use of Boolean algebra, the use of a Karnaugh map, and the use of tabulation techniques.

The algebraic simplification of logic functions is made easier by the use of the logic rules which follow:

1. $A + \bar{A} = 1$
2. $A + A = A$
3. $AA = A$
4. $A\bar{A} = 0$
5. $A + 0 = A$
6. $A + 1 = 1$
7. $A \cdot 1 = A$
8. $A \cdot 0 = 0$
9. $AB = BA$ $\left.\begin{array}{l} \\ \\ \end{array}\right\}$ Commutative law
10. $A + B = B + A$
11. $A(B + C) = AB + AC$ $\left.\begin{array}{l} \\ \\ \end{array}\right\}$ Distributive law
12. $A + BC = (A + B)(A + C)$
13. $A + B + C = (A + B) + C$
    $= A + (B + C)$ $\left.\begin{array}{l} \\ \\ \\ \\ \end{array}\right\}$ Associative law
14. $ABC = A(BC) = (AB)C$
15. $A(B + \bar{B}) = A$
16. $A + AB = A$
17. $A(A + B) = A$
18. $A + \bar{A}B = A + B$
19. $B(A + \bar{B}) = AB$
20. $(A + B)(B + C)(C + \bar{A}) = (A + B)(C + \bar{A})$
21. $\overline{A + B} = \bar{A}\bar{B}$
22. $\overline{AB} = \bar{A} + \bar{B}$
23. $AB + BC + \bar{A}C = AB + \bar{A}C$
24. $\bar{A} + AB = \bar{A} + B$
25. $AB + AC + \bar{B}C = AB + \bar{B}C$

Rules 21 and 22 are known as *De Morgan's* rules and they can be extended to deal with three, or more, input variables. Thus: $F = \overline{A + B + C} = \bar{A}\bar{B}\bar{C}$, and $F = \overline{ABC} = \bar{A} + \bar{B} + \bar{C}$.

EXAMPLE 2.4

Simplify $F = \bar{A}(B + \bar{C})(A + \bar{B} + C)\bar{A}\bar{B}\bar{C}$.

Solution

Multiplying out gives

$$F = (AB + B\bar{B} + BC + A\bar{C} + \bar{B}\bar{C} + C\bar{C})\bar{A}\bar{A}\bar{B}\bar{C}$$

Rules 3 and 4 give

$$F = (AB + BC + A\bar{C} + \bar{B}\bar{C})\bar{A}\bar{B}\bar{C}$$

Multiplying out again and applying rules 3 and 4 gives

$$F = \bar{A}\bar{B}\bar{C} \ (Ans.)$$

EXAMPLE 2.5

Simplify $F = (A + B)(\overline{AB} + C) + AB$.

Solution

Rule 22 gives

$$F = (A + B)(\bar{A} + \bar{B} + C) + AB$$

Multiplying out and using rule 4 gives

$$
\begin{aligned}
F &= A\bar{B} + AC + \bar{A}B + BC + AB \\
&= A(B + \bar{B}) + AC + \bar{A}B + BC \\
&= A(1 + C) + \bar{A}B + BC && \text{(from rule 1)} \\
&= A + \bar{A}B + BC && \text{(from rule 6)} \\
&= A + B + BC && \text{(from rule 18)} \\
&= A + B(1 + C) && \text{(from rule 11)} \\
&= A + B \ (Ans.) && \text{(from rule 6)}
\end{aligned}
$$

With some practice it is possible to write several of the above steps down at once.

EXAMPLE 2.6

Simplify the equation $F = ABC + ABD + \bar{A}B\bar{C} + CD + B\bar{D}$.

Solution

$$
\begin{aligned}
F &= ABC + ABD + \bar{A}B\bar{C} + CD + B\bar{D} \\
&= ABC + \bar{A}B\bar{C} + CD + B(\bar{D} + AD) \\
&= ABC + \bar{A}B\bar{C} + CD + B(\bar{D} + A) && \text{(from rule 18)} \\
&= AB(1 + C) + \bar{A}B\bar{C} + CD + B\bar{D}
\end{aligned}
$$

$$\begin{aligned}
&= AB + \bar{A}B\bar{C} + CD + B\bar{D} &&\text{(from rule 6)}\\
&= B(A + \bar{A}\bar{C}) + CD + B\bar{D}\\
&= AB + B\bar{C} + CD + B\bar{D} &&\text{(from rule 18)}\\
&= AB + B\bar{C} + CD + BC + B\bar{D} &&\text{(from rule 23)}\\
&= AB + B(C + \bar{C}) + CD + B\bar{D}\\
&= AB + B + CD + B\bar{D} &&\text{(from rule 1)}\\
&= B(1 + A + \bar{D}) + CD\\
&= B + CD \ (Ans.) &&\text{(from rule 6)}
\end{aligned}$$

EXAMPLE 2.7

Simplify the equations:

(a) $F = \overline{A\bar{B}} + \overline{A(\bar{A} + C)}$.

(b) $F = A\bar{B}(A + C) + \bar{A}B\overline{(A + \bar{B} + \bar{C})}$.

Solution

(a) $F = \bar{A} + B + \bar{A} + \overline{\bar{A} + C} = \bar{A} + B + \bar{A} + A\bar{C} = \bar{A} + B + A\bar{C}$

$\quad = \bar{A} + B + \bar{C} \ (Ans.)$ $\hfill$ (rule 24)

(b) $F = A\bar{B} + \overline{A + C} + \bar{A}B(\bar{A}BC) = \bar{A} + B + \bar{A}\bar{C} + \bar{A}BC = \bar{A}(1 + \bar{C} + BC) + B$

$\quad = \bar{A} + B \ (Ans.)$

Sometimes an equation can be further simplified if extra redundant terms are added. Consider the Boolean equation

$$F = AB\bar{C} + ABC + A\bar{B}C + \bar{A}BC$$

Simplifying,

$$F = AB(C + \bar{C}) + A\bar{B}C + \bar{A}BC = AB + A\bar{B}C + \bar{A}BC$$

Adding redundant terms ABC and ABC to the original equation gives

$$F = AB\bar{C} + ABC + A\bar{B}C + ABC + \bar{A}BC + ABC$$

Simplifying,

$$F = AB(C + \bar{C}) + AC(B + \bar{B}) + BC(A + \bar{A}) = AB + AC + BC$$

Conversion of a sum-of-products equation into the equivalent product-of-sums form

The conversion of an equation in product-of-sums form into the equivalent sum-of-products form is easily accomplished by merely multiplying out (see Examples 2.4 and 2.5). The reverse process, namely converting an equation from its sum-of-products form into its equivalent product-of-sums form, is more difficult. Consider the equation

$$F = ABC + \bar{A}B\bar{C}$$

The first step is to obtain the complement of the equation

$$\bar{F} = \overline{ABC + \bar{A}B\bar{C}}$$
$$= \overline{ABC} \cdot \overline{\bar{A}B\bar{C}}$$
$$= (\bar{A} + \bar{B} + \bar{C})(A + \bar{B} + C)$$

Multiplying out and then simplifying gives

$$\bar{F} = A\bar{B} + \bar{A}C + A\bar{B} + \bar{B} + \bar{B}C + A\bar{C} + \bar{B}C$$
$$= \bar{B}(1 + \bar{A} + A + C + \bar{C}) + \bar{A}C + A\bar{C}$$
$$= \bar{B} + \bar{A}C + A\bar{C}$$

Applying De Morgan's rule again:

$$F = \overline{\bar{B} + \bar{A}C + A\bar{C}} = B(\overline{\bar{A}C})(\overline{A\bar{C}})$$

or

$$F = B(A + \bar{C})(\bar{A} + C)$$

EXAMPLE 2.8

Write the SOP expression $F = A\bar{B} + BC\bar{D} + \bar{B}CD$ in its POS form.

Solution

$$\bar{F} = \overline{A\bar{B} + BC\bar{D} + \bar{B}CD} = \overline{A\bar{B}} \cdot \overline{BC\bar{D}} \cdot \overline{\bar{B}CD} = (\bar{A} + B)(\bar{B} + C + D)(B + C + \bar{D})$$
$$= BD + BC + \bar{A}C + \bar{A}\bar{B}\bar{D}$$
$$F = \overline{BD} \cdot \overline{BC} \cdot \overline{\bar{A}C} \cdot \overline{\bar{A}\bar{B}\bar{D}} = (\bar{B} + \bar{D})(\bar{B} + \bar{C})(A + \bar{C})(A + B + D) \ (Ans.)$$

The algebraic method of simplifying Boolean equations possesses the disadvantages that:

(a) It can be very time consuming.
(b) It requires considerable practice and experience before the most appropriate approach and/or rule can be selected quickly.

Usually it is better to employ the mapping method described later.

Canonical form

The canonical form of a Boolean equation is one in which *each* term contains each of the input variables *once* only. If two equations are written down in their canonical forms, they can be compared term by term and this fact will be utilized in Chapter 3 when a tabular method of simplifying Boolean equations is described. Also, the canonical form of a sum-of-products equation is required when the equation is to be implemented using a multiplexer (see p. 116).

EXAMPLE 2.9

Write the equation $\bar{A}B + \bar{C}B$ in its canonical form.

Solution

$$F = \bar{A}B(C + \bar{C}) + \bar{C}B(A + \bar{A})$$
$$= \bar{A}BC + \bar{A}B\bar{C} + AB\bar{C} + \bar{A}B\bar{C}$$
$$= \bar{A}BC + \bar{A}B\bar{C} + AB\bar{C} \ (Ans.)$$

EXAMPLE 2.10

Write the equation $(\bar{A} + B)(\bar{C} + B)$ in its canonical form.

Solution

$$F = (\bar{A} + B + C)(\bar{A} + B + \bar{C})(A + B + \bar{C})(\bar{A} + B + \bar{C})$$
$$= (\bar{A} + B + C)(\bar{A} + B + \bar{C})(A + B + \bar{C}) \ (Ans.)$$

EXAMPLE 2.11

Express the equation $F = A(\bar{B} + C)$ in:

(a) SOP form.
(b) POS canonical form.

Solution

(a) Multiplying out: $F = A\bar{B} + AC$
    $= A\bar{B}(C + \bar{C}) + AC(B + \bar{B}) = A\bar{B}C + A\bar{B}\bar{C} + ABC + A\bar{B}C$
    $= A\bar{B}\bar{C} + A\bar{B}C + ABC \ (Ans.)$
(b) For A: $(A + B + C)(A + B + \bar{C})(A + \bar{B} + C)(A + \bar{B} + \bar{C})$
    For $\bar{B} + C$: $(A + \bar{B} + C)(\bar{A} + \bar{B} + C)$

Therefore, $F = (A + B + C)(A + B + \bar{C})(A + \bar{B} + C)(A + \bar{B} + \bar{C})(\bar{A} + \bar{B} + C) \ (Ans.)$

EXAMPLE 2.12

Write the equation $F = AB + \bar{C}\bar{D}$ in its canonical form.

Solution

$$F = AB(C + \bar{C})(D + \bar{D}) + \bar{C}\bar{D}(A + \bar{A})(B + \bar{B})$$
$$= ABCD + ABC\bar{D} + AB\bar{C}D + AB\bar{C}\bar{D} + A\bar{B}\bar{C}\bar{D} + \bar{A}B\bar{C}\bar{D} + \bar{A}\bar{B}\bar{C}\bar{D} \ (Ans.)$$

Duality

Every sum-of-products equation has a dual product-of-sums equation, and vice versa, and this duality provides another method which can be used in the simplification of Boolean equations.

The dual of an equation is simply obtained by merely replacing every AND symbol ($\cdot$) by the OR symbol (+) and vice versa. For example, the dual of $F = (\bar{A} + B + C)(A + D)$ is $F' = \bar{A}BC + AD$.

EXAMPLE 2.13

Use duality to simplify Example 2.5.

Solution

$$F = (A + B)(\bar{A} + \bar{B} + C) + AB$$
$$F' = (AB + \bar{A}\bar{B}C)(A + B)$$
$$= AB$$

Therefore, $F = A + B$ (*Ans.*)

Clearly, in this case, the algebra involved is much simpler than previously.

EXAMPLE 2.14

Repeat Example 2.6 using duality.

Solution

$$F = ABC + ABD + \bar{A}B\bar{C} + CD + B\bar{D}$$
$$F' = (A + B + C)(A + B + D)(\bar{A} + B + \bar{C})(C + D)(B + \bar{D})$$
$$= (A + B + CD)(\bar{A} + B + \bar{C})(C + D)(B + \bar{D})$$
$$= (A\bar{C} + \bar{A}CD + B)(C + D)(B + \bar{D}) = (A\bar{C}D + \bar{A}CD + BC + BD)(B + \bar{D})$$
$$= BD + BC$$

Therefore, $F = (B + D)(B + C) = B + CD$ (as before) (*Ans.*)

Complement of *F*

The complement $\bar{F}$ of a Boolean equation can be obtained by replacing each input variable by its complement in the corresponding dual equation.

Thus if $F = AB + C$, then $F' = (A + B)C$ and so

$$\bar{F} = (\bar{A} + \bar{B})\bar{C} = \bar{A}\bar{C} + \bar{B}\bar{C}$$

EXAMPLE 2.15

Use duality to find the complement of the equation

$$F = ABC + \bar{A}D$$

Check your answer using De Morgan's rules.

Solution

$$F' = (A + B + C)(\bar{A} + D)$$

so

$$\bar{F} = (\bar{A} + \bar{B} + \bar{C})(A + \bar{D})$$
$$= A\bar{B} + A\bar{C} + \bar{A}\bar{D} + \bar{B}\bar{D} + \bar{C}\bar{D} \ (Ans.)$$

Using De Morgan's rule

$$\bar{F} = \overline{ABC + \bar{A}D}$$
$$= \overline{ABC} \cdot \overline{\bar{A}D} = (\bar{A} + \bar{B} + \bar{C})(A + \bar{D}) \text{ (as before) } (Ans.)$$

EXAMPLE 2.16

Find the complement of $F = ABC + \bar{A}B\bar{C} + \bar{A}\bar{B}C$:

(a) Directly.
(b) Using duality.

Solution

(a) $\bar{F} = \overline{ABC + \bar{A}B\bar{C} + \bar{A}\bar{B}C} = (\overline{ABC})(\overline{\bar{A}B\bar{C}})(\overline{\bar{A}\bar{B}C})$
    $= (\bar{A} + \bar{B} + \bar{C})(A + \bar{B} + C)(A + B + \bar{C})$
    $= (\bar{A}\bar{B} + \bar{A}C + A\bar{B} + \bar{B} + \bar{B}C + A\bar{C} + \bar{B}C)(A + B + \bar{C})$
    $= (\bar{A}C + A\bar{C} + \bar{B})(A + B + \bar{C})$
    $= \bar{A}BC + A\bar{C} + A\bar{B} + \bar{B}\bar{C} \ (Ans.)$

(b) $F' = (A + B + C)(\bar{A} + B + \bar{C})(\bar{A} + \bar{B} + C)$
    $\bar{F} = (\bar{A} + \bar{B} + \bar{C})(A + \bar{B} + C)(A + B + \bar{C})$ (as before) $(Ans.)$

Decimal representation of SOP and POS equations

A canonical Boolean equation in either its SOP or POS form may be written as:

SOP: $F = \Sigma(DCBA)$

POS: $F = \Pi(DCBA)$

Each term in the equation can be represented by its equivalent decimal number. Consider the equation

$$F = AB\bar{C}\bar{D} + \bar{A}\bar{B}\bar{C}D + \bar{A}B\bar{C}D$$

Taking A as the least significant number and D as the most significant number, $AB\bar{C}\bar{D}$ = 3, $\bar{A}\bar{B}\bar{C}D$ = 8 and $\bar{A}B\bar{C}D$ = 10. Hence, $F = \Sigma(3, 8, 12)$.

For a POS equation the true form of a variable is given the decimal value 0 and the inverted form of a variable is given a decimal value, e.g. $A = 0$ $\bar{A} = 1$, $B = 0$ $\bar{B} = 2$, $C = 0$ $\bar{C} = 4$, etc.

The POS equation $F = (A + B + \bar{C} + \bar{D})(\bar{A} + \bar{B} + \bar{C} + \bar{D})(\bar{A} + \bar{B} + C + D)$ can be written as

$$F = \Pi(3, 12, 15)$$

EXAMPLE 2.17

(a) Write down the Boolean equation represented by $F = \Sigma(0, 1, 2, 4, 6, 7)$.
(b) Convert the equation into its POS form.
(c) Express the POS equation in the form $F = \Pi(CBA)$.

Solution

(a) $F = \bar{A}\bar{B}\bar{C} + A\bar{B}\bar{C} + \bar{A}B\bar{C} + \bar{A}\bar{B}C + \bar{A}BC + ABC = \bar{A}B\bar{C} + \bar{A}\bar{B}C + BC + \bar{B}\bar{C}$

Add redundant terms $\bar{A}BC$ and $\bar{A}\bar{B}\bar{C}$ to obtain

$F = \bar{A}B + \bar{A}\bar{B} + BC + \bar{B}\bar{C}$
   $= \bar{A} + BC + \bar{B}\bar{C}$ (*Ans.*)

(b) $\bar{F} = \overline{\bar{A} + BC + \bar{B}\bar{C}} = A \cdot \overline{BC} \cdot \overline{\bar{B}\bar{C}} = A(\bar{B} + \bar{C})(B + C)$
   $= A(\bar{B}C + B\bar{C}) = A\bar{B}C + AB\bar{C}$

$F = \overline{A\bar{B}C + AB\bar{C}} = (\overline{A\bar{B}C})(\overline{AB\bar{C}}) = (\bar{A} + B + \bar{C})(\bar{A} + \bar{B} + C)$ (*Ans.*)

(c) $F = \Pi(3, 5)$ (*Ans.*)

Note that the numbers in the $\Pi(CBA)$ equation are the numbers that are missing from the $\Sigma(CBA)$ equation. This is always the case and it gives an easy way of converting from SOP to POS or vice versa.

EXAMPLE 2.18

Convert $F = (\bar{A} + B + C)(A + \bar{B} + C)(A + B + \bar{C})$ into its SOP form.

Solution

$F = \Pi(1, 2, 4)$. Hence, the equivalent POS equation is $F = \Sigma(0, 3, 5, 6, 7)$, or $F = \bar{A}\bar{B}\bar{C} + AB\bar{C} + A\bar{B}C + \bar{A}BC + ABC$ (*Ans.*)

EXAMPLE 2.19

(a) Reduce the Boolean equation $F = \Sigma(9, 10, 11, 12, 13, 14, 15)$ to its simplest form.
(b) Convert the equation into its canonical POS form.

Solution

(a) $F = \bar{A}B\bar{C}D + AB\bar{C}D + \bar{A}BCD + A\bar{B}CD + \bar{A}BCD + ABCD + A\bar{B}\bar{C}D$
$= \bar{A}BD + A\bar{B}D + ABD + \bar{A}\bar{B}CD = BD + \bar{B}D(A + \bar{A}C) = BD + \bar{B}D(A + C)$
$= BD + AD + CD$ (*Ans.*)

(b) $F = \Pi(0, 1, 2, 3, 4, 5, 6, 7, 8)$
$= (\bar{A} + B + C + D)(A + \bar{B} + C + D)(\bar{A} + \bar{B} + C + D)(A + B + \bar{C} + D)(\bar{A} + B$
$+ \bar{C} + D)(A + \bar{B} + \bar{C} + D) + (\bar{A} + \bar{B} + \bar{C} + D)(A + B + C + \bar{D})(A + B$
$+ C + D)$ (*Ans.*)

It would be very tedious to simplify answer (b) algebraically but the equation can be simplified using a Karnaugh map as shown in Example 3.7 on p. 72.

PRACTICAL EXERCISE 2.3

Aim: to determine whether the two canonical forms of an equation, although looking different, produce the same output for the same input variables.
Components and equipment: two 74HC32 quad 2-input OR gates, two 74HC11 triple 3-input AND gates and one 74HC04 hex inverter, two LEDs and two 270 Ω resistors. Breadboard. Power supply.
Procedure:

(a) The logic function to be implemented is $F = \Sigma(2, 4, 5, 7)$. Build the circuit shown in Fig. 2.6.

Fig. 2.6

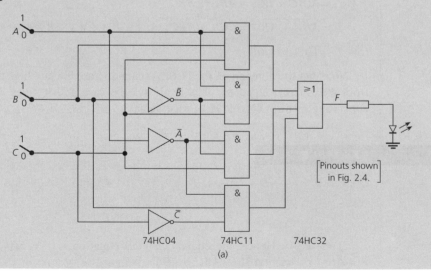

74HC04   74HC11   74HC32

(a)

Fig. 2.6 *(Cont'd)*

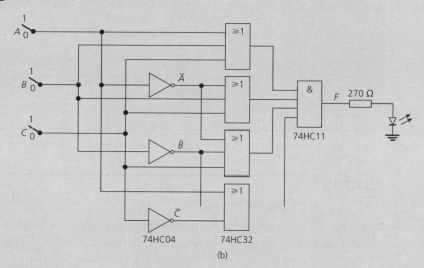

(b)

(b) Apply each of the possible combinations of the three variables A, B and C to the input terminals of the circuit. Each time note the logical state of the LED. Write down the truth table for the circuit.
(c) Obtain the POS version of the function and then build a circuit to implement it.
(d) Repeat procedure (b).
(e) The SOP and POS canonical circuits obtained from the same truth table always give identical outputs when driven by the same inputs. Compare the two truth tables and confirm that this was so in this exercise.

PRACTICAL EXERCISE 2.4

Aim: to design and build a circuit that will have a HIGH output only when its BCD input number is 7, 8 or 9.
Components and equipment: either (i) one 74LS21 or 74HC21 dual 4-input AND gate, one 74LS32 or 74HC32 quad 2-input OR gate, and one 74LS04 or 74HC04, or 74AC04 hex inverter, one LED and one 270 Ω resistor. Breadboard. Power supply.
   The truth table for the required circuit is given in Table 2.9.
   The Boolean expression for the circuit is $F = ABC\bar{D} + \bar{A}\bar{B}CD + A\bar{B}CD$, and this reduces to $F = ABC\bar{D} + \bar{B}CD$.
   If the can't happen conditions are used, $F = D + ABC$.

Table 2.9

A	0	1	0	1	0	1	0	1	0	1	0	1	0	1	0	1
B	0	0	1	1	0	0	1	1	0	0	1	1	0	0	1	1
C	0	0	0	0	1	1	1	1	0	0	0	0	1	1	1	1
D	0	0	0	0	0	0	0	0	1	1	1	1	1	1	1	1
F	0	0	0	0	0	0	0	1	1	1	×	×	×	×	×	×

Codes

In digital circuitry, numbers are represented by a sequence of the bits 1 and 0. The number of *words* or codewords which can be formed from *n* bits is equal to 2^n. For example, 4 bits can give 16 different words. By allocating a different value or *weighting* to each of the 4 bits, a large number of different codes can be generated. However, the number of codes actually employed is relatively few and only the more important of them will be discussed here.

The basic 4-bit binary code is well known but it is repeated in Table 2.10 so that it can be compared with the other codes which are given later.

Table 2.10 *Four-bit binary code*

0	0000	1	0001	2	0010	3	0011
4	0100	5	0101	6	0110	7	0111
8	1000	9	1001	10	1010	11	1011
12	1100	13	1101	14	1110	15	1111

Binary-coded decimal

A binary-coded decimal (BCD) code uses 4 bits to represent a single decimal number between 0 and 9. Numbers greater than 9 are not represented and, when required, combinations of BCD numbers are employed. A large number of different weightings can potentially be allocated to each of the 4 bits of a BCD word but, in practice, only a few of the possible weightings are used. Here, five different weightings, namely 8421, excess-3 (XS3), 5421, 2421 and 84-2-1 will be considered. The differences between these five codes are shown in Table 2.11. It should be noted that the first ten steps of the 8421 code are identical with those of the pure binary code. Note also that the XS3 code is actually 8421 with decimal 3 (or 0011) added to each codeword.

Table 2.11

	8421	XS3	5421	2421	84-2-1
0	0000	0011	0000	0000	0000
1	0001	0100	0001	0001	0111
2	0010	0101	0010	0010	0110
3	0011	0110	0011	0011	0101
4	0100	0111	0100	0100	0100
5	0101	1000	0101	0101	1011
6	0110	1001	0110	0110	1010
7	0111	1010	0111	0111	1001
8	1000	1011	1011	1110	1000
9	1001	1100	1101	1111	1111

EXAMPLE 2.20

Represent the decimal number 38 in each of the BCD codes given in Table 2.11.

Solution

(a)	8421	0011	1000	
(b)	XS3	0110	1011	
(c)	5421	0011	1011	
(d)	2421	0011	1110	
(e)	84-2-1	0101	1000	(*Ans.*)

The most frequently employed of these BCD codes is the 8421 version, partly because of its close similarity to pure binary. The 2421 code is sometimes employed in digital-to-analogue converters because it allows the use of a smaller range of resistance values.

Two features of a code that may influence its choice for a particular application are, first, any *self-complementing* property and, second, any *reflective* property it may possess.

Some codes possess the advantage of being self-complementing. This means that their logic and arithmetic complements are identical. For example, the 9s complement of an XS3, or a 2421 BCD number, is the same as its logical complement. Some examples taken from Table 2.11 are:

(a) Decimal 5 = 100 in XS3   9s complement = 4 = 0111
                                Logical complement = 0111
(b) Decimal 7 = 1010 in XS3   9s complement = 2 = 0101
                                Logical complement = 0101

This property can be of advantage in circuits which employ decimal arithmetic.

Table 2.12 *Gray code*							
0	1	2	3	4	5	6	7
0000	0001	0011	0010	0110	0111	0101	0100
8	9	10	11	12	13	14	15
1100	1101	1111	1110	1010	1011	1001	1000

Gray code

A code is said to be *reflective* if two numbers which are equally spaced either side of the centre numbers 7 and 8 differ in only 1 bit. Several examples exist but the most important, which is also an example of a *unit distance code*, is known as the *Gray code*.

A unit distance code is one that changes in only one bit from one codeword to the next adjacent codeword. The Gray code is given in Table 2.12. The Gray code finds particular application in conjunction with rotational encoders which convert the angular position of a shaft into an equivalent binary number. The code may seem to be difficult to remember but a useful, easily remembered means of converting from pure binary into Gray is available.

First, put a 0 in front of the most significant bit. Then, carry out the exclusive-OR logical operation on adjacent bits starting from the left.

EXAMPLE 2.21

Convert the pure binary numbers 1100 and 0111 into the Gray code.

Solution

(a) 0 1 1 0 0
     \/\/\/\/
      1 0 1 0

(b) 0 0 1 1 1
     \/\/\/\/
      0 1 0 0

Alphanumeric codes

An alphanumeric code is one that includes all the letters of the alphabet, punctuation marks and arithmetic signs as well as numbers. One such code is known as the Murray code and is widely employed in telegraphy systems. The code most often used in conjunction with computers and microprocessors is known as the *American Standard Code for Information Interchange* (ASCII) and it is listed in Table 2.13.

Each character requires 7 bits. Some of the characters shown in the table are various control characters used in data systems; one example is CR which stands for carriage return.

Table 2.13 ASCII

b_7			0	0	0	0	1	1	1	1
	b_6		0	0	1	1	0	0	1	1
		b_5	0	1	0	1	0	1	0	1
b_4 b_3 b_2 b_1										
0 0 0 0			NUL	DLE	SP	0	@	P	'	p
0 0 0 1			SOH	DC1	!	1	A	Q	a	q
0 0 1 0			STX	DC2	"	2	B	R	b	r
0 0 1 1			ETX	DC3	#	3	C	S	c	s
0 1 0 0			EOT	DC4	$	4	D	T	d	t
0 1 0 1			ENQ	NAK	%	5	E	U	e	u
0 1 1 0			ACK	SYN	&	6	F	V	f	v
0 1 1 1			BEL	ETB	'	7	G	W	g	w
1 0 0 0			BS	CAN	(	8	H	X	h	x
1 0 0 1			HT	EM	)	9	I	Y	i	y
1 0 1 0			LF	SUB	*	:	J	Z	j	z
1 0 1 1			VT	ESC	+	;	K	[	k	{
1 1 0 0			FF	FS	,	<	L	\	l	l
1 1 0 1			CR	GS	–	=	M	]	m	}
1 1 1 0			SO	RS	.	>	N		‖	~
1 1 1 1			SI	US	/	?	O	—	o	DEL

From the table

$$\text{decimal } 0 = 0110000 = 48_{10}$$
$$A = 1000001 = 65_{10}$$
$$j = 1101010 = 106_{10} \text{ and so on.}$$

Code converters

Many instances arise for a decimal number to be encoded into the corresponding binary or binary-coded decimal (or some other code) number. Similarly, it is often necessary to decode a number from pure binary or BCD into decimal.

There are no MSI code converter ICs available in the 74HC and other 74CMOS logic families other than the 74HC42 BCD-to-decimal converter. If a converter is required it will have to be built using SSI devices (or a PLD).

Decimal-to-binary-coded decimal converter

A decimal-to-BCD converter will have nine input lines representing, respectively, the decimal integers 0, 1, 2, 3, 4, 5, 6, 7, 8, 9, and four output lines representing 2^0, 2^1, 2^2 and 2^3, respectively. The presence of any one of the nine decimal numbers is indicated by a high level (logic 1 voltage) on the appropriate input line and a low level (logic 0

Table 2.14 *Decimal-to-BCD truth table*

Decimal		0	1	2	3	4	5	6	7	8	9
BCD outputs	2^0	0	1	0	1	0	1	0	1	0	1
	2^1	0	0	1	1	0	0	1	1	0	0
	2^2	0	0	0	0	1	1	1	1	0	0
	2^3	0	0	0	0	0	0	0	0	1	1

voltage) on all of the remaining input lines. Decimal zero is represented by *all* the input lines being at the low level.

The converter is required to generate the logic 1 state on the appropriate output lines to produce the binary equivalent of the decimal input number. The truth table of a decimal-to-BCD converter is given in Table 2.14.

From Table 2.14 the Boolean equations describing the operation of the circuit are:

$$2^0 = 1 + 3 + 5 + 7 + 9$$

$$2^1 = 2 + 3 + 6 + 7$$

$$2^2 = 4 + 5 + 6 + 7$$

$$2^3 = 8 + 9$$

The design requires one 5-input, two 4-input and one 2-input OR gates and could be implemented using three 74HC32A quad 2-input OR gate ICs.

The design of a BCD-to-decimal converter can also start from the truth table given in Table 2.14. Let $2^0 = A$, $2^1 = B$, $2^2 = C$ and $2^3 = D$. Then

$$0 = \bar{A}\bar{B}\bar{C}\bar{D} \qquad 1 = A\bar{B}\bar{C}\bar{D} \qquad 2 = \bar{A}B\bar{C}\bar{D} \qquad 3 = AB\bar{C}\bar{D} \qquad 4 = \bar{A}\bar{B}C\bar{D}$$

$$5 = A\bar{B}C\bar{D} \qquad 6 = \bar{A}BC\bar{D} \qquad 7 = ABC\bar{D} \qquad 8 = \bar{A}\bar{B}\bar{C}D \qquad 9 = A\bar{B}\bar{C}D$$

These equations could be implemented directly using four inverters and ten AND gates but MSI devices, e.g. the 74HC42 BCD-to-decimal converter, are available.

Binary-to-Gray converters

Table 2.15 shows the binary and the Gray code equivalents of the decimal numbers 0 through to 15. From the truth table,

$$G_1 = A\bar{B}\bar{C}\bar{D} + \bar{A}B\bar{C}\bar{D} + A\bar{B}C\bar{D} + \bar{A}BC\bar{D} + A\bar{B}\bar{C}D + \bar{A}B\bar{C}D + A\bar{B}CD + \bar{A}BCD$$

$$G_2 = \bar{A}B\bar{C}\bar{D} + AB\bar{C}\bar{D} + \bar{A}\bar{B}C\bar{D} + A\bar{B}C\bar{D} + \bar{A}B\bar{C}D + AB\bar{C}D + \bar{A}\bar{B}CD + A\bar{B}CD$$

$$G_3 = \bar{A}\bar{B}C\bar{D} + A\bar{B}C\bar{D} + \bar{A}BC\bar{D} + ABC\bar{D} + \bar{A}\bar{B}CD + A\bar{B}CD + \bar{A}BCD + ABCD$$

$$G_4 = \bar{A}\bar{B}\bar{C}D + A\bar{B}\bar{C}D + \bar{A}B\bar{C}D + AB\bar{C}D + \bar{A}\bar{B}CD + A\bar{B}CD + \bar{A}BCD + ABCD$$

Simplifying,

Table 2.15 Binary-to-Gray code truth table

Decimal number	Binary				Gray			
	D	C	B	A	G_4	G_3	G_2	G_1
0	0	0	0	0	0	0	0	0
1	0	0	0	1	0	0	0	1
2	0	0	1	0	0	0	1	1
3	0	0	1	1	0	0	1	0
4	0	1	0	0	0	1	1	0
5	0	1	0	1	0	1	1	1
6	0	1	1	0	0	1	0	1
7	0	1	1	1	0	1	0	0
8	1	0	0	0	1	1	0	0
9	1	0	0	1	1	1	0	1
10	1	0	1	0	1	1	1	1
11	1	0	1	1	1	1	1	0
12	1	1	0	0	1	0	1	0
13	1	1	0	1	1	0	1	1
14	1	1	1	0	1	0	0	1
15	1	1	1	1	1	0	0	0

$$G_1 = A\bar{B}\bar{D} + \bar{A}B\bar{D} + AB D + \bar{A}BD = A\bar{B} + \bar{A}B$$

$$G_2 = \bar{A}B\bar{C} + AB\bar{C} + \bar{A}\bar{B}C + ABC = BC + \bar{B}C$$

$$G_3 = \bar{B}C\bar{D} + BC\bar{D} + B\bar{C}D + \bar{B}\bar{C}D = C\bar{D} + \bar{C}D$$

$$G_4 = \bar{B}\bar{C}D + B\bar{C}D + \bar{B}CD + BCD = \bar{C}D + CD = D$$

The binary-to-Gray converter could be implemented using three exclusive-OR gates.

Gray-to-binary converter

A Gray-to-binary converter can be designed by reading the truth table of Table 2.15 the other way around. Thus

$$A = G_1\bar{G}_2\bar{G}_3\bar{G}_4 + \bar{G}_1G_2\bar{G}_3\bar{G}_4 + G_1G_2G_3\bar{G}_4 + \bar{G}_1\bar{G}_2G_3\bar{G}_4$$
$$+ G_1\bar{G}_2G_3G_4 + \bar{G}_1G_2G_3G_4 + G_1G_2\bar{G}_3G_4 + \bar{G}_1\bar{G}_2\bar{G}_3G_4$$

This equation cannot be simplified

$$B = G_1G_2\bar{G}_3\bar{G}_4 + \bar{G}_1G_2\bar{G}_3\bar{G}_4 + G_1\bar{G}_2G_3\bar{G}_4 + \bar{G}_1\bar{G}_2G_3\bar{G}_4$$
$$+ G_1G_2G_3G_4 + \bar{G}_1G_2G_3G_4 + G_1\bar{G}_2\bar{G}_3G_4 + \bar{G}_1\bar{G}_2\bar{G}_3G_4$$

$$C = \bar{G}_1G_2G_3\bar{G}_4 + G_1G_2G_3\bar{G}_4 + G_1\bar{G}_2G_3\bar{G}_4 + \bar{G}_1\bar{G}_2G_3\bar{G}_4$$
$$+ \bar{G}_1G_2\bar{G}_3G_4 + G_1G_2\bar{G}_3G_4 + G_1\bar{G}_2\bar{G}_3G_4 + \bar{G}_1\bar{G}_2\bar{G}_3G_4$$

$$D = G_4$$

Simplifying,

$$C = G_3\bar{G}_4 + \bar{G}_3G_4$$

$$B = G_2G_3G_4 + \bar{G}_2G_3\bar{G}_4 + G_2\bar{G}_3\bar{G}_4 + \bar{G}_2\bar{G}_3G_4$$

The equations for A, B, and C can be implemented directly but considerable simplification results if it is noticed that

$$B = \bar{G}_2C + G_2\bar{C}$$

and

$$A = G_1\bar{B} + \bar{G}_1B$$

Aim: to build and test (a) a binary-to-Gray code converter, (b) a Gray-to-binary code converter.
Components and equipment: one 74HC86 quad 2-input exclusive-OR gate IC, four LEDs and four 270 Ω resistors. Power supply.
Procedure:

(a) Build the binary-to-Gray code converter shown in Fig. 2.7.
(b) Apply, in sequence, the binary equivalents to decimal 0 through to 15 to the input terminals of the circuit. Each time note the logical states of the four LEDs.
(c) Write down the truth table for the circuit and use it to check whether the circuit converts an input binary word into an output Gray code word.
(d) Study the equations for the Gray-to-binary code converter and decide how the circuit can be modified to act in the reverse manner. Then apply, in sequence, the Gray code equivalents of decimal 0 through to 15 to the input terminals of the modified circuit. Each time note the logical states of the LEDs and hence obtain the truth table of the circuit.
(e) Did the modified circuit work correctly as a Gray-to-binary converter?

Fig. 2.7

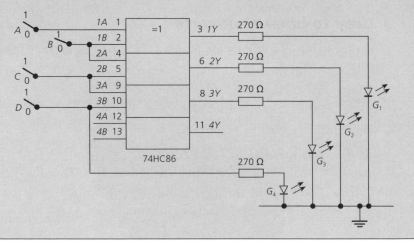

Fig. 2.8 *Multi-level circuit*

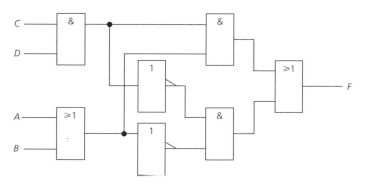

Multi-level logic

The minimization of a logical function using either Boolean algebra or a Karnaugh map results in a two-level circuit that has minimum propagation delay. However, it may not employ the least possible number of gates.

The logic function $F = \bar{A}\bar{B}\bar{C} + \bar{A}\bar{B}\bar{D} + ACD + BCD$ requires nine gates: four 3-input AND gates, one 4-input OR gate, and four inverters for its implementation. This means one 74HC11, one 74HC32 and one 74HC04 ICs.

Re-writing the equation:

$$F = CD(A + B) + \bar{A}\bar{B}(\bar{C} + \bar{D})$$
$$= CD(A + B) + (\overline{A + B})(\overline{CD})$$
$$= XY + \bar{X}\bar{Y}$$

This form of the equation can be implemented using the circuit given in Fig. 2.8. Its implementation requires the use of seven gates: three 2-input AND gates, two 2-input OR gates and two inverters. Again, the same three ICs are needed even though there are two fewer gates employed. This circuit has three levels and hence its propagation delay will be longer. However, in VLSI designs the main requirement is to reduce the number of *literals*, i.e. variables in both complemented and true form, involved in a circuit, and a multi-level solution will achieve that aim.

Deriving the best multi-level circuit can be a complex procedure and a number of techniques are available, including software such as MisII, for example.

EXERCISES

2.1 Show that:
  (a) $A + \bar{A}B = AB$.
  (b) $AB + \bar{B}C + AB\bar{D} + A\bar{B}C + ABCE = AB + \bar{B}C$.
  (c) $AC + \bar{B}\bar{C}D + B\bar{C}\bar{D} + \bar{A}\bar{B}D + \bar{A}\bar{B}\bar{C}D + AB\bar{C}\bar{D} = AC + CD + \bar{C}\bar{D} + \bar{C}B$.

2.2 (a) Express $F = \overline{\overline{A + BC} + (A + \overline{C})B}$ in sum-of-products form.

(b) Express $F = AB + AC + B\overline{C}$ in product-of-sums form.

2.3 Simplify:

(a) $F = \overline{(\overline{A}C + D)(\overline{A} + \overline{C})\overline{B}D}$.

(b) $F = \overline{(AD + \overline{CD})(\overline{AC + B})}$.

2.4 Simplify:

(a) $F = \overline{A}\overline{B}C + A\overline{B}\overline{D} + BCD + B\overline{C}$.

(b) $F = A(\overline{B} + C)\overline{D} + B(\overline{C} + D)$.

2.5 Simplify:

(a) $F = \overline{A}B \cdot \overline{A} + C$.

(b) $F = \overline{A}B + \overline{B}C + A\overline{C} + A\overline{B}C$.

(c) $F = (A + \overline{BC})(\overline{A}BC)$.

(d) $F = \overline{A}(C + \overline{D}) + \overline{C}(A + \overline{B})$.

2.6 Simplify:

(a) $F = (\overline{\overline{A}} + \overline{B})(\overline{C} + \overline{AB})$.

(b) $F = \overline{A}B\overline{C} + \overline{A}\overline{B}C + AB\overline{C} + \overline{A}B\overline{C}$.

2.7 Simplify:

(a) $F = \overline{(A\overline{B}CD)(A\overline{B})} + BC$.

(b) $F = \overline{A}\overline{B}CD + A\overline{B}\overline{C}D + \overline{A}\overline{B}\overline{C}D + \overline{A}\overline{B}CD + \overline{A}B\overline{C}D + A\overline{B}\overline{C}D + ABCD$.

(c) $F = (\overline{A} + B)(C + \overline{D})$.

2.8 Simplify $F = \overline{(A\overline{B}C)(\overline{A}\overline{B})} + BC$.

2.9 Simplify $F = (A + B)(A + \overline{A}B)C + \overline{A}B + ABC + \overline{A(\overline{B} + \overline{C})}$.

2.10 Convert the equation $F = (A + B + C)(A + B + \overline{C})(A + \overline{B} + C)(\overline{A} + B + C)(\overline{A} + \overline{B} + C)$ into its SOP form using two different methods.

2.11 Determine the complement of $F = \overline{BC}(A + \overline{CD})$

using: (a) De Morgan's rules.

(b) Duality.

2.12 Design a car alarm system to give an alarm if:

(a) The ignition (I) is turned ON and the driver's door (D) is open.

(b) The headlamps (H) or the sidelamps (S) are ON and the driver's door is open.

(c) The headlamps are ON and the engine (E) is not running.

2.13 Show that $\overline{A}\overline{B}\overline{C} + A\overline{B}\overline{C} + \overline{A}B\overline{C} + \overline{A}\overline{B}C + \overline{A}BC + ABC = \overline{A} + BC + \overline{B}\overline{C}$.

2.14 A digital circuit has four inputs A, B, C and D and two outputs F_1 and F_2. Output F_1 is to go HIGH whenever three or more inputs are HIGH. Output F_2 is to go HIGH if any two, but not three, inputs are HIGH.

(a) Obtain the truth table for the circuit.

(b) Obtain the SOP expressions.

(c) Simplify the SOP equations.

2.15 Design a combinational logic circuit to have a HIGH output only when the input 4-bit number is between decimal 4 and 10.

2.16 A house has two lights to illuminate the stairs leading from the hall to the upstairs landing. The lights can be switched OFF and ON by either one of two switches, one in the hall and one on the landing. The lights are to be OFF when both switches are either ON or OFF together, and the lights are to be ON when one switch is ON and other is OFF. Design the system.

2.17 Simplify the equations:
(a) $F = (A + B)BC + A$.
(b) $F = (A + B)\bar{B}C + B + \bar{B}\bar{C}$.
(c) $F = (A + \bar{B})(B + C)\bar{A}$.
(d) $F = (A + B)(\bar{A} + BC) + \bar{A}B + CD$.

2.18 Simplify:
(a) $F = \bar{B}C(A + \bar{A}) + AB\bar{C} + BC(A + \bar{A})$.
(b) $F = (A + B)(\bar{A} + C)$.
(c) $F = A + ABC + \bar{A}BC + \bar{A}B + A\bar{D} + AD$.
(d) $F = AC + C(D + BC + AB)(ACD + BD)$.

2.19 What is meant by duality? Find the complements of the equations:
(a) $F = ABC + AB$ using:
  (i) duality
  (ii) De Morgan's rules.
(b) $F = (AB + C)(ABD + BC)$ using:
  (i) duality
  (ii) De Morgan's rules.

2.20 (a) Convert from POS form to SOP form:
  (i) $\Pi(1, 5, 8, 14, 15)$
  (ii) $\Pi(0, 2, 6, 7)$.
(b) Convert from SOP form to POS form:
  (i) $\Sigma(1, 2, 3, 4, 5)$
  (ii) $\Sigma(2, 4, 5, 12, 14)$.

2.21 (a) Write, in decimal form, the equations:
  (i) $F = \bar{A}\bar{B}\bar{C}D + A\bar{B}C\bar{D} + AB\bar{C}\bar{D} + ABC$
  (ii) $A\bar{B}C + A\bar{B} + B\bar{C}$.
(b) Write, in decimal form, the equations:
  (i) $F = (A + B + C + D)(\bar{A} + \bar{B} + \bar{C} + \bar{D})(\bar{A} + B + \bar{C} + D)(A + \bar{B} + C + D) + (A + \bar{B} + C + \bar{D})$
  (ii) $F = (A + \bar{B} + C)(\bar{A} + B)(\bar{B} + \bar{C})$.

3 Mapping and tabulation methods of simplifying Boolean equations

After reading this chapter you should be able to:

- Plot a sum-of-products Boolean equation on a Karnaugh map.
- Plot a product-of-sums Boolean equation on a Karnaugh map.
- Use a Karnaugh map to reduce a complex Boolean equation to its simplest form.
- Use a Karnaugh map to convert an SOP equation into its POS equivalent form.
- Make use of don't care conditions to simplify Boolean equations.
- Map a Boolean equation that is given in its decimal form.
- Use a Karnaugh map to simplify Boolean equations with five variables.
- Use the Quine–McCluskey tabular method of simplifying a Boolean equation.
- Understand what is meant by static and dynamic hazards in a combinational logic circuit.
- Know some simple ways of eliminating static hazards.

The algebraic method of simplifying Boolean equations has the disadvantages that:

(a) It can be very time consuming.
(b) It requires considerable practice and experience before the most appropriate approach and/or rule can be selected quickly.

The *Karnaugh map* provides a convenient method for the simplification of Boolean equations with up to four input variables and it can be used with five input variables. When the number of input variables is six or more, it is easier to employ a tabular method of simplification, such as that devised by Quine and McCluskey. This technique is rather lengthy, and hence error-prone, but it can be programmed by solution on a computer. Also some software packages are available for the simplification of complex Boolean equations, e.g. *Espresso*.

The Karnaugh map

The Karnaugh map provides a convenient method of simplifying sum-of-products Boolean equations in which the function to be simplified is displayed diagrammatically on a map of *cells*. Each cell maps one term of the function. The number of cells is equal to 2^n, where n is the number of input variables. Thus, if the equation to be simplified has three variables A, B and C, then $n = 3$ and $2^3 = 8$. The rows and columns of the Karnaugh map can be labelled as shown for 4-, 8- and 16-cell maps (the labels in the cells are not normally given). Note that only one variable is changed if a move is made between adjacent rows and columns. A 1 written in a cell indicates the presence in the function being mapped of the term represented by that cell. A 0 written in a cell means that that term is not present in the mapped function.

A B	0	1
0	$\bar{A}\bar{B}$	$A\bar{B}$
1	$\bar{A}B$	AB

AB C	00	01	11	10
0	$\bar{A}\bar{B}\bar{C}$			$A\bar{B}\bar{C}$
1		$\bar{A}BC$	ABC	

AB CD	00	01	11	10
00	$\bar{A}\bar{B}\bar{C}\bar{D}$			
01				
11			$ABCD$	
10				$A\bar{B}C\bar{D}$

To simplify a Boolean equation using a Karnaugh map adjacent cells containing 1 are looped together. This step eliminates any terms of the form $A\bar{A}$. In this context adjacent means:

(a) Side-by-side in the horizontal and vertical directions (but *not* diagonal).
(b) The right-hand and left-hand sides, and the top and bottom, of the map.
(c) The four corner cells in a four-variable map.

Cells may only be looped together in twos, fours, or eights. As few groups as possible should be formed; but groups may overlap one another and may contain only one cell. The larger the number of 1s looped together in a group the simpler is the product term that the group represents. If the cells that contain 0 are looped together, instead of the cells containing 1, the complement of the expression is obtained.

EXAMPLE 3.1

Use a Karnaugh map to simplify the Boolean equation $F = \bar{A}B + \bar{A}B\bar{C} + AB\bar{C} + A\bar{B}\bar{C}$.

Solution

The Karnaugh mapping of the equation is

AB C	00	01.	11	10
0	1	1	1	1
1	0	1	0	0

The cells that contain 1 can be looped together in one group of four cells and one group of two cells as shown in the map below

AB C	00	01	11	10
0	(1	1)	1	1)
1	0	(1)	0	0

From this mapping, $F = \bar{A}B + \bar{C}$ (*Ans.*)

The result can be checked using Boolean algebra:

$$F = \bar{A}B + \bar{A}\bar{B}\bar{C} + AB\bar{C} + A\bar{B}\bar{C} = \bar{A}(B + \bar{B}\bar{C}) + A\bar{C}(B + \bar{B}) = \bar{A}B + C(A + \bar{A}) = \bar{A}B + \bar{C}.$$

Clearly, little has been gained by the use of the mapping method in this instance, but for more complex Boolean equations the mapping method is much easier to perform.

EXAMPLE 3.2

(a) Use a Karnaugh map to simplify the equation $F = ACD + \bar{A}BCD + \bar{B}\bar{C}\bar{D} + \bar{A}\bar{B}C\bar{D} + \bar{A}\bar{B}CD + A\bar{B}C\bar{D}$.

Determine $\bar{F}$ by:

(b) The use of De Morgan's rules.
(c) The use of duality.
(d) By looping the cells that contain 0.

Solution

(a) The mapping of the function is

AB CD	00	01	11	10
00	1)	0	0	1
01	0	0	0	0
11	1	1	1	1
10	1)	0	0	1

Looping adjacent cells in two groups of four simplifies the equation to

$$F = CD + \bar{B}\bar{D} \ (Ans.)$$

(b) $\bar{F} = \overline{CD + \bar{B}\bar{D}} = (\overline{CD})(\overline{\bar{B}\bar{D}}) = (\bar{C} + \bar{D})(B + D)$
   $= B\bar{C} + B\bar{D} + \bar{C}D = B\bar{D} + \bar{C}D \ (Ans.)$

(c) $F' = (C + D)(\bar{B} + \bar{D}) = \bar{B}C + \bar{B}D + C\bar{D}$

   $\bar{F} = B\bar{C} + B\bar{D} + \bar{C}D = B\bar{D} + \bar{C}D \ (Ans.)$

(d) Looping the 0 cells, as shown by the map below, gives

CD \ AB	00	01	11	10
00	1	0	0	1
01	0	0	0	0
11	1	1	1	1
10	1	0	0	1

$$F = \bar{C}D + B\bar{D} \ (Ans.)$$

EXAMPLE 3.3

(a) Simplify the equation $F = ABCD + \bar{A}BCD + A\bar{C}D + A\bar{C}\bar{D} + \bar{A}B\bar{C}$.
    Obtain $\bar{F}$ using:
(b) De Morgan's rules.
(c) Duality.
(d) By looping the 0 cells.

Solution

(a) Looping the 1 squares, $F = BD + A\bar{C} + B\bar{C} \ (Ans.)$

CD \ AB	00	01	11	10
00	0	1	1	1
01	0	1	1	1
11	0	1	1	0
10	0	0	0	0

(b) $\bar{F} = \overline{BD + A\bar{C} + B\bar{C}}$
   $= \overline{BD} \ \overline{A\bar{C}} \ \overline{B\bar{C}} = (\bar{B} + \bar{D})(\bar{A} + C)(\bar{B} + C)$
   $= (\bar{A}\bar{B} + \bar{B}C + \bar{A}\bar{D} + C\bar{D})(\bar{B} + C)$

$$= \bar{A}\bar{B} + \bar{B}C + \bar{A}\bar{B}D + \bar{B}C\bar{D} + \bar{A}\bar{B}C + \bar{B}C + \bar{A}C\bar{D} + C\bar{D}$$
$$= \bar{A}\bar{B} + \bar{B}C + C\bar{D} \ (Ans.)$$

(c) $F' = (B + D)(A + \bar{C})(B + \bar{C})$

$$\bar{F} = (\bar{B} + \bar{D})(\bar{A} + C)(\bar{B} + C) = (\bar{A}\bar{B} + \bar{A}\bar{D} + \bar{B}C + C\bar{D})(\bar{B} + C)$$
$$= AB + BC + CD \ (Ans.)$$

(d) Looping the 0 squares, $\bar{F} = \bar{A}\bar{B} + \bar{B}C + C\bar{D} \ (Ans.)$

EXAMPLE 3.4

The Boolean equation which describes the operation of a digital logic circuit is

$$F = \overline{\bar{A} + B + \bar{C}} + \overline{A + \bar{B} + D} + \overline{(A + \bar{B} + \bar{C})(A + C + D)} + \bar{A}\bar{B}\bar{C}\bar{D} + \bar{A}\bar{B}C$$

Use a Karnaugh map to reduce the expression to its simplest form.

Solution

Applying De Morgan's rules,

$$F = A\bar{B}C + \bar{A}B\bar{D} + \overline{A + \bar{B} + \bar{C}} + \overline{A + C + D} + \bar{A}\bar{B}\bar{C}\bar{D} + \bar{A}\bar{B}C$$
$$= A\bar{B}C + \bar{A}B\bar{D} + \bar{A}\bar{B}C + \bar{A}\bar{C}\bar{D} + \bar{A}\bar{B}\bar{C}\bar{D} + \bar{A}\bar{B}C$$

This expression is mapped on a 16-cell Karnaugh map

CD \ AB	00	01	11	10
00	1	1	0	0
01	0	0	0	0
11	1	1	0	1
10	1	1	0	1

From the looped 1 cells, $F = \bar{A}C + \bar{B}C + \bar{A}\bar{D} \ (Ans.)$
Alternatively, looping the 0 cells gives $\bar{F} = A\bar{C} + \bar{C}D + AB$, and

$$F = \overline{A\bar{C} + \bar{C}D + AB} = \overline{A\bar{C}} \cdot \overline{\bar{C}D} \cdot \overline{AB} = (\bar{A} + C)(C + \bar{D})(\bar{A} + \bar{B})$$
$$= (\bar{A}C + \bar{A}\bar{D} + C)(\bar{A} + \bar{B}) = \bar{A}C + \bar{B}C + \bar{A}\bar{D} \ (Ans.)$$

Product-of-sums equations

The Karnaugh map can also be used to simplify an equation of the product-of-sums form. Either the equation can be multiplied out into the equivalent sum-of-product form, or each term can be mapped separately and then the individual maps can be combined by ANDing them. Note that for corresponding squares in each map

$$1\ 1 = 1 \qquad 1\ 0 = 0 \qquad 0\ 0 = 0$$

EXAMPLE 3.5

Repeat Example 2.5 using a Karnaugh map.

Solution

Applying De Morgan's rules, $F = (A + B)(\bar{A} + \bar{B} + C) + AB$.

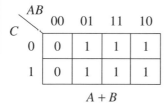

C \ AB	00	01	11	10
0	0	1	1	1
1	0	1	1	1

$A + B$

C \ AB	00	01	11	10
0	1	1	0	1
1	1	1	1	1

$\bar{A} + \bar{B} + C$

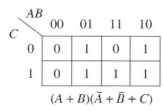

C \ AB	00	01	11	10
0	0	1	0	1
1	0	1	1	1

$(A + B)(\bar{A} + \bar{B} + C)$

Mapping AB gives F.

C \ AB	00	01	11	10
0	0	1	1	1
1	0	1	1	1

F

Looping the cells that contain 1 gives $F = A + B$ (*Ans.*)

EXAMPLE 3.6

Simplify the Boolean equation $F = (\bar{A} + B + \bar{C})(\bar{A} + B + D)(\bar{B} + \bar{C} + \bar{D})(\bar{B} + C)$.

Solution

The mappings of each of the terms are shown below

CD \ AB	00	01	11	10
00	1	1	1	1
01	1	1	1	1
11	1	1	1	0
10	1	1	1	0

$\bar{A} + B + \bar{C}$

CD \ AB	00	01	11	10
00	1	1	1	0
01	1	1	1	1
11	1	1	1	1
10	1	1	1	0

$\bar{A} + B + D$

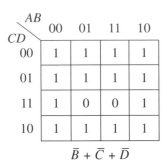

$\bar{B} + \bar{C} + \bar{D}$

CD\AB	00	01	11	10
00	1	1	1	1
01	1	1	1	1
11	1	0	0	1
10	1	1	1	1

$\bar{B} + C$

CD\AB	00	01	11	10
00	1	0	0	1
01	1	0	0	1
11	1	1	1	1
10	1	1	1	1

To obtain the overall mapping each cell in an individual map must be ANDed with the corresponding cells in all the other maps. This means that if any cell in a map contains a 0 then the corresponding cell in the composite map also contains 0. The composite map shown has had its 1 cells looped to give

F

CD\AB	00	01	11	10
00	1	0	0	0
01	1	0	0	1
11	1	0	0	0
10	1	1	1	0

$$F = \bar{A}\bar{B} + \bar{B}\bar{C}D + BC\bar{D} \ (Ans.)$$

EXAMPLE 3.7

Use a Karnaugh map to simplify the POS equation $F = (A + B + C + D)(\bar{A} + B + C + D)(A + \bar{B} + C + D)(\bar{A} + \bar{B} + C + D)(A + B + \bar{C} + D)(\bar{A} + B + \bar{C} + D)(A + \bar{B} + \bar{C} + D)(\bar{A} + \bar{B} + \bar{C} + D)(A + B + C + \bar{D})$.

Solution

Plotting each sum term on a separate map

$A + B + C + D$

CD\AB	00	01	11	10
00	0	1	1	1
01	1	1	1	1
11	1	1	1	1
10	1	1	1	1

$\bar{A} + B + C + D$

CD\AB	00	01	11	10
00	1	1	1	0
01	1	1	1	1
11	1	1	1	1
10	1	1	1	1

$A + \bar{B} + C + D$

CD\AB	00	01	11	10
00	1	0	1	1
01	1	1	1	1
11	1	1	1	1
10	1	1	1	1

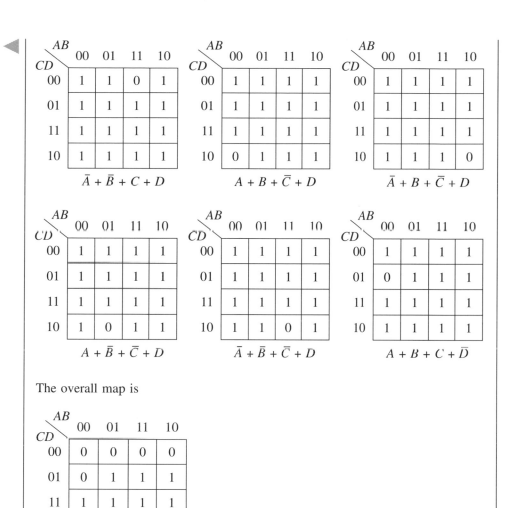

The overall map is

AB\CD	00	01	11	10
00	0	0	0	0
01	0	1	1	1
11	1	1	1	1
10	0	0	0	0

F

Looping the 1 cells: $F = AD + BD + CD = D(A + B + C)$ (*Ans.*)

Looping the 0 cells: $F' = \bar{D} + \bar{A}\bar{B}\bar{C}$, so $F = D(A + B + C)$ (*Ans.*)

Numerical method

For each sum term in the POS equation write 0 if the input variable is in its true form, and write 1 if the variable is in its complemented form. Then map the 0s in the corresponding cells in the Karnaugh map. Consider Example 3.7 again. The equation to be mapped becomes:

$$F = (ABCD) = (0000)(1000)(0100)(1100)(0010)(1010)(0110)(1110)(0001)$$

Putting a 0 into each of the numbered cells gives

CD\AB	00	01	11	10
00	0	0	0	0
01	0	1	1	1
11	1	1	1	1
10	0	0	0	0

Note that this map is the same as that obtained in a much longer way in Example 3.7.

EXAMPLE 3.8

Use a Karnaugh map to simplify $F = (\bar{A} + B + \bar{C} + D)(A + \bar{B} + C + \bar{D})(A + B + C + \bar{D})(\bar{A} + \bar{B} + C + D)(\bar{A} + \bar{B} + \bar{C} + D)(A + B + C + D)(\bar{A} + B + C + D)(A + \bar{B} + C + D)$. Give the solution in:

(a) SOP form.
(b) POS form.

Solution

$$F = (1010)(0101)(0001)(1100)(1110)(0000)(1000)(0100)$$

The mapping is

CD\AB	00	01	11	10
00	0	0	0	0
01	0	0	1	1
11	1	1	1	1
10	1	1	0	0

(a) Looping the 1 cells: $F = AD + CD + \bar{A}C$ (*Ans.*)
(b) Looping the 0 cells: $F' = \bar{C}\bar{D} + \bar{A}\bar{C} + A\bar{D}$, and
$F = (C + D)(A + C)(\bar{A} + D)$ (*Ans.*)

Converting SOP equations into POS equations

A Karnaugh map can be used to convert a sum-of-products equation into its corresponding product-of-sums form.

Looping 1 cells

The sum-of-products equation should be mapped in the usual way and then the 0 cells should be looped to obtain the minimal equation for F. The application of De Morgan's rules will then give the required POS equation.

Looping 0 cells

The use of De Morgan's rules can be avoided and the required equation taken straight from the map if, for each grouping of 0 cells, the complement of the map labelling is used together with duality. Thus, from the 0 cells in the first map, $F' = \bar{B}C$

C \ AB	00	01	11	10
0	1	1	1	1
1	0	1	1	0

But the complement of the map labelling is to be used plus duality. Hence, $F' = \bar{B}C$ should be read as $F = B + C$.

Similarly, from the second map, $F' = \bar{A}\bar{C} + AC$ and using duality and taking complements, $F = (A + C)(\bar{A} + \bar{C})$

C \ AB	00	01	11	10
0	0	0	1	1
1	1	1	0	0

Obtain the product-of-sums form of the equation $F = \bar{A}\bar{B} + \bar{B}C + C\bar{D}$ using:

(a) De Morgan's rules.
(b) A Karnaugh map.

Solution

(a) $\bar{F} = \overline{\bar{A}\bar{B} + \bar{B}C + C\bar{D}} = \overline{\bar{A}\bar{B}} \cdot \overline{\bar{B}C} \cdot \overline{C\bar{D}}$
$= (A + B)(B + \bar{C})(\bar{C} + D) = B\bar{C} + BD + A\bar{C}$

$F = \overline{B\bar{C} + BD + A\bar{C}} = \overline{B\bar{C}} \cdot \overline{BD} \cdot \overline{A\bar{C}}$
$= (\bar{B} + C)(\bar{B} + \bar{D})(\bar{A} + C)$ *(Ans.)*

(b) The mapping of F is

	AB			
CD	00	01	11	10
00	1	0	0	0
01	1	0	0	0
11	1	0	0	1
10	1	1	1	1

From the looped 0 cells, $F' = B\bar{C} + BD + A\bar{C}$, so that $F = (\bar{B} + C)(\bar{B} + \bar{D})(\bar{A} + C)$ (*Ans.*)

EXAMPLE 3.10

Obtain the minimal product-of-sums form of $F = AB + \bar{A}CD + \bar{A}\bar{B}C\bar{D} + \bar{A}BC\bar{D}$.

Solution

The mapping is

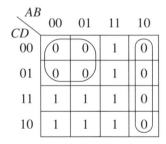

Looping the 0 squares gives $F = (\bar{A} + B)(A + C)$ (*Ans.*)

Decimal representation of terms

It is sometimes convenient to express the various terms of a Boolean equation in their decimal form. Thus, referring to Table 3.1, assuming A to be the least significant bit, the decimal mapping is as shown on the next page.

The decimal form of a sum-of-products Boolean equation can be written as

$$F = \Sigma(ABCD)$$

Thus, $F = \Sigma(3, 7, 10, 12)$ represents the equation

$$F = AB\bar{C}\bar{D} + ABC\bar{D} + \bar{A}B\bar{C}D + \bar{A}\bar{B}CD$$

Table 3.1 *Decimal representation of Boolean equations*

Decimal number	0	1	2	3	4	5	6	7	8	9	10	11	12	13	14	15
A	0	1	0	1	0	1	0	1	0	1	0	1	0	1	0	1
B	0	0	1	1	0	0	1	1	0	0	1	1	0	0	1	1
C	0	0	0	0	1	1	1	1	0	0	0	0	1	1	1	1
D	0	0	0	0	0	0	0	0	1	1	1	1	1	1	1	1

AB / CD	00	01	11	10
00	0	2	3	1
01	8	10	11	9
11	12	14	15	13
10	4	6	7	5

EXAMPLE 3.11

Map the function $F = \Sigma(4, 6, 10, 11, 14, 15)$ and then simplify.

Solution

AB / CD	00	01	11	10
00	0	0	0	0
01	0	1	1	0
11	0	1	1	0
10	1	1	0	0

From the map, $F = BD + \bar{A}C\bar{D}$ (*Ans.*)

EXAMPLE 3.12

Simplify the equations:

(a) $F = \Sigma(0, 1, 2, 3, 7, 8, 9, 12, 14)$.
(b) $\Pi(0, 1, 2, 3, 7, 8, 9, 12, 14)$.

Solution

(a) The mapping is

CD \ AB	00	01	11	10
00	1	1	1	1
01	1	0	0	1
11	1	1	0	0
10	0	0	1	0

From the map, $F = \bar{C}\bar{D} + \bar{B}\bar{C} + \bar{A}CD + AB\bar{D}$ (*Ans.*)

(b) The mapping is

CD \ AB	00	01	11	10
00	0	0	0	0
01	0	1	1	0
11	0	0	1	1
10	1	1	0	1

From the map, $F = B\bar{C}D + ACD + \bar{A}C\bar{D} + \bar{B}C\bar{D}$ (*Ans.*)

'Don't care' conditions

The truth tables of some logic circuits contain certain combinations of input variables for which the output F is unimportant and so they can be either 1 or 0. Such combinations are said to be 'don't care' conditions or states. When a Boolean function is mapped, any don't care terms are represented by a × and can be looped in with *either* the 1 cells *or* the 0 cells in the simplification of the function. A × cell should be looped with a group of 1 cells if the looping then gives a greater reduction in the plotted equation. Suppose, for example, that the mapping of a particular function is

CD \ AB	00	01	11	10
00	1	1	×	1
01	0	1	1	1
11	0	×	1	1
10	0	1	1	×

Looping the 1 squares only gives

$$F = AD + ABC + \bar{A}B\bar{D} + \bar{A}B\bar{C} + \bar{B}C\bar{D}$$

If the don't care squares are looped together with the 1 squares, the function can be reduced to the much simpler result:

$$F = A + B + \bar{C}\bar{D}$$

EXAMPLE 3.13

Use a Karnaugh map to simplify the equation $F = \Sigma(0, 2, 4, 5, 6, 7, 8)$ with don't cares $\times = \Sigma(12, 13, 14, 15)$.

Solution

The Karnaugh map is

CD\AB	00	01	11	10
00	1	1	0	0
01	1	0	0	0
11	×	×	×	×
10	1	1	1	1

From the map $F = C + \bar{A}\bar{B} + \bar{A}\bar{D}$ (*Ans.*)

EXAMPLE 3.14

A logic circuit is to produce an output at logic 1 whenever its BCD input word represents an odd number. Use a Karnaugh map to simplify the Boolean equation that describes the operation of the circuit:

(a) Without using don't cares.
(b) Using the don't cares.

Solution

$F = \Sigma(1, 3, 5, 7, 9)$ and $\times = \Sigma(10, 11, 12, 13, 14, 15)$

The mapping is

CD \ AB	00	01	11	10
00	0	0	1	1
01	0	×	×	1
11	×	×	×	×
10	0	0	1	1

(a) From the map: $F = A\bar{D} + A\bar{B}\bar{C}$ (*Ans.*)
(b) $F = A$ (*Ans.*)

Karnaugh map for more than four variables

The number of cells in a Karnaugh map must be equal to 2^n, where n is the number of input variables.

Five input variables

Five input variables A, B, C, D and E will require $2^5 = 32$ cells. These can be obtained by drawing two 16-square maps side-by-side, one of which represents E while the other represents $\bar{E}$

CD \ AB	00	01	11	10
00	0	2	3	1
01	8	10	11	9
11	12	*Z* 14	15	13
10	4	6	7	5

$\bar{E}$

CD \ AB	00	01	11	10
00	16	18	19	17
01	24	26	27	25
11	28	*Z* 30	31	29
10	20	22	23	21

E

The 32-cell map is used to simplify a five-input variable Boolean equation in a similar manner to that previously described. Note that the similarly positioned cells in each map, e.g. the cells marked Z, are considered to be adjacent for looping purposes.

EXAMPLE 3.15

Simplify the Boolean equation $F = \Sigma(2, 5, 6, 9, 10, 14, 15, 16, 18, 20, 22, 26, 29, 30, 31)$.

Solution

The mapping is

AB\CD	00	01	11	10
00	0	1	0	0
01	0	1	0	1
11	0	1	1	0
10	0	1	0	1

$\bar{E}$

AB\CD	00	01	11	10
00	1	1	0	0
01	0	1	0	0
11	0	1	1	1
10	1	1	0	0

E

From the map:

(a) Cells 5 and 9 cannot be looped: $A\bar{B}C\bar{D}\bar{E} + \bar{A}B\bar{C}D\bar{E}$.
(b) Cells 29 and 31 can be looped: $ACDE$.
(c) Cells 14, 15, 30 and 31 can be looped: BCD.
(d) Cells 16, 18, 20 and 22 can be looped: $\bar{A}\bar{D}E$.
(e) Cells 2, 6, 10, 14, 18, 22, 26 and 30 can be looped: $\bar{A}B$.

Hence, $F = \bar{A}B + \bar{A}\bar{D}E + BCD + ACDE + A\bar{B}C\bar{D}\bar{E} + \bar{A}B\bar{C}D\bar{E}$ (*Ans.*)

EXAMPLE 3.16

Simplify the equations mapped on the Karnaugh map below.

AB\CD	00	01	11	10
00	1	1	0	1
01	1	1	0	1
11	0	0	0	0
10	0	0	0	0

$\bar{E}$

AB\CD	00	01	11	10
00	1	1	0	1
01	1	1	0	1
11	1	0	0	0
10	1	0	0	0

E

Solution

Looping the cells that contain 1:

(a) Cells 0, 2, 8, 10, 16, 18, 24, 26 give $\bar{A}\bar{C}$.
(b) Cells 0, 1, 8, 9, 16, 17, 24, 25 give $\bar{B}\bar{C}$.
(c) Cells 16, 20, 24, 28 give $\bar{A}\bar{B}E$.

Therefore, $F = \bar{A}\bar{C} + \bar{B}\bar{C} + \bar{A}\bar{B}E$ (*Ans.*)
    Alternatively, looping the 0 cells gives:

(a) Cells 3, 7, 11, 15, 19, 23, 27 and 31: $\bar{A} + \bar{B}$.
(b) Cells 4, 5, 6, 7, 12, 13, 14 and 15: $\bar{C} + \bar{E}$.
(c) Cells 5, 7, 13, 15, 21, 23, 29 and 31: $\bar{A} + \bar{C}$.
(d) Cells 6, 7, 14, 15, 22, 23, 30 and 31: $\bar{B} + \bar{C}$.

Therefore, $F = (\bar{A} + \bar{B})(\bar{C} + \bar{E})(\bar{A} + \bar{C})(\bar{B} + \bar{C})$ (*Ans.*)

Six input variables

The use of a Karnaugh map can also be extended to six input variables but it becomes rather unwieldy since four maps are required. Each map has the usual A, B, C and D labelling plus individual labelling of the 16-cell maps of EF, $E\bar{F}$, $\bar{E}F$ and $\bar{E}\bar{F}$. Simplification of a six-variable equation is cumbersome and error-prone and so some alternative approach is desirable. A tabular method of solution may be used or various computer packages are available for solving problems with five or more variables, such as Espresso. The logic converter icon in the Electronics Workbench software package can simplify Boolean equations with up to eight input variables and then, on the click of a button, give either the AND/OR or the NAND gate implementation of the simplified equation. [Note, however, that it takes A as the most significant variable and D as the least significant.]

Tabular simplification of Boolean equations

The Karnaugh map provides a convenient method for the simplification of Boolean equations with up to four input variables and can be used with five input variables. When the number of input variables is in excess of this number, it is generally easier to employ a tabular method of simplification. The method to be described is that of Quine and McCluskey.

An *implicant* is an input variable, or a group of variables, that may be combined together in a Karnaugh map. A prime implicant is an implicant that cannot be combined with another implicant to eliminate an input variable from a simplified equation.

Electronic Workbench uses the Quine–McCluskey method to simplify Boolean equations.

Quine–McCluskey tabular reduction

The procedure to be followed when simplifying a Boolean equation using the Quine–McCluskey method is as follows:

(a) Write down the equation to be simplified in its *canonical* form.
(b) Write all the canonical terms in a column (column A), starting with those terms that include the *most complemented* input variables.

(c)   In column *B* write, for each term in (b), a 1 for an uncomplemented input variable and 0 for a complemented input variable.

(d)   Divide column *B* into *groups* containing, in order, no 1s, a single 1, two 1s, three 1s, and so on.

(e)   An attempt is now made to 'match' each term within a group with another group. Here *matching* means that the two terms differ from one another by only *one* input variable. For example, the terms 0001 and 1001 would match. Each term that is matched with another term is ticked. A term should be matched as many times as possible.

(f)   The *matched terms* are then entered in column *C* but the two different input variables have cancelled out and are hence omitted. A dash is entered in this position.

(g)   When column *C* has been completed, it is divided into further groups as before and another set of matched terms is found. Again, variables which cancel out are indicated by a dash.

(h)   These matched terms are entered in column *D* and so on.

(i)   When no more matchings can be found the table is inspected for all the ticked items. These are the prime implicants and they may give the simplest form of the equations but it is possible that some further simplification can be achieved.

(j)   A *prime implicant table* (or chart) is drawn up and it is used to locate any common terms among the prime implicants and thereby allows a further reduction in the function.

The rules for constructing the prime implicant table are:

(a)   Use each term (minterm) in the original equation as a column label.

(b)   Use the prime implicants as row labels.

(c)   Enter a tick in each position in the table that corresponds with a minterm that is covered by a particular prime implicant.

(d)   Examine the table for the *essential prime implicants*. A prime implicant is one that provides the *only* cover for one, or more, minterms.

(e)   Other minterms may be covered by two, or more, essential prime implicants and then a choice must be made as to which one(s) to discard.

EXAMPLE 3.17

Table 3.2 gives the prime implicant table (or chart) that has been obtained by the Quine–McCluskey reduction of a Boolean equation. Obtain the minimum coverage for *F*.

Table 3.2

	0	1	2	3	4	5	6	7	8	9	10	11	12	13	14	15
$A\bar{B}C$						✓								✓		
$B\bar{C}$		✓	✓								✓	✓				
BCD															✓	✓
BD						✓						✓			✓	✓

Solution

The first two terms contain essential prime implicants and must be retained. One of the other terms must also be retained to cover 14 and 15; choosing the last term gives $F = A\bar{B}C + B\bar{C} + BD$ (*Ans.*)

EXAMPLE 3.18

Solve the equation $F = AB + B\bar{C} + CD + \bar{B}D$ using a tabular method.

Solution

This is, of course, an equation that could easily be simplified using a Karnaugh map.
Step A Writing the given equation in its canonical form:

$$F = AB(C + \bar{C})(D + \bar{D}) + B\bar{C}(A + \bar{A})(D + \bar{D})$$
$$+ CD(A + \bar{A})(B + \bar{B}) + \bar{B}D(A + \bar{A})(C + \bar{C})$$
$$= ABCD + AB\bar{C}D + ABC\bar{D} + AB\bar{C}\bar{D} + \bar{A}B\bar{C}D + \bar{A}B\bar{C}\bar{D}$$
$$+ A\bar{B}CD + \bar{A}BCD + \bar{A}\bar{B}CD + A\bar{B}\bar{C}D + \bar{A}\bar{B}\bar{C}D$$

Step B These terms are now listed in column A of Table 3.3.
Step C Column B is then produced by writing 1 for each uncomplemented input variable and 0 for each complemented input variable in column A.
Step D Column B is then divided into four groups by drawing horizontal lines. In each group the number of 1s is the same.
Step E The matching process can now begin. Terms 1 and 4 differ only in their second bit from the left and are matched together and ticked. 0-01 is then entered into column C. Term 1 can also be matched with term 5 and so 00-1 is entered into column C.

Table 3.3 *Tabular simplification of Boolean equation*

	A	B	C	D	E
1	$\bar{A}\bar{B}\bar{C}D$	0 0 0 1 ✓	1 0 - 0 1 ✓	1 0 - - 1 ✓	- - - 1
2	$\bar{A}B\bar{C}\bar{D}$	0 1 0 0 ✓	2 0 0 - 1 ✓	2 - - 0 1 ✓	- - - 1 → D
3	$AB\bar{C}\bar{D}$	1 1 0 0 ✓	3 - 0 0 1 ✓	3 - 0 - 1 ✓	- - - 1
4	$\bar{A}B\bar{C}D$	0 1 0 1 ✓	4 - 1 0 0 ✓	4 - 1 0 -	→ $B\bar{C}$
5	$\bar{A}\bar{B}CD$	0 0 1 1 ✓	5 0 1 0 - ✓	5 1 1 - -	→ AB
6	$A\bar{B}\bar{C}D$	1 0 0 1 ✓	6 1 1 0 - ✓	6 - 1 - 1 ✓	
7	$AB\bar{C}D$	1 1 0 1 ✓	7 1 1 - 0 ✓	7 - - 1 1 ✓	
8	$ABC\bar{D}$	1 1 1 0 ✓	8 - 1 0 1 ✓	8 1 - - 1 ✓	
			9 0 1 - 1 ✓		
9	$\bar{A}BCD$	0 1 1 1 ✓	10 0 - 1 1 ✓	*Note:* the same	
			11 - 0 1 1 ✓	combination is	
10	$A\bar{B}CD$	1 0 1 1 ✓	12 1 - 0 1 ✓	NOT written	
			13 1 0 - 1 ✓	down twice.	
11	$ABCD$	1 1 1 1 ✓	14 1 1 - 1 ✓		
			15 1 1 1 - ✓		
			16 - 1 1 1 ✓		
			17 1 - 1 1 ✓		

Similarly, other matched terms are 1 and 6, 2 and 3, 2 and 4, 3 and 7, 3 and 8, 4 and 7, 4 and 9, 5 and 10, 6 and 7, 6 and 10, 7 and 11, 8 and 11, 9 and 11, and, finally, 10 and 11.

The matching process is now repeated with the entries in column C. Thus:

terms 1 and 10 give   0--1 entered in column D
terms 2 and 9 give   0--1 entered in column D
      3 and 8 give   --01
      3 and 11 give   -0-1
      4 and 8 give   -10-
      5 and 6 give   -10-
      6 and 15 give   11--
      7 and 14 give   11--
      8 and 16 give   -1-1
      9 and 14 give   -1-1
      10 and 17 give --11
      11 and 16 give --11
      12 and 17 give 1--1
      13 and 14 give 1--1

Two of the column D terms cannot be matched and are therefore *prime implicants*. The terms in column D that can be matched all lead to the same result, i.e. ---1.

Hence the prime implicants of the function are

$$D + B\overline{C} + AB$$

This may be the simplest form possible but a check can be carried out by drawing up a table of the prime implicants. The first step in obtaining the prime implicant table is to write down all the possible combinations for each term. Therefore,

- - BA		- $\overline{C}$B-		D- - -	
0011	3	0010	2	1000	8
0111	7	0011	3	1001	9
1011	11	1010	10	1010	10
1111	15	1011	11	1011	11
				1100	12
				1101	13
				1110	14
				1111	15

Hence, the prime implicant table is given in Table 3.4. The table shows that all three terms must be retained to cover all the required decimal numbers and this means that no further simplification is possible. Hence, $F = AB + B\overline{C} + D$ (*Ans.*)

The accuracy of this result can easily be checked by means of a Karnaugh map since there are only four input variables. It can also be checked using the logic converter in Electronics Workbench. [Enter $BC + B\overline{D} + \overline{C}D + DE$ which gives the simplified result as $\overline{C}D + DE + B$. This, on changing A to LSB and E to MSB, gives the same result as the answer to Example 3.17.]

Table 3.4 *Prime implicant table*

	1	2	3	4	5	6	7	8	9	10	11	12	13	14	15
AB			✓				✓				✓				✓
$B\bar{C}$		✓	✓							✓	✓				
D								✓	✓	✓	✓	✓	✓	✓	✓

Table 3.5

Decimal number	Equation number	A	B	Equation number	C	Equation number	D	E
2	1	$\bar{A}B\bar{C}\bar{D}\bar{E}$	01000	1	01-00	1	01--0	01---✓
16	2	$\bar{A}\bar{B}\bar{C}\bar{D}E$	00001	2	010-0	2	01-0-	
5	3	$A\bar{B}\bar{C}\bar{D}\bar{E}$✓	10100	3	0100-	3	010--	
6	4	$\bar{A}B\bar{C}D\bar{E}$	01100	4	0-001✓	4	0--01✓	
9	5	$A\bar{B}\bar{C}D\bar{E}$✓	10010	5	00-01	5	011--	
10	6	$\bar{A}B\bar{C}D\bar{E}$	01010	6	011-0	6	01-1-	
18	7	$\bar{A}B\bar{C}\bar{D}E$	01001	7	0110-	7	01--1	
20	8	$\bar{A}\bar{B}C\bar{D}E$	00101	8	01-10	8	-111-✓	
14	9	$\bar{A}BCD\bar{E}$	01110	9	0101-			
22	10	$\bar{A}BC\bar{D}E$	01101	10	01-01			
26	11	$\bar{A}B\bar{C}DE$	01011	11	010-1			
15	12	$ABCD\bar{E}$	11110	12	-1110			
29	13	$A\bar{B}CDE$	10111	13	0111-			
30	14	$\bar{A}BCDE$	01111	14	011-1			
31	15	$ABCDE$	11111	15	01-11			
				16	1111-			
				17	1-111✓			
				18	-1111			

EXAMPLE 3.19

Solve Example 3.15 using the Quine–McCluskey method.

Solution

The equation is $F = \bar{A}B\bar{C}\bar{D}\bar{E} + \bar{A}\bar{B}\bar{C}\bar{D}E + A\bar{B}\bar{C}D\bar{E} + \bar{A}BC\bar{D}\bar{E} + A\bar{B}\bar{C}D\bar{E} + \bar{A}B\bar{C}D\bar{E} + \bar{A}B\bar{C}\bar{D}E + \bar{A}\bar{B}C\bar{D}E + \bar{A}BCD\bar{E} + \bar{A}BC\bar{D}E + \bar{A}B\bar{C}DE + ABCD\bar{E} + A\bar{B}CDE + \bar{A}BCDE + ABCDE$.

These terms have been entered into Table 3.5 in order of the number of complemented input variables starting with 1.

Column B shows the same equations written with a 1 for each true variable and a 0 for each complemented variable. The column is split into groups that contain terms having:

(a) Single 1s.
(b) Two 1s.
(c) Three 1s.
(d) Four 1s.
(e) Five 1s.

The terms in column B must now be matched.

Terms 1 and 4: 01-00 is entered into column C; 1 and 6: 010-0 is entered into column C; 1 and 7: 0100- is entered into column C.

2 and 7: 0-001   2 and 8: 00-01
4 and 9: 011-0   4 and 10: 0110-
6 and 9: 01-10   6 and 11: 0101-
7 and 10: 01-01   7 and 11: 010-1
9 and 12: -1110   9 and 14: 0111-
10 and 14: 011-1   11 and 14: 01-11
12 and 15: 1111-   13 and 15: 1-111
14 and 15: -1111

In column C the matched terms are: 1 and 8: 01--0 entered into column D; 1 and 10: 01-0-; 2 and 6: 01--0; 2 and 11: 010--; 3 and 7: 01-0-; 3 and 9: 010--; 5 and 10: 0--01; 6 and 14: 011--; 7 and 13: 011--; 8 and 15: 01-1-; 9 and 13: 01-1-; 10 and 15: 01--1; 11 and 14: 01--1; 12 and 18: -111-; 13 and 16: -111-.

For column D the matched terms are: 1 and 7: 01--- entered into column E; 2 and 6: 01---; 3 and 5: 01---.

The prime implicants are the terms that have a tick ✓, i.e.

$$F = \bar{A}B + A\bar{B}\bar{C}D\bar{E} + \bar{A}DE + A\bar{B}C\bar{D}\bar{E} + BCD + ACDE$$

The possible combinations for $\bar{A}B$--- are 01000   2   For $\bar{A}$--$\bar{D}E$ are

01001	18	00001	16
01010	10	01001	18
01011	26	00101	20
01100	6	01101	22
01101	22		
01110	14		
01111	30		

For $A\bar{B}\bar{C}D\bar{E}$ 9   For   -BCD-   For $A\bar{B}C\bar{D}\bar{E}$ 5   For A-CDE

01110	14	10111	29
11110	15	11111	31
01111	30		
11111	31		

The prime implicants are shown in Table 3.6.

All terms are necessary. Therefore, $F = \bar{A}B + \bar{A}DE + BCD + ACDE + A\bar{B}\bar{C}D\bar{E} + A\bar{B}C\bar{D}\bar{E}$ (Ans.)

Table 3.6

	0	1	2	3	4	5	6	7	8	9	10	11	12	13	14	15
$\overline{A}B$			✓				✓				✓				✓	
$\overline{A}\overline{D}E$																
$ACDE$																
BCD															✓	✓
$A\overline{B}\overline{C}\overline{D}\overline{E}$										✓						
$A\overline{B}C\overline{D}\overline{E}$						✓										

	16	17	18	19	20	21	22	23	24	25	26	27	28	29	30	31
$\overline{A}B$			✓				✓				✓				✓	
$\overline{A}\overline{D}E$	✓		✓		✓		✓									
$ACDE$													✓		✓	✓
BCD															✓	✓
$A\overline{B}\overline{C}\overline{D}\overline{E}$																
$A\overline{B}C\overline{D}\overline{E}$																

Espresso

The use of the Quine–McCluskey method becomes more and more difficult as the number of input variables is increased. *Espresso* is a software package that can be employed to simplify a complex two-level Boolean equation. Suppose that the equation to be simplified is $F = \overline{A}B\overline{C} + AB\overline{C} + A\overline{B} + \overline{A}CD + \overline{A}\overline{B}C + \overline{A}BC$.

The Karnaugh mapping of this equation is

AB CD	00	01	11	10
00	0	1	1	1
01	0	1	1	1
11	1	1	0	1
10	1	1	0	1

The Espresso program starts by expanding all implicants into their maximum size. This is shown in the next map and the resulting equation is $F = \overline{A}C + \overline{A}B + A\overline{C} + A\overline{B}$.

AB CD	00	01	11	10
00	0	1	1	1
01	0	1	1	1
11	1	1	0	1
10	1	1	0	1

Next, the program extracts an irredundant cover to obtain the next mapping. This is (clearly) not the best solution.

AB
$\begin{array}{c|cccc} & 00 & 01 & 11 & 10 \\ \hline 00 & 0 & 1 & 1 & 1 \\ 01 & 0 & 1 & 1 & 1 \\ 11 & 1 & 1 & 0 & 1 \\ 10 & 1 & 1 & 0 & 1 \end{array}$

CD

Espresso then reduces the prime implicants to the smallest size that still covers the Boolean equation being simplified. This step is shown by the map. Now, $F = A\bar{C} + A\bar{B}C' + \bar{A}B\bar{C} + \bar{A}C$.

This result is again expanded to give the penultimate map

AB
$\begin{array}{c|cccc} & 00 & 01 & 11 & 10 \\ \hline 00 & 0 & 1 & 1 & 1 \\ 01 & 0 & 1 & 1 & 1 \\ 11 & 1 & 1 & 0 & 1 \\ 10 & 1 & 1 & 0 & 1 \end{array}$

BC

From this map $F = \bar{A}C + A\bar{C} + B\bar{C} + A\bar{B}$. The irredundant cover is again extracted to give the final map

AB
$\begin{array}{c|cccc} & 00 & 01 & 11 & 10 \\ \hline 00 & 0 & 1 & 1 & 1 \\ 01 & 0 & 1 & 1 & 1 \\ 11 & 1 & 1 & 0 & 1 \\ 10 & 1 & 1 & 0 & 1 \end{array}$

CD

Finally, $F = \bar{A}C + A\bar{B} + B\bar{C}$.

The program will repeat the sequence, reduce irredundant cover and expand until the result obtained is no better than the one previously arrived at. The input data to the Espresso program is entered in the form of the truth table of the required circuit.

Race hazards

A *race hazard* in a combinational logic circuit is an unwanted transient that produces a spike or *glitch* at the output of the circuit. To put it alternatively: a hazard is said to exist if a single input change can result in the occurrence of one or more unwanted momentary outputs. The hazard usually results because of the existence of two or more paths through the circuit which introduce unequal time delays. The hazard may be either *static* or *dynamic*.

Static hazards

A *static hazard* is said to exist in a combinational logic circuit when the change of a single input variable from 0 to 1, or from 1 to 0, causes a glitch at the output when no change should occur.

A hazard may be produced, for example, by the use of an inverter to obtain the complement of an input variable. Referring to Fig. 3.1(a), if the variable A is applied to the inverter to produce $\bar{A}$, the action is not instantaneous. Hence, if A is changed from 0 to 1, some time delay is inevitably introduced in $\bar{A}$ becoming available. This means that for some short time $A + \bar{A}$ is *not* equal to 1 and, consequently, a false output is generated for a short time (see Fig. 3.1(b)).

Consider a circuit which produces an output $F = \bar{A}B + AC$. The use of an inverting stage to obtain $\bar{A}$ means that $\bar{A}$ must always have a small time delay relative to A at the output of the circuit. When $B = C = 1$, then F should be of the form $\bar{A} + A = 1$. Suppose that A changes from 1 to 0. Then $\bar{A}$ takes a short time to change from 0 to 1, and hence for a short while $A + \bar{A}$ will be equal to 0, producing a short, unwanted 0 spike at the output (Fig. 3.2). Similarly, when A changes from 0 to 1, a short period of time will exist during which *both* A and $\bar{A}$ will be a logical 1 but this will not affect the output.

A static hazard exists in a circuit if the input variables can alter to produce a change between adjacent cells in the Karnaugh mapping of the functions that are *not* looped together. Thus, in the mapping for $\bar{A}B + AC$, a hazard exists since the cells marked with an × are not looped together. A hazard can be removed by ensuring that such adjacent cells *are* included within a loop. Clearly this will mean the introduction of one, or perhaps more, redundant terms.

Fig. 3.1 *(a) Circuit producing a race hazard and (b) waveform for (a)*

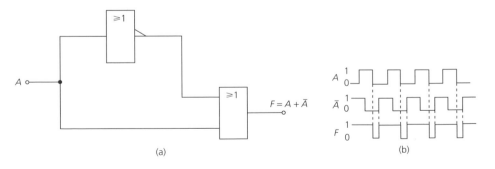

(a)   (b)

Fig. 3.2 *Circuit producing F = AB + AC gives unwanted output when A changes from 1 to 0*

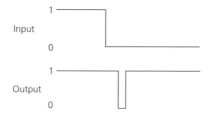

AB\C	00	01	11	10
0	0	1	0	0
1	0	1 ✕	1 ✕	1

If the squares marked ✕ are looped together, the extra term will be BC, so that $F = \bar{A}B + AC + BC$. Now, when $B = C = 1$, the output will be 1 regardless of the instantaneous state of A.

EXAMPLE 3.20

The mapping of a Boolean equation is shown overleaf. Obtain:

(a) The minimal solution.
(b) The hazard-free solution of the equation.

Solution

(a) The minimal solution is shown in Map A and is

$$F = \bar{B}C + A\bar{C}\bar{D} + \bar{A}\bar{C}D \ (Ans.)$$

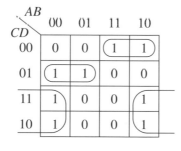   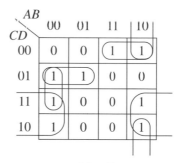

Map A   Map B

(b) The hazard-free solution is obtained from Map B and is

$$F = \bar{B}C + A\bar{C}\bar{D} + \bar{A}\bar{C}D + \bar{A}\bar{B}D + A\bar{B}\bar{D} \ (Ans.)$$

AB\CD	00	01	11	10
00	0	0	1	1
01	1	1	0	0
11	1	0	0	1
10	1	0	0	1

Adding a redundant term to an equation does not always eliminate all the static hazards and sometimes it may be necessary to employ an alternative technique instead, or perhaps, as well. A static hazard may also be overcome by adding extra time delay into a circuit in the appropriate place(s). The insertion into a circuit of a 2-input OR (or AND) gate with its input terminals connected together will add the propagation delay of the OR gate to the non-inverting path, but will not otherwise affect the circuit. Figure 3.3(a) shows a simple circuit in which the inputs to a 2-input OR gate are A and $\bar{A}$, the complemented value being obtained by the use of an inverter. The output of the OR gate should, of course, always be at the logic 1 level, but when the input changes from 1 to 0 the propagation delay of the inverter will ensure that there will be a short time during which the inputs are both at 0. For this time the output of the OR gate will be at 0, Fig. 3.3(b). The static hazard can be eliminated by putting an OR gate in series with the non-inverting input to the OR gate as shown in Fig. 3.3(b).

Fig. 3.3 *(a) Circuit with static hazard and (b) elimination of hazard*

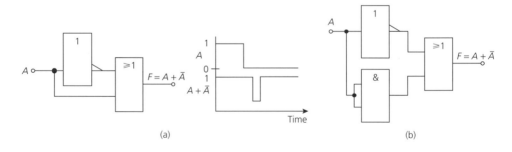

(a)

(b)

Fig. 3.4 *Illustrating a dynamic hazard*

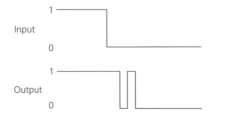

Dynamic hazards

A *dynamic hazard* is said to exist when the output of a circuit changes two or more times when it should have changed once only in response to an input change. An example of a dynamic hazard is shown in Fig. 3.4.

PRACTICAL EXERCISE 3.1

Aim: to investigate the presence of a static hazard and its elimination.
Components and equipment: two 74HC32 quad 2-input OR gate ICs, one 74HC08 2-input AND gate IC, one 74HC04 hex inverter IC, one LED and one 270 Ω resistor. Breadboard. Power supply.
Procedure:

(a) Build the circuit shown in Fig. 3.5.
(b) Apply each combination of binary 1 and 0 voltage levels to the inputs *A*, *B* and *C* and determine the truth table for the circuit.
(c) Connect the pulse generator to the *C* input terminal of the circuit and connect the CRO to the output of the circuit. Set the frequency of the pulse generator to 100 Hz and the CRO timebase to the appropriate setting. Observe the output waveform and determine if a static hazard exists.
(d) Draw the Karnaugh map of the Boolean equation that describes the circuit's operation and decide how the hazard can be eliminated. Modify the circuit to include the extra term that is necessary.
(e) Repeat procedure (c).
(f) Another way of removing the hazard is to increase the time delays of the inputs *A* and *B*. In series with both inputs connect a 2-input AND gate that has both inputs connected together. Repeat procedure (c).

See ade 3.1.ewb file in Electronics Workbench.

Fig. 3.5

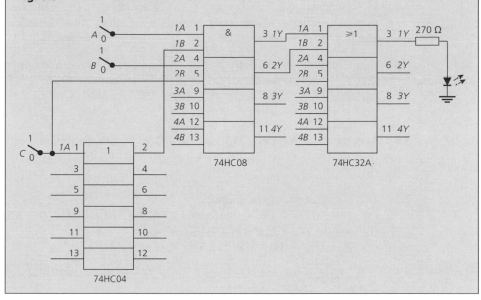

3.1 Use the Quine–McCluskey tabular method to simplify the equation

$$F = \bar{A}\bar{B}CD + A\bar{B}CD + A\bar{B}\bar{C}D + \bar{A}BCD + \bar{A}B\bar{C}D + \bar{A}B\bar{C}\bar{D}$$
$$+ \bar{A}\bar{B}C\bar{D} + A\bar{B}C\bar{D} + A\bar{B}\bar{C}\bar{D}$$

3.2 What is meant by the terms: prime implicant, input variable, minterm and maxterm? Table 3.7 shows a prime implicant chart which has been produced by a Quine–McCluskey reduction of a Boolean equation. Obtain the minimum coverage for the function F.

Table 3.7

	0	1	2	3	4	5	6	7	8	9	10	11	12	13	14	15
$\bar{A}\bar{B}\bar{C}$	✓		✓											✓		
AB			✓								✓					
BCD	✓		✓								✓	✓				

3.3 Simplify, using a Karnaugh map:
(a) $F = \bar{A}\bar{B} + \bar{A}B + A\bar{B}$.
(b) $F = AB + \bar{A}BC + B\bar{C}$.
(c) $F = (\bar{A} + \bar{B})(\bar{A} + B)(A + B)$.
(d) $F = (AB + \bar{B}\bar{C})(\bar{A}B + C)$.

3.4 Plot the function $F = AC\bar{D} + \bar{B}C\bar{D} + \bar{B}CD + \bar{A}B\bar{C} + \bar{A}CD + ABCD$ on a Karnaugh map and indicate where static hazards exist. State how the hazards could be eliminated.

3.5 Simplify, both algebraically and by mapping, each of the following:
(a) $F = AB + \bar{A}C + BC$.
(b) $F = \bar{A}B\bar{D} + \bar{A}BC + A\bar{B}C\bar{D} + \bar{A}BD$.
(c) $F = \bar{A}B\bar{C}D + A\bar{B}CD + \bar{A}\bar{B}C\bar{D} + \bar{A}C$.

3.6 Map the expression $F = \Sigma(3, 8, 12, 14, 15)$. Obtain the minimal expressions for:
(a) F.
(b) $\bar{F}$.

3.7 Use a Karnaugh map to simplify:
(a) $F = \bar{A}\bar{B}CD + A\bar{B}C\bar{D} + \bar{A}B\bar{C}\bar{D} + AB\bar{C}D + \bar{A}\bar{B}\bar{C}D + A\bar{B}\bar{C}D + ABCD$.
(b) $F = \bar{B}CD + C\bar{D} + \bar{A}BC\bar{D} + \bar{A}\bar{B}C$.
(c) $F = A\bar{B}\bar{C} + A\bar{D} + \bar{A}\bar{B}C\bar{D} + \bar{C}D + \bar{A}C\bar{D} + \bar{A}BC\bar{D}$.

3.8 Use a Karnaugh map to simplify $F = \bar{A}\bar{B}CD + BCDE + A\bar{B}\bar{C}\bar{D}\bar{E} + CD\bar{E} + \bar{A}\bar{B}\bar{C}\bar{D}E + A\bar{B}CD + A\bar{B}C\bar{E} + \bar{A}\bar{B}C\bar{D}E$.

3.9 Solve exercise 3.8 using the Quine–McCluskey tabulation method.

3.10 Use a Karnaugh map to simplify the Boolean POS equation $F = (\bar{A} + B + C + D)(A + B + C + D)(\bar{A} + \bar{B} + \bar{C} + D)(A + B + \bar{C} + D)(\bar{A} + B + \bar{C} + D)(A + \bar{B} + C + D)(\bar{A} + \bar{B} + \bar{C} + D)(A + \bar{B} + \bar{C} + D)$. Give the result in both SOP and POS form.

3.11 Simplify the equation $F = \prod(0, 1, 2, 3, 5, 6, 7, 8, 9)$ using:
(a) Boolean algebra.
(b) A Karnaugh map.

3.12 For the equation $F = AB + BCD + \bar{B}\bar{C}\bar{D}$:
(a) Obtain the complement.
(b) Obtain the POS equivalent equation.
(c) Multiply out the POS equation to return to the original equation.

3.13 Use the Quine–McCluskey method to solve $F = \Sigma(0, 2, 3, 5, 7, 8, 10, 13, 15)$.

3.14 Use a Karnaugh map to simplify $F = A\bar{B}D + BD + \bar{A}CD$. Give the result in both the SOP and POS forms.

3.15 The 2-bit binary numbers BA and DC are applied to a circuit that has three outputs F, G and H. One of the outputs is to go HIGH to indicate:
(a) $BA < CD$.
(b) $BA > CD$.
(c) $BA = CD$.
Design the circuit.

3.16 The output of a circuit with inputs A, B, C and D is required to:
(a) Go LOW when $A = B = 1$.
(b) Go HIGH when $A = 1$ and $B = 0$.
(c) Be equal to C when $A = B = 0$.
(d) Be equal to D when $A = 0$ and $B = 1$.
Design the circuit.

3.17 Two 2-bit numbers BA and CD are multiplied together to produce a 4-bit number. Design the circuit.

3.18 Map the function $F = (A + \bar{B} + C + \bar{D})(\bar{A} + B + \bar{C} + D)(\bar{A} + \bar{B} + \bar{C} + D)(A + B + \bar{C} + \bar{D})(A + B + C + \bar{D})(\bar{A} + \bar{B} + \bar{C} + \bar{D})(A + \bar{B} + \bar{C} + \bar{D})(\bar{A} + B + \bar{C} + \bar{D})$. Hence, obtain the minimal:
(a) SOP expression.
(b) POS expression.

3.19 (a) Write the function $F = A\bar{B}\bar{C}\bar{D} + \bar{A}\bar{B}C\bar{D} + \bar{A}BC\bar{D} + ABC\bar{D}$ in decimal form and hence determine the POS form of the expression.
(b) Find $\bar{F}$ by looping the 0 cells.

3.20 Show that a glitch (static hazard) occurs in the circuit shown in Fig. 3.6 when $A = 0$, $B = 1$ and C changes from 0 to 1, or C changes from 1 to 0. Draw the output waveform.

Fig. 3.6

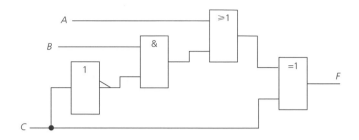

3.21 Table 3.8 shows a prime implicant table. Obtain the minimum coverage for F.

Table 3.8																
	0	1	2	3	4	5	6	7	8	9	10	11	12	13	14	15
$\bar{A}B$		✓					✓				✓				✓	
BD											✓	✓			✓	✓
$A\bar{B}\bar{C}$		✓								✓						
$A\bar{B}\bar{D}$		✓				✓										
$\bar{A}\bar{C}\bar{D}$	✓		✓													
$A\bar{C}D$									✓		✓					
$\bar{B}\bar{C}\bar{D}$	✓	✓														

3.22 Two inputs A and B are applied to the inputs $1A$ and $1B$ of a 74HC08 quad 2-input AND gate as shown in Fig. 3.7. Input A always arrives 5 ns before input B. Draw the output waveform of the gate if:

Fig. 3.7

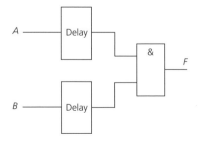

(a) Delay A is 10 ns and delay B is 20 ns.
(b) Delay A is 20 ns and delay B is 10 ns.

3.23 Design a logic circuit that will change the input denary numbers given in Table 3.9 into the given output denary numbers. Both input and output numbers are represented in binary code.

Table 3.9

Input denary numbers	0	1	2	3	4	5	6	7
Output denary numbers	0	6	4	7	0	3	7	7

3.24 Design a circuit to give the product of two binary numbers A and B where A is in the range 0–3 and B is in the range 0–5.

3.25 Distinguish between weighted and unit-distance codes, giving an example of each. Show how a unit-distance code can be generated using a Karnaugh map. Devise a circuit to convert from the unit-distance code shown in Table 3.10 to the first ten states of a binary count. Simplify the circuits as far as possible, assuming that the states shown are the only permissible ones.

Table 3.10

	0	1	2	3	4	5	6	7	8	9
A	0	0	0	0	0	1	1	1	1	1
B	0	1	1	1	0	0	1	1	1	0
C	0	0	1	1	1	1	1	1	0	0
D	1	1	1	0	0	0	0	1	1	1

3.26 Design a circuit that will convert the pure binary number $ABCD$ into the excess-three coded number $WXYZ$.

4 NAND/NOR and multiplexer logic

After reading this chapter you should be able to:

- Use NAND gates only to implement a circuit represented by an SOP Boolean expression.
- Use NOR gates only to implement a circuit represented by a POS Boolean expression.
- State why it is desirable to use either NAND gates or NOR gates only.
- Design a combinational logic circuit from its truth table.
- Use a multiplexer to generate a given logic function.
- Understand the data sheet of a multiplexer.

The implementation of a combination logic circuit usually requires the use of more than one kind of gate and often this results in uneconomic and inefficient use of ICs. If, for example, only one 2-input NAND gate is required it would demand the use of a quad 2-input NAND gate IC and this would leave three of the gates unused. [Note, however, the recent introduction of single gate ICs into the AHC and AHCT families, e.g. the 74AHC1G00 is a single 2-input NAND gate.] To avoid the use of too many partly used ICs a combinational logic circuit can be built using NAND gates only or, alternatively, using only NOR gates. This is possible because both the NAND gate and the NOR gate are able to be connected to perform any of the other logic functions. An alternative, that will often result in fewer ICs being required, is to implement a logic function using either a *multiplexer* or a *programmable logic device* (PLD).

NAND/NOR logic

The AND logical function can be obtained using two NAND gates connected as shown in Fig. 4.1(a); the left-hand gate provides the NAND function and this is inverted by the right-hand gate to give the AND function. Similarly, the OR logical function can be obtained by connecting two NOR gates in the way shown in Fig. 4.1(b). NAND gates can also be used to give the OR logical function and NOR gates to give the AND function; the necessary circuits are not quite as easy to see but they can both be derived using De Morgan's rules.

Fig. 4.1 *Implementation of (a) the AND function using NAND gates and (b) the OR function using NOR gates*

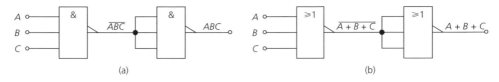

(a) (b)

Fig. 4.2 *Implementation of (a) the AND function using NOR gates and (b) the OR function using NAND gates*

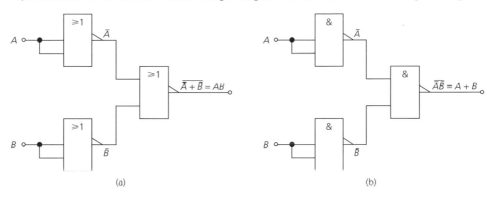

(a) (b)

Fig. 4.3 *Four-input AND function implemented using NOR gates*

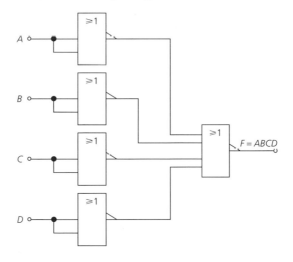

Rule 22 is $\overline{AB} = \overline{A} + \overline{B}$ and hence $AB = \overline{\overline{A} + \overline{B}}$. The right-hand side of this equation is easily implemented using NOR gates as shown in Fig. 4.2(a).

The other De Morgan rule (No. 21) is $\overline{A + B} = \overline{A}\overline{B}$ and hence $A + B = \overline{\overline{A}\overline{B}}$ and this can be implemented by NAND gates connected as shown in Fig. 4.2(b).

This principle can be extended to three, four, or more input variables. For example, the function $F = ABCD = \overline{\overline{A} + \overline{B} + \overline{C} + \overline{D}}$ can be built using the arrangement shown in Fig. 4.3.

Clearly, more gates are needed to implement the AND/OR functions using NAND/NOR gates but very often the apparent increase in the number of gates required is not as great as may be anticipated. This is because consecutive stages of inversion are redundant (since $\bar{\bar{A}} = A$) and need not be provided.

NAND gates

The NAND gate only implementation of a circuit can be obtained by first drawing the AND/OR/NOT circuit that gives the required Boolean equation and then replacing each AND or OR gate by its NAND or OR gate equivalent circuit. This method, although accurate, is rather time consuming and it is better to employ either one of two alternative methods that are available.

OR/AND gate method

To implement a sum-of-products equation using NAND gates only:

(a) Take the final gate(s) at the output as OR.
(b) Take the *even* levels of gate, numbered from the output, as AND.
(c) Take the *odd* levels of gate, numbered from the output, as OR.
(d) Any input variables entering the circuit at an *odd* gate level, numbered from the output, must be inverted.

The rules are illustrated in Fig. 4.4.

Fig. 4.4 *Rules for the implementation of an SOP logic function using NAND gates only*

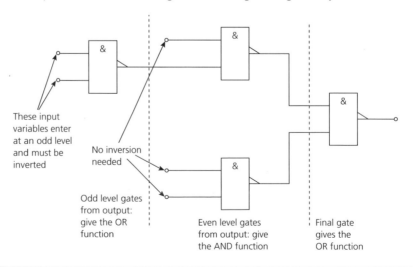

These input variables enter at an odd level and must be inverted

No inversion needed

Odd level gates from output: give the OR function

Even level gates from output: give the AND function

Final gate gives the OR function

EXAMPLE 4.1

Implement, using NAND gates only, the function $F = \bar{A}C + \bar{B}C + A\bar{C}D$.

Solution

Using the OR/AND rule, Fig. 4.5 can be drawn directly. To check:

$$F = \overline{(\overline{\bar{A}C})(\overline{\bar{B}C})(\overline{A\bar{C}D})} = \bar{A}C + \bar{B}C + A\bar{C}D$$

The method can also be employed to determine the function performed by an existing circuit that uses only NAND gates (*Ans.*)

Fig. 4.5

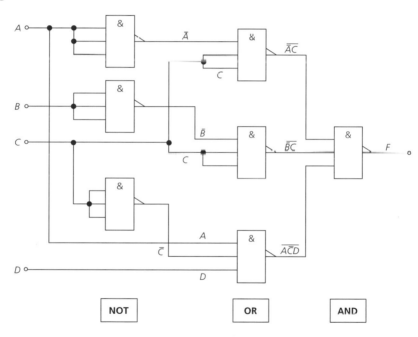

EXAMPLE 4.2

Implement $F = A(BC + D) + \bar{B}C$ using NAND gates only.

Solution

The AND/OR/NOT gate version of the circuit that will implement the given equation is shown in Fig. 4.6(a). The circuit has been divided into four levels moving backwards from the output terminal. Using the OR/AND rules, the NAND gate version of the circuit can be drawn (see Fig. 4.6(b)) (*Ans.*)

Fig. 4.6

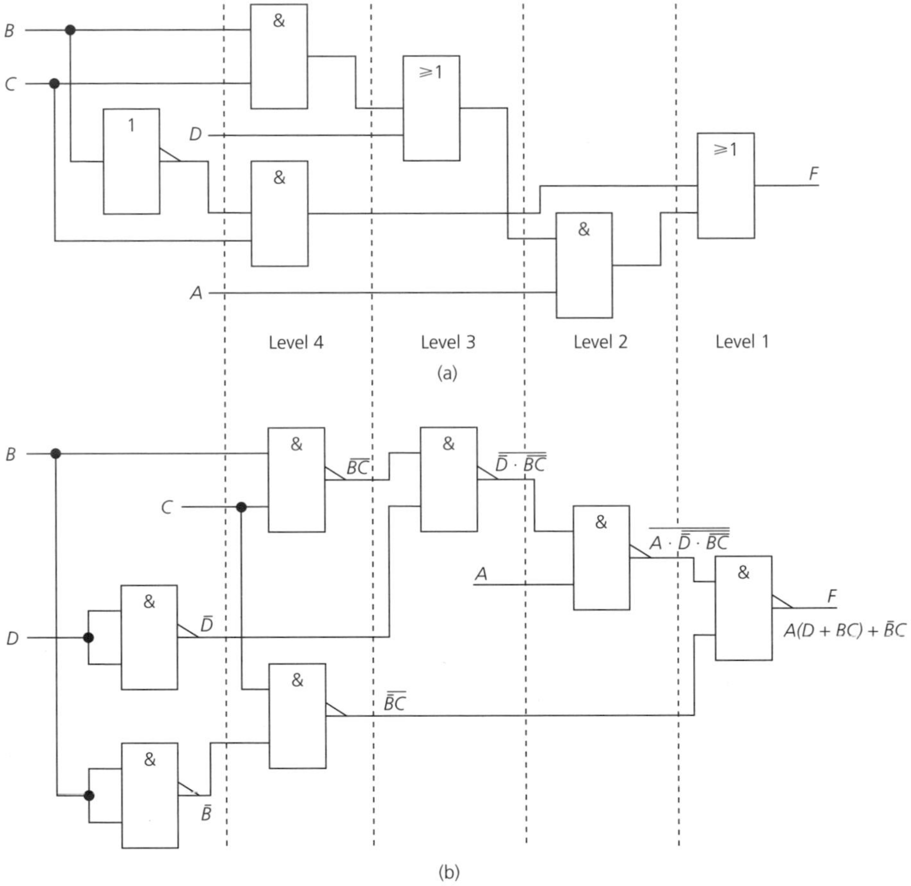

B
C
D
A
&
≥1
1
F
Level 4
Level 3
Level 2
Level 1
(a)
$A(D + BC) + \bar{B}C$
(b)

Double inversion method

This method of obtaining the NAND gate only version of a circuit consists of *doubly* inverting the equation to be implemented. Thus, for the equation $F = \bar{A}C + \bar{B}C + A\bar{C}D$ (again),

$$\bar{F} = \overline{\bar{A}C + \bar{B}C + A\bar{C}D} = (\overline{\bar{A}C})(\overline{\bar{B}C})(\overline{A\bar{C}D})$$

and

$$F = \overline{(\overline{\bar{A}C})(\overline{\bar{B}C})(\overline{A\bar{C}D})}$$

This equation shows immediately that, if the complements of input variables *A*, *B* and *C* are available, four NAND gates are required. The implementation is shown in Fig. 4.5.

Implement the function $F = A\bar{B}CD + AC\bar{D} + \bar{A}\bar{B}C\bar{D} + \bar{A}BD + AB\bar{D}$ using 74CMOS NAND gates only.

Solution

This would be a straightforward application of one of the two rules except that there is no 5-input NAND gate in any of the 74CMOS logic families. Doubly inverting:

$$F = \overline{\overline{A\bar{B}CD + AC\bar{D} + \bar{A}\bar{B}C\bar{D} + \bar{A}BD + AB\bar{D}}}$$
$$= \overline{(A\bar{B}CD + AC\bar{D}) \cdot (\bar{A}\bar{B}C\bar{D}) \cdot (\bar{A}BD) \cdot (AB\bar{D})}$$
$$= \overline{WXYZ} = \bar{W} + \bar{X} + \bar{Y} + \bar{Z}$$

$\bar{X}$, $\bar{Y}$ and $\bar{Z}$ can all be implemented by a single NAND gate. $\bar{W}$ needs to be rearranged.

$$\bar{W} = \overline{A\bar{B}CD + AC\bar{D}} = \overline{A\bar{B}CD} \cdot \overline{AC\bar{D}} = \overline{\overline{A\bar{B}CD} \cdot \overline{AC\bar{D}}}$$

and this equation can be implemented by 4-input NAND gates. The required circuit is given in Fig. 4.7 (*Ans.*)

Fig. 4.7

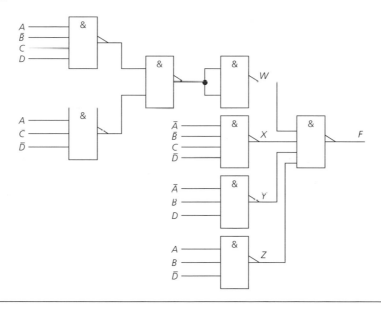

Aims:

(a) To confirm that the circuit shown in Fig. 4.5 implements a given logic function.
(b) To determine how to obtain a 3-input NOR gate using a 2-input NOR gate IC.

Components and equipment: one quad 74HC00 NAND gate IC, one 74HC10 triple 3-input NAND gate IC, one 74HC02 quad 2-input NOR gate IC, one LED and one 270 Ω resistor. Breadboard. Power supply.

Procedure:

(a) Build the circuit shown in Fig. 4.8.
(b) Connect each of the 15 possible combinations of 1 and 0 to the four inputs to the circuit. Each time note the logical state of the LED. Hence write down the truth table of the circuit.
(c) From the truth table determine the Boolean equation that describes the operation of the circuit. Compare it with the equation given in Example 4.1.
(d) If each NAND gate in the circuit is replaced by a NOR gate the dual of the implemented function should be obtained. However, the 74HC logic family does not contain a 3-input NOR gate. Investigate how three 2-input NOR gates may be connected to act as one 3-input NOR gate.
(e) Build the circuit shown in Fig. 4.8 using NOR gates and then determine
  (i) the truth table
  (ii) the Boolean equation for the new circuit.

Fig. 4.8

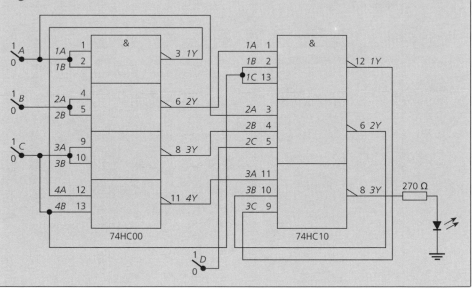

EXAMPLE 4.4

(a) Map the Boolean equation $F = \Sigma(0, 1, 2, 4, 6, 8, 9, 14, 15)$ and minimize it.
(b) State where a static hazard exists and modify the minimized equation to eliminate it.
(c) Implement the circuit using NAND gates only.

Solution

(a) The mapping is

CD \ AB	00	01	11	10
00	1	1	0	1
01	1	0	0	1
11	0	1	1	0
10	1	1	0	0

From the map: $F = \bar{B}\bar{C} + \bar{A}\bar{D} + BCD$ (*Ans.*)

(b) The term BCD is not linked to either of the other two terms and so a static hazard exists. It can be eliminated by adding a term that links BCD to $\bar{A}\bar{D}$, i.e. by adding $\bar{A}BC$ (*Ans.*)

(c) $F = \bar{B}\bar{C} + \bar{A}\bar{D} + \bar{A}BC + BCD$
$\bar{F} = \overline{\bar{B}\bar{C} + \bar{A}\bar{D} + \bar{A}BC + BCD} = \overline{\bar{B}\bar{C}} \cdot \overline{\bar{A}\bar{D}} \cdot \overline{\bar{A}BC} \cdot \overline{BCD}$
$F = \overline{\overline{\bar{B}\bar{C}} \cdot \overline{\bar{A}\bar{D}} \cdot \overline{\bar{A}BC} \cdot \overline{BCD}}$

The required circuit is given in Fig. 4.9. The circuit could be built using three ICs – one 74HC02 hex inverter, one 74HC10 triple 3-input NAND gate and one 74HC20 dual 4-input NAND gate (*Ans.*)

Fig. 4.9

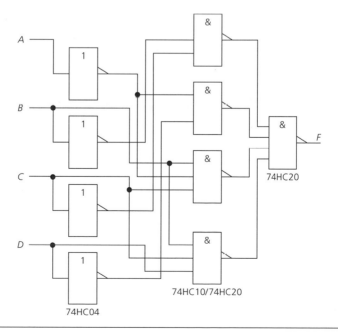

Fig. 4.10 *Rules for the implementation of a POS logic function using NOR gates only*

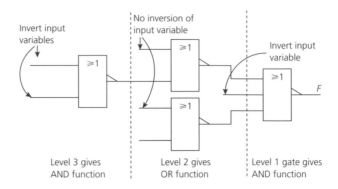

Invert input variables

No inversion of input variable

Invert input variable

Level 3 gives AND function

Level 2 gives OR function

Level 1 gate gives AND function

NOR gates

AND/OR method

To implement a product-of-sums equation using NOR gates only:

(a) Take the final gate at the output as AND.
(b) Take the *even* levels of gate, numbered from the output, as OR.
(c) Take the *odd* levels of gate, numbered from the output, as AND.
(d) Any input variables entering at an *odd* level must be inverted.

The method is illustrated in Fig. 4.10.

EXAMPLE 4.5

Implement, using NOR gates only, the function $F = (\bar{A} + C)(B + \bar{C})(A + D)$.

Solution

Using the AND/OR method gives the circuit shown in Fig. 4.11 (*Ans.*)

Fig. 4.11

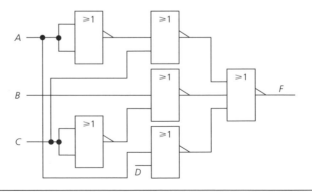

Fig. 4.12

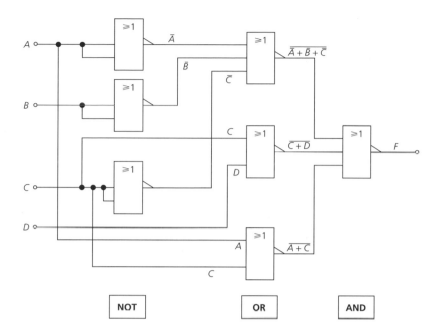

Double inversion method

The equation that is to be implemented using NOR gates only must be put into POS form; then double invert the equation. Thus, for the logic function $F = (\bar{A} + \bar{B} + \bar{C})(C + D)(A + C)$, doubly inverting gives:

$$F = \overline{\overline{(\bar{A} + \bar{B} + \bar{C})} + \overline{(C + D)} + \overline{(A + C)}}$$

The implementation of this equation requires the use of two 2-input NOR gates and two 3-input NOR gates. Two other NOR gates are connected as inverters to allow A and C to be obtained and hence the circuit shown in Fig. 4.12 requires the use of two ICs – one quad 2-input NOR gate and one triple 3-input NOR gate.

EXAMPLE 4.6

Map the function $\Sigma(2, 6, 7, 8, 12, 13)$. Obtain the minimum expression by:

(a) Looping the 1 cells.
(b) Looping the 0 cells.
(c) Draw the circuits that implement (a) and (b) using NOR gates only.
(d) State which circuit is the better and re-draw it using the 74HC02 quad 2-input and 74HC27 triple 3-input NOR gates.

Solution

The mapping is

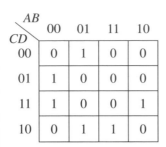

CD \ AB	00	01	11	10
00	0	1	0	0
01	1	0	0	0
11	1	0	0	1
10	0	1	1	0

(a) Looping the 1 cells gives: $F = BC\bar{D} + \bar{A}\bar{B}D + \bar{A}B\bar{D} + \bar{B}CD$ (Ans.)

(b) Looping the 0 cells gives: $\bar{F} = A\bar{C} + BD + \bar{B}\bar{D}$ and, therefore, $F = (\bar{A} + C)(\bar{B} + \bar{D})(B + D)$ (Ans.)

(c) Applying De Morgan's rules to answer (a), $F = (\bar{B} + \bar{C} + D)(A + B + \bar{D})(A + \bar{B} + D)(B + \bar{C} + \bar{D})$. This equation is implemented by the circuit shown in Fig. 4.13(a). Answer (b) can be implemented directly to give the circuit in

Fig. 4.13

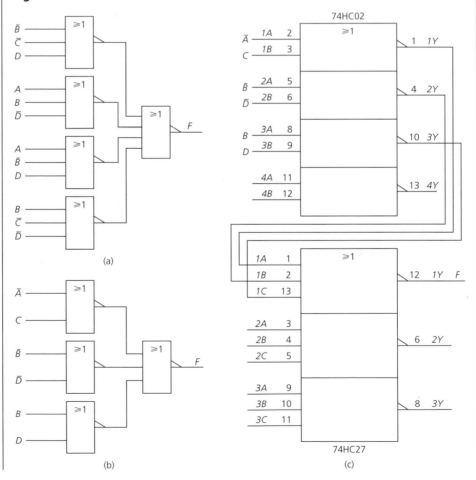

(a)

(b)

(c)

Fig. 4.13(b). Clearly, circuit (b) is the better since it requires the use of only one IC.

(d) Figure 4.13(c) shows the circuit implemented using the 74HS02 quad 2-input NOR gate IC.

Designing a circuit from a truth table

In the design of a combinational logic circuit, the truth table of the required logical operation should be written down and then used to obtain an expression for the output F of the circuit. If the expression is required in the sum-of-products form, each 1 appearing in the output column of the truth table must be represented by a term in the Boolean equation describing the circuit. Every such term must contain each input variable that is in the logical 1 state and the complement of each input variable that is in the logical 0 state. This expression can then be simplified, using one of the methods presented earlier, before it is implemented by the suitable interconnection of a number of gates.

If a product-of-sums expression for the output signal is required, each 0 in the output column of the truth table must be represented by a term in the Boolean equation. This term must contain each input variable that is at 0 and the complement of each variable that is at 1. Consider the truth table of a circuit that is given in Table 4.1.

The sum-of-products expression describing the circuit is

$$F = \bar{A}\bar{B}\bar{C} + \bar{A}\bar{B}C + \bar{A}BC + ABC$$

The product-of-sums expression is

$$F = (A + B + C)(A + \bar{B} + C)(\bar{A} + \bar{B} + C)(\bar{A} + B + \bar{C})$$

[Both expressions reduce to $F = \bar{A}\bar{B} + BC$.]

Table 4.1

A	0	1	0	1	0	1	0	1
B	0	0	1	1	0	0	1	1
C	0	0	0	0	1	1	1	1
F	1	0	0	0	1	0	1	1

EXAMPLE 4.7

Design a circuit that will indicate whether a 4-bit number is either odd and greater than 8 or even and less than 5. Assume decimal 0 to be an even number. Implement the circuit using:

Table 4.2

A	0	1	0	1	0	1	0	1	0	1	0	1	0	1	0	1
B	0	0	1	1	0	0	1	1	0	0	1	1	0	0	1	1
C	0	0	0	0	1	1	1	1	0	0	0	0	1	1	1	1
D	0	0	0	0	0	0	0	0	1	1	1	1	1	1	1	1
F	1	0	1	0	1	0	0	0	0	1	0	1	0	1	0	1

Fig. 4.14

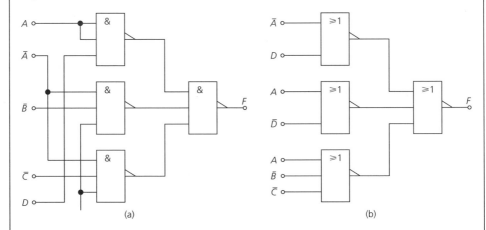

(a)   (b)

(a)   NAND gates only.
(b)   NOR gates only.

Solution

Table 4.2 gives the truth table for the required circuit. From the table,

$$F = \bar{A}\bar{B}\bar{C}\bar{D} + \bar{A}B\bar{C}\bar{D} + \bar{A}\bar{B}C\bar{D} + A\bar{B}\bar{C}D + AB\bar{C}D + A\bar{B}CD + ABCD$$

The mapping is

$\frac{AB}{CD}$	00	01	11	10
00	1	1	0	0
01	0	0	1	1
11	0	0	1	1
10	1	0	0	0

From the map, $F = AD + \bar{A}\bar{B}\bar{D} + \bar{A}\bar{C}\bar{D}$ or $F = (\bar{A} + D)(A + \bar{D})(A + \bar{B} + \bar{C})$. The required circuits are shown in Fig. 4.14(a) and (b) (*Ans.*)

EXAMPLE 4.8

A circuit is required that has ten inputs, numbered 0 through to 9, and one output. The output is to be HIGH whenever any one or more of the inputs numbered 1, 3, 7 or 9 are HIGH. Design the circuit using:

(a) NAND gates only.
(b) NOR gates only.

Solution

The truth table for the circuit is shown in Table 4.3.

From the table, $F = A\bar{B}\bar{C}\bar{D} + AB\bar{C}\bar{D} + ABC\bar{D} + A\bar{B}\bar{C}D$ and its mapping is, with don't cares 10 through to 15,

CD \ AB	00	01	11	10
00	0	0	1	1
01	0	×	×	1
11	×	×	×	×
10	0	0	1	0

From the looped 1 cells, $F = A\bar{C} + AB$ and this is implemented by the circuit shown in Fig. 4.15(a). From the looped 0 cells, $F' = \bar{A} + \bar{B}C$ and hence $F = A(B + \bar{C})$. This equation is implemented by the circuit given in Fig. 4.15(b) (*Ans.*)

Table 4.3

Input	0	1	2	3	4	5	6	7	8	9
A	0	1	0	1	0	1	0	1	0	1
B	0	0	1	1	0	0	1	1	0	0
C	0	0	0	0	1	1	1	1	0	0
D	0	0	0	0	0	0	0	0	1	1
Output F	0	1	0	1	0	0	0	1	0	1

Fig. 4.15

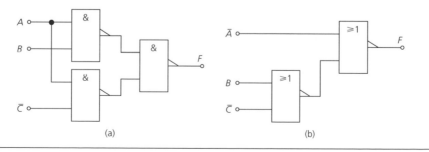

(a)   (b)

Aim: to build the circuits designed in Example 4.8 and confirm their correct operation. Components and equipment: one 74HC00 quad 2-input NAND gate IC, one 74HC02 quad 2-input NOR gate IC, one LED and one 270 Ω resistor. Breadboard. Power supply.

Procedure:

(a) Build the circuit shown in Fig. 4.16(a).

(b) Apply, in turn, the binary equivalents of the decimal numbers 0 through to 9 to the inputs of the circuit. Take A as the least significant bit. Each time note the logical state of the LED and, by comparing with Table 4.3, confirm the correct operation of the circuit.

(c) Build the circuit shown in Fig. 4.16(b) and repeat procedure (b).

Fig. 4.16

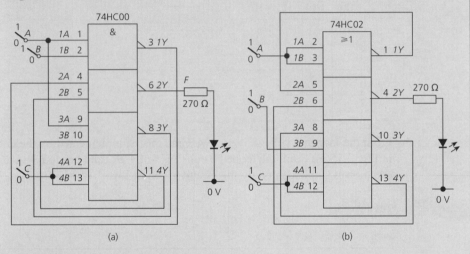

(a)   (b)

Aim: to implement the circuit designed in Example 4.8 using:

(a) NAND gates only.

(b) NOR gates only.

Components and equipment: one 74HC10 triple 3-input NAND gate IC, one 74HC00 quad 2-input NAND gate IC, one 74HC04 hex inverter, one LED and one 270 Ω resistor. Breadboard. Power supply.

Procedure:

(a) Build a circuit using NAND gates only to implement the SOP form of the Boolean equation describing the operation of the circuit.

(b) Apply the input variable combinations given in Table 4.3 and each time note the logical state of the LED. Hence, confirm the correct operation of the circuit.

(c) Build a circuit using NOR gates only to implement the POS form of the Boolean equation for the circuit and then repeat procedure (c).

Logic design using multiplexers

The availability of integrated circuit devices known as multiplexers provides an alternative to the use of gates for the implementation of logic functions. The equation to be implemented must first be expressed in its canonical sum-of-products form.

A multiplexer, or data selector, is a circuit that has several inputs and one output (sometimes an inverted output is also provided). The signal applied to any one of the inputs can be selected to appear at the output terminal. The routing of the desired data input is determined by the binary signal applied to the select inputs of the circuit. Multiplexers are employed for a variety of purposes including: the selection of data, the routing of data, parallel-to-serial conversion, and – of interest in this chapter – the generation of logic functions. A multiplexer can be used instead of several logic gates to implement a logic function. The advantage of using a multiplexer in this way is that a single IC is able to implement a logic function that otherwise might require several gate ICs. [A *programmable array logic* (PAL) device (see p. 275) might provide an even better solution.]

Multiplexers

A number of different multiplexers are available in the various logic families and two of the most commonly employed are the 74151 8-to-1 line multiplexer and the 74153 dual 4-to-1 line multiplexer.

74HC153 dual 4-to-1 line multiplexer

Figure 4.17 gives a part of the data sheet of the 74HC153. The data inputs $1C_0$ through to $1C_3$, and $2C_0$ through to $2C_3$ (denoted as 1C0, 1C3, 2C0, 2C3 etc. in the data sheet) are used to transfer data to the output of the circuit. The select inputs A and B are used to select a particular data input. The address of a data input is placed on the select inputs to select any one of the $2^n = 4$ data inputs. The strobe or enable inputs $1\overline{G}$ and $2\overline{G}$ are used to disable either multiplexer; the pin must be held LOW for the multiplexer to be enabled. When a multiplexer is disabled its output is held LOW and this is shown by the function table. The Boolean equation that describes the operation of the circuit is

$$F = \overline{A}\overline{B}D_0 + A\overline{B}D_1 + \overline{A}BD_2 + ABD_3 \tag{4.1}$$

The logic symbol for the device gives the pin numbers for each of the inputs and outputs. The $1\overline{G}$ input is labelled EN to denote enable and this label also applies to the lower multiplexer. The G symbol in the control box indicates AND dependency and shows that the output(s) depend upon the AND relationship between the select inputs and each of the data inputs 0 through to 3. This means that each data input is ANDed with one particular select word; when this word exists at the select inputs the data input is routed to the output.

The tables for recommended operating conditions, electrical characteristics, and switching characteristics are similar to those discussed earlier in Chapter 1 and have the same meanings.

SN54HC153, SN74HC153
DUAL 4-LINE TO 1-LINE DATA SELECTORS/MULTIPLEXERS

SCLS112B – DECEMBER 1982 – REVISED MAY 1997

- **Permit Multiplexing from n Lines to One Line**
- **Perform Parallel-to-Serial Conversion**
- **Strobe (Enable) Line Provided for Cascading (N Lines to n Lines)**
- **Package Options Include Plastic Small-Outline (D), Thin Shrink Small-Outline (PW), and Ceramic Flat (W) Packages, Ceramic Chip Carriers (FK), and Standard Plastic (N) and Ceramic (J) 300-mil DIPs**

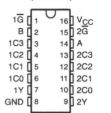

SN54HC153 . . . J OR W PACKAGE
SN74HC153 . . . D, N, OR PW PACKAGE
(TOP VIEW)

description

Each of these data selectors/multiplexers contains inverters and drivers to supply full binary decoding data selection to the AND-OR gates. Separate strobe ($\overline{G}$) inputs are provided for each of the two 4-line sections.

The SN54HC153 is characterized for operation over the full military temperature range of –55°C to 125°C. The SN74HC153 is characterized for operation from –40°C to 85°C.

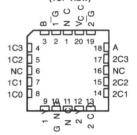

SN54HC153 . . . FK PACKAGE
(TOP VIEW)

NC – No internal connection

FUNCTION TABLE

INPUTS							OUTPUT
SELECT†		DATA				$\overline{G}$	Y
B	A	C0	C1	C2	C3		
X	X	X	X	X	X	H	L
L	L	L	X	X	X	L	L
L	L	H	X	X	X	L	H
L	H	X	L	X	X	L	L
L	H	X	H	X	X	L	H
H	L	X	X	L	X	L	L
H	L	X	X	H	X	L	H
H	H	X	X	X	L	L	L
H	H	X	X	X	H	L	H

† Select inputs A and B are common to both sections.

LOGIC SYMBOL

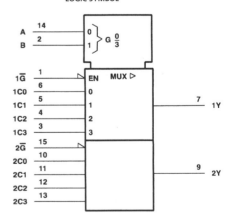

Fig. 4.17 *(Cont'd)*

absolute maximum ratings over operating free-air temperature range[†]

Supply voltage range, V_{CC} .... −0.5 V to 7 V
Input clamp current, I_{IK} ($V_I < 0$ or $V_I > V_{CC}$) (see Note 1) .... ±20 mA
Output clamp current, I_{OK} ($V_O < 0$ or $V_O > V_{CC}$) (see Note 1) .... ±20 mA
Continuous output current, I_O ($V_O = 0$ to V_{CC}) .... ±35 mA
Continuous current through V_{CC} or GND .... ±70 mA
Package thermal impedance, θ_{JA} (see Note 2): D package .... 113°C/W
                                                            N package .... 78°C/W
                                                            PW package .... 149°C/W
Storage temperature range, T_{stg} .... −65°C to 150°C

[†] Stresses beyond those listed under "absolute maximum ratings" may cause permanent damage to the device. These are stress ratings only, and functional operation of the device at these or any other conditions beyond those indicated under "recommended operating conditions" is not implied. Exposure to absolute-maximum-rated conditions for extended periods may affect device reliability.

NOTES: 1. The input and output voltage ratings may be exceeded if the input and output current ratings are observed.
        2. The package thermal impedance is calculated in accordance with JESD 51, except for through-hole packages, which use a trace length of zero.

recommended operating conditions

			SN54HC153			SN74HC153			UNIT
			MIN	NOM	MAX	MIN	NOM	MAX	
V_{CC}	Supply voltage		2	5	6	2	5	6	V
V_{IH}	High-level input voltage	$V_{CC} = 2$ V	1.5			1.5			V
		$V_{CC} = 4.5$ V	3.15			3.15			
		$V_{CC} = 6$ V	4.2			4.2			
V_{IL}	Low-level input voltage	$V_{CC} = 2$ V	0		0.5	0		0.5	V
		$V_{CC} = 4.5$ V	0		1.35	0		1.35	
		$V_{CC} = 6$ V	0		1.8	0		1.8	
V_I	Input voltage		0		V_{CC}	0		V_{CC}	V
V_O	Output voltage		0		V_{CC}	0		V_{CC}	V
t_t	Input transition (rise and fall) time	$V_{CC} = 2$ V	0		1000	0		1000	ns
		$V_{CC} = 4.5$ V	0		500	0		500	
		$V_{CC} = 6$ V	0		400	0		400	
T_A	Operating free-air temperature		−55		125	−40		85	°C

electrical characteristics over recommended operating free-air temperature range (unless otherwise noted)

PARAMETER	TEST CONDITIONS		V_{CC}	$T_A = 25$°C			SN54HC153		SN74HC153		UNIT
				MIN	TYP	MAX	MIN	MAX	MIN	MAX	
V_{OH}	$V_I = V_{IH}$ or V_{IL}	$I_{OH} = -20$ μA	2 V	1.9	1.998		1.9		1.9		V
			4.5 V	4.4	4.499		4.4		4.4		
			6 V	5.9	5.999		5.9		5.9		
		$I_{OH} = -6$ mA	4.5 V	3.98	4.3		3.7		3.84		
		$I_{OH} = -7.8$ mA	6 V	5.48	5.8		5.2		5.34		
V_{OL}	$V_I = V_{IH}$ or V_{IL}	$I_{OL} = 20$ μA	2 V		0.002	0.1		0.1		0.1	V
			4.5 V		0.001	0.1		0.1		0.1	
			6 V		0.001	0.1		0.1		0.1	
		$I_{OL} = 6$ mA	4.5 V		0.17	0.26		0.4		0.33	
		$I_{OL} = 7.8$ mA	6 V		0.15	0.26		0.4		0.33	
I_I	$V_I = V_{CC}$ or 0		6 V		±0.1	±100		±1000		±1000	nA
I_{CC}	$V_I = V_{CC}$ or 0,	$I_O = 0$	6 V			8		160		80	μA
C_i			2 V to 6 V		3	10		10		10	pF

Fig. 4.17 *(Cont'd)*

switching characteristics over recommended operating free-air temperature range, C_L = 50 pF (unless otherwise noted) (see Figure 1)

PARAMETER	FROM (INPUT)	TO (OUTPUT)	Vcc	$T_A = 25°C$			SN54HC153		SN74HC153		UNIT
				MIN	TYP	MAX	MIN	MAX	MIN	MAX	
t_{pd}	A or B	Y	2 V		90	150		225		190	ns
			4.5 V		21	30		45		38	
			6 V		17	26		38		32	
	Data (Any C)	Y	2 V		73	126		189		158	
			4.5 V		17	28		42		35	
			6 V		14	23		35		29	
	$\overline{G}$	Y	2 V		38	95		150		125	
			4.5 V		11	19		28		24	
			6 V		9	16		24		20	
t_t		Y	2 V		20	60		90		75	ns
			4.5 V		8	12		18		15	
			6 V		6	10		15		13	

Fig. 4.18 *Logic symbol for the 74HC151 8-to-1 line multiplexer*

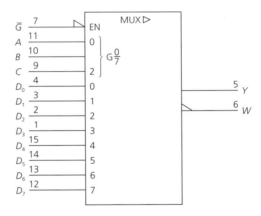

74HC151 8-to-1 line multiplexer

Figure 4.18 shows the logic symbol for the 74HC151 8-to-1 multiplexer and Table 4.4 gives its function table. The data sheet for this device is similar to the 153's data sheet: the figures in the operating conditions and electrical characteristics tables are identical; however, the figures in the switching characteristics table are different.

The Boolean equation describing the operation of an 8-to-1 line multiplexer is

$$F = \bar{A}\bar{B}\bar{C}D_0 + A\bar{B}\bar{C}D_1 + \bar{A}B\bar{C}D_2 + AB\bar{C}D_3 + \bar{A}\bar{B}CD_4$$
$$+ A\bar{B}CD_5 + \bar{A}BCD_6 + ABCD_7 \tag{4.2}$$

Implementing Boolean equations with multiplexers

When the number of variables in the Boolean equation is equal to the number of multiplexer select inputs the multiplexer can implement a function directly from the truth table. The variables are connected to the select inputs and each data input is connected to

Table 4.4 74HC151 function table

Inputs				Outputs	
Select			Strobe		
C	B	A	$\bar{G}$	Y	W
X	X	X	H	L	H
L	L	L	L	D0	$\overline{D0}$
L	L	H	L	D1	$\overline{D1}$
L	H	L	L	D2	$\overline{D2}$
L	H	H	L	D3	$\overline{D3}$
H	L	L	L	D4	$\overline{D4}$
H	L	H	L	D5	$\overline{D5}$
H	H	L	L	D6	$\overline{D6}$
H	H	H	L	D7	$\overline{D7}$

$D_0, D_1 \ldots D_7$ = the level of the respective D input.

either 1 or 0; logic 1 is connected to each data input that corresponds to a combination of input variables for which the truth table gives $F = 1$, and 0 is connected to all other data inputs. Alternatively, the equation to be implemented can be compared term-by-term with the Boolean equation for the multiplexer.

EXAMPLE 4.9

Implement $F = AB\bar{C} + A\bar{B}C + \bar{A}\bar{B}\bar{C} + \bar{A}BC$ on an 8-to-1 multiplexer.

Solution

(a) Comparing with equation (4.2), $D_0 = D_3 = D_4 = D_5 = 1$, and $D_1 - D_2 - D_6 = D_7 = 0$.

(b) The truth table of the function is given in Table 4.5.

Table 4.5

A	0	1	0	1	0	1	0	1
B	0	0	1	1	0	0	1	1
C	0	0	0	0	1	1	1	1
F	1	0	0	1	1	1	0	0

The columns in which $F = 1$ give the data inputs that must be connected to 1. The required connections are shown in Fig. 4.19 (*Ans.*)

Fig. 4.19   **Fig. 4.20**

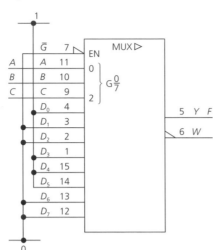

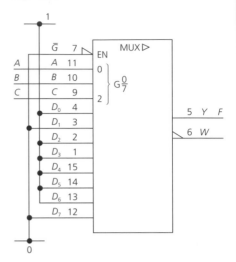

EXAMPLE 4.10

Implement, using an 8-input multiplexer, the Boolean equation $F = \bar{A} + B\bar{C} + \bar{B}C$.

Solution

Writing the equation in its canonical form:

$$F = \bar{A}(B + \bar{B})(C + \bar{C}) + (A + \bar{A})B\bar{C} + (A + \bar{A})\bar{B}C$$
$$= \bar{A}BC + \bar{A}\bar{B}C + \bar{A}B\bar{C} + \bar{A}\bar{B}\bar{C} + AB\bar{C} + A\bar{B}C$$

Comparing with equation (4.2) only the terms $A\bar{B}\bar{C}$ and ABC are missing giving $D_1 = D_7 = 0$. All the other terms are present so $D_0 = D_2 = D_3 = D_4 = D_5 = D_6 = 1$. The circuit is shown implemented by a 74HC151 in Fig. 4.20 (*Ans.*)

Select inputs one less than number of variables

A multiplexer with n select inputs can also be employed to implement a Boolean equation that has $n + 1$ variables. The most significant input variable and its complement are then applied to some of the data inputs. Suppose that the logic function $F(A, B, C, D)$ is to be implemented, where D is the most significant bit. The design procedure is as follows: write down the truth table of the logic function. Each combination of A, B and C must occur for both $D = 0$ and for $D = 1$. If

- $F = 0$ both times an ABC combination occurs, then the logic 0 voltage must be connected to the data input selected by that combination.

- $F = 1$ both times a particular ABC combination occurs, connect the logic 1 voltage to the selected data inputs.
- F is different for the two particular combinations of A, B and C and $F = D$ each time, then connect D to the selected data input.
- F is different for the two combinations of A, B and C and $F = \bar{D}$, then $\bar{D}$ must be connected to the selected data input.

Sometimes it may be necessary to employ a 4-to-1 line multiplexer to implement a 4-variable logic function and it will then be necessary to rearrange the function. Consider the logic function

$$F = AB\bar{C}\bar{D} + \bar{A}\bar{B}C\bar{D} + \bar{A}\bar{B}CD + A\bar{B}\bar{C}D + \bar{A}B\bar{C}D + A\bar{B}CD + \bar{A}BCD + ABCD$$

Rearranging:

$$\begin{aligned}
F &= \bar{C}\bar{D}(AB) + C\bar{D}(\bar{A}\bar{B}) + \bar{C}D(\bar{A}\bar{B} + A\bar{B} + \bar{A}B) + CD(A\bar{B} + \bar{A}B + AB) \\
&= \bar{C}\bar{D}(AB) + C\bar{D}(\bar{A}\bar{B}) + \bar{C}D(\bar{B} + \bar{A}B) + CD(B + A\bar{B}) \\
&= \bar{C}\bar{D}(AB) + C\bar{D}(\bar{A}\bar{B}) + \bar{C}D(\bar{A} + \bar{B}) + CD(A + B)
\end{aligned}$$

$\bar{C}\bar{D}$ selects input D_0; hence, AB is connected to D_0.
$C\bar{D}$ selects input D_1; hence, $\bar{A}\bar{B}$ is connected to D_1.
$\bar{C}D$ selects input D_2; hence, $\bar{A} + \bar{B}$ is connected to D_2.
CD selects input D_3; hence, $A + B$ is connected to D_3.

The required circuit, shown in Fig. 4.21, can be implemented using one quad 2-input NAND gate, one hex inverter and, of course, a 4-to-1 multiplexer.

Fig. 4.21

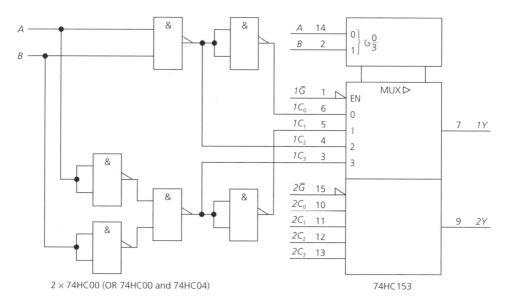

2 × 74HC00 (OR 74HC00 and 74HC04)   74HC153

EXAMPLE 4.11

Implement the function $F = AB\bar{C}D + A\bar{B}C\bar{D} + \bar{A}\bar{B}\bar{C}\bar{D} + AB\bar{C} + \bar{A}BC$ on an 8-to-1 line multiplexer.

Solution

The truth table is given in Table 4.6.

Table 4.6

A	0	1	0	1	0	1	0	1	0	1	0	1	0	1	0	1
B	0	0	1	1	0	0	1	1	0	0	1	1	0	0	1	1
C	0	0	0	0	1	1	1	1	0	0	0	0	1	1	1	1
D	0	0	0	0	0	0	0	0	1	1	1	1	1	1	1	1
F	1	0	0	0	0	1	0	0	0	1	0	1	0	0	1	0
	D_0		D_1		D_2		D_3		D_4		D_5		D_6		D_7	

When $ABC = 000$, $F = 1$ when $D = 0$ and $F = 0$ when $D = 1$. Hence, $F = \bar{D}$. Connect D_0 to $\bar{D}$.
When $ABC = 100$, $F = 0$ when $D = 0$ and $F = 1$ when $D = 1$. Hence, $F = D$. Connect D_1 to D.
When $ABC = 010$, $F = 0$ when $D = 0$ and $F = 0$ when $D = 1$. Hence, $F = 0$. Connect D_2 to 0.
When $ABC = 110$, $F = 0$ when $D = 0$ and $F = 1$ when $D = 1$. Hence, $F = D$. Connect D_3 to D.
When $ABC = 001$, $F = 0$ when $D = 0$ and $F = 0$ when $D = 1$. Hence, $F = 0$. Connect D_4 to 0.
When $ABC = 101$, $F = 1$ when $D = 0$ and $F = 0$ when $D = 1$. Hence, $F = \bar{D}$. Connect D_5 to $\bar{D}$.
When $ABC = 011$, $F = 0$ when $D = 0$ and $F = 1$ when $D = 1$. Hence, $F = D$. Connect D_6 to D.
When $ABC = 111$, $F = 0$ when $D = 0$ and $F = 0$ when $D = 1$. Hence, $F = 0$. Connect D_7 to 0 (*Ans.*)

PRACTICAL EXERCISE 4.4

Aim: to show how a multiplexer may be used to generate a logic function when:

(a)   The number of variables is equal to the number of select inputs.
(b)   The number of variables is one more than the number of select inputs.

Components and equipment: one 74HC153 dual 4-to-1 line multiplexer, one LED and one 270 Ω resistor. Breadboard. Power supply.

EXERCISES

4.1 What is meant by duality in Boolean algebra? Find the complements of the equations:
(a) $F = C(AB + \bar{A}\bar{B})$
(b) $F = AC(A + B)(\bar{A} + \bar{B})$
by
     (i) the use of duality
     (ii) using a Karnaugh map.
Implement the complements using either NAND or NOR gates only.

4.2 What are race hazards in a logic circuit? Plot the function $F = ABCD + \bar{A}\bar{C}\bar{D} + A\bar{B}D + A\bar{C}\bar{D} + \bar{A}C\bar{D}$ on a Karnaugh map. Thence obtain:
(a) The minimal solution.
(b) The simplest race-free solution.
Implement both (a) and (b) using either NAND or NOR gates only.

4.3 The function $F = C(A + \bar{A}B + \bar{A}\bar{B}D)$ is to be implemented using:
(a) NAND gates.
(b) NOR gates only.
Simplify the equation and put it into suitable form for implementation.

4.4 The function $F = ABC + \bar{B}C + \bar{A}\bar{C}D + A\bar{B}C + \bar{A}BCD$ is to be implemented:
(a) Simplify the equation using a Karnaugh map and implement using the 74HC00 only.
(b) Obtain $\bar{F}$ from the map and invert to get F.
Implement the result using the 74HC02 and one 74HC27.

4.5 Implement the function $F = (\bar{A} + B + \bar{C})(AB + \bar{A}\bar{B} + AC) + ABC$ using:
(a) NAND gates only.
(b) NOR gates only.

4.6 A circuit is required, using the minimum number of NAND gates only, that has two outputs F_1 and F_2, when

$$F_1 = \bar{B}C\bar{D} + \bar{A}\bar{B}CD + \bar{A}\bar{B}\bar{C}D$$
$$F_2 = A\bar{B}C\bar{D} + \bar{A}BD + \bar{A}BC\bar{D}$$

Design the circuit. Assume the complements of the input variables are available.

A	0	0	0	0	1	1	1	1
B	0	0	1	1	0	0	1	1
C	0	1	0	1	0	1	0	1
F	1	1	1	0	0	0	0	1

4.7 (a) Map the function $\Sigma(1, 3, 5, 7, 9, 10, 11)$ with don't care conditions 2 and 8. Obtain two simplified Boolean equations, one suitable for NAND gate, the other for NOR gate implementation.

(b) Obtain the Boolean equations describing the circuit whose truth table is given in Table 4.7. Simplify the equation and then implement it using
(i) NAND gates only
(ii) NOR gates only.

4.8 Implement the function $F = ACD + \bar{A}BCD + \bar{B}\bar{C}D + \bar{A}BC\bar{D} + \bar{A}BCD + A\bar{B}C\bar{D}$ using:
(a) A 4-input multiplexer.
(b) An 8-input multiplexer.

4.9 Implement the function $F = A\bar{B}C + \bar{A}B\bar{C}$ using NOR gates only.

4.10 Implement the function $F = A\bar{B}C + \bar{A}B\bar{C} + A\bar{B}\bar{C} + \bar{A}\bar{B}C$ using a 4-input multiplexer.

4.11 Figure 4.16 shows the data sheet for the 74HC153 dual 4-to-1 line multiplexer. Determine:
(a) The function of the device.
(b) The purpose of each pin.
(c) The minimum values for
(i) the logic 1 input voltage
(ii) the logic 1 output voltage.
(d) The maximum values for
(i) the logic 0 input voltage
(ii) the logic 0 output voltage.
(e) The propagation time from
(i) select input
(ii) data input
(iii) enable input to the output.
Take V_{CC} as 4.5 V.

4.12 A light is to be switched ON and OFF from any one of three different points.
(a) Write down the truth table for the required circuit and derive the SOP equation for it.
(b) Obtain the POS equation for the required circuit.

4.13 (a) Implement the logic function $F = A\bar{B}\bar{C}\bar{D} + \bar{A}B\bar{C}\bar{D} + A\bar{B}C\bar{D} + \bar{A}BC\bar{D} + AB\bar{C}\bar{D} + \bar{A}\bar{B}\bar{C}D + A\bar{B}\bar{C}D + \bar{A}\bar{B}CD + A\bar{B}CD + ABCD$ on a 74HC151 8-to-1 line multiplexer.

(b) Simplify the equation and determine how many NAND gates would be required to implement the circuit.

4.14 Use an 8-to-1 line multiplexer to implement the function $F = AC + B\bar{C} + \bar{A}BC$.

4.15 Figure 4.22 shows the logic symbols of two multiplexers:

(a) For each state what kind of multiplexer it is.

(b) Should the enable input be HIGH or LOW to disable the circuit?

(c) Explain the meanings of the labels G1 and $G\frac{0}{7}$.

(d) Is/are the output(s) active-HIGH or LOW?

(e) Figure 4.23 gives the logic symbol for a demultiplexer. Explain the meanings of the labels in the box marked EN.

Fig. 4.22

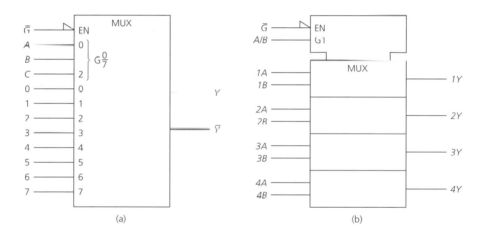

(a) (b)

Fig. 4.23

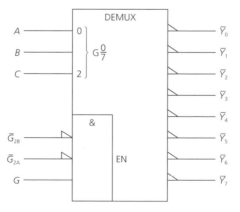

4.16 Implement the logic function $F = A\bar{B}\bar{C} + AB\bar{C} + A\bar{B}C + \bar{A}BC$ on an 8-to-1 multiplexer.

4.17 Simplify the equation $F = ABC + \bar{B}\bar{C} + ACD + \bar{A}B\bar{C} + AB\bar{C}D$ and put it into suitable form for implementation by:

(a) NAND gates only.

(b) NOR gates only.

Draw the circuit using

    (i) the 74HC00

    (ii) the 74HC02.

5 MSI devices

After reading this chapter you should be able to:

- Understand the advantages to be gained by implementing a combinational logic circuit using either NAND gates only, or NOR gates only.
- Implement the OF function using NAND gates only.
- Implement the AND function using NOR gates only.
- Understand the rules for implementing an SOP equation using NAND gates only.
- Understand the double-inversion method of obtaining the NAND gate version of a combination logic circuit.
- Use NAND gates only to implement a given combinational logic circuit.
- Understand the rules for implementing a POS equation using NOR gates only.
- Understand the double-inversion method of obtaining the NOR gate version of a combinational logic circuit.
- Use NOR gates only to implement a given combinational logic circuit.
- Design a combinational logic circuit from a truth table.
- Use a multiplexer to implement Boolean equations.
- Understand the use of multi-level logic.

A wide variety of digital circuits can be obtained in medium-scale integration (MSI), large-scale integration (LSI), and very large-scale integration (VSLI) packages. The vast majority of digital circuit applications can be carried out using one or more of these integrated circuits (ICs), but small-scale integration (SSI) devices are no longer employed for other than 'glue' logic, interfacing, and delay purposes. MSI circuits include counters, decoders, encoders, and multiplexers. LSI devices include memories, display drivers and keyboard encoders, while VSLI circuits include large-capacity memories, microprocessors, floppy disc controllers and various peripheral devices such as the asynchronous communications interface adaptor (ACIA).

Most MSI devices are available in all the TTL logic families and often in the 74HC logic family also, but the other 74CMOS logic families offer far fewer devices. Table 5.1 gives three examples.

Table 5.1

Data selector/multiplexer	AS	ALS	LS	AHC	AHCT	HC	HCT	AC	ACT	LV	LVC
dual 4-to-1 '153	✓	✓	✓			✓					
Decoder/demultiplexer											
3-to-8 '138	✓	✓	✓			✓	✓			✓	✓
Synchronous counter											
4-bit binary '161	✓		✓			✓					

Multiplexers/data selectors

A *multiplexer*, or *data selector*, is a circuit that is able to accept data from any one of a number of sources and then output the data on to a single line. The circuit may be employed to multiplex data, to select data, or to generate logic functions. A multiplexer has both data and select input terminals and a single output terminal. The data that is to be routed to the single output terminal is applied to the data input terminals, each of which has its own unique address. For example, data input 2 has the address 0010 in a 4-to-1 line multiplexer. The digital word applied to the input select terminals addresses the multiplexer to determine which of the data inputs is switched to the common output. The operation of a multiplexer is detailed by its function table which is given in its data sheet (see Fig. 4.17 for the 74HC153). Three multiplexers in the 74HC logic family are the 74HC151 8-to-1 data selector/multiplexer, the 74HC153 dual 4-to-1 data selector/ multiplexer and the 74HC157 quad 2-to-1 data selector/multiplexer.

The 74HC153 has two select inputs A and B, and four data inputs C_0 through to C_3. When the strobe input $\overline{G}$ is HIGH the circuit is disabled and the output Y is HIGH. When $\overline{G}$ is LOW the circuit is enabled and the output Y takes up the same logical state as the selected data input.

When a 74HC151 is used as a data selector it is required to transmit the data applied to any one of its eight inputs to the single output terminal. If, say, only four data inputs are to be used then only two select inputs will need to be active. Select input C can then be connected to $\overline{G}$ and thence to earth. The output can be taken from either the HIGH Y or the LOW $\overline{W}$ output terminal.

A multiplexer may be used to generate logic functions and this topic has been covered in the previous chapter. A typical multiplexer application is shown in Fig. 5.1. Multi-digit LED displays are usually multiplexed in order to reduce the cost of components and to reduce power dissipation. The bus between the register and the decoder/ driver, and between the decoder/driver and the seven-segment display, are common conductors. When two rectangular waveforms at frequencies f and $2f$ are applied to the inputs of the 2-to-4 line multiplexer the inputs follow the cycle 00, 01, 10, 11, 00 and so on. The multiplexer outputs therefore go LOW in sequence. A LOW at the input to a register enables both that register and the associated seven-segment display. The data stored in a register is then passed to its decoder/driver and the decoded signal drives the

Fig. 5.1 *Multiplexed seven-segment display*

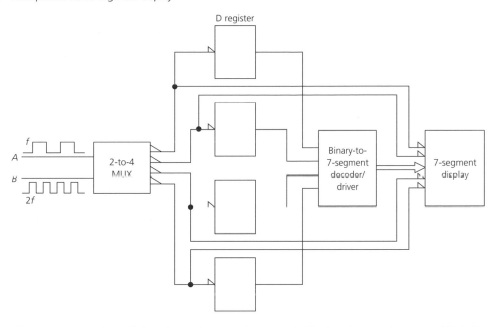

display. The action of the circuit is to activate each display in turn but, provided the frequency f is at least 1 kHz, the visible display will appear to be quite steady.

Decoders/demultiplexers

The function of a demultiplexer is to route input data to any one of several output terminals. This operation is the opposite to that of a multiplexer. The basic function of a decoder is to detect the presence of a specific digital word at its input terminals and to indicate the presence of that word at the output terminal(s). A decoder/driver has n input lines to accept n bits and between 1 and 2^n output lines that indicate the presence of one, or more, n-bit digital words at the input.

Decoders have several applications: they may be employed to indicate when a specified input digital word has occurred, be employed to enable another circuit, or used as an address decoder in a microprocessor system. The operation of a decoder is similar to that of a demultiplexer in that only one output is active at a time. A demultiplexer routes input data to the active output but a decoder does not; it merely indicates the presence at the input terminals of a particular codeword. MSI devices are decoder/demultiplexers and can be used for either purpose.

Decoders/demultiplexers in the 74CMOS logic families include the 74HC138 3-to-8 line decoder and the 74HC139 dual 2-to-4 line decoder. Figure 5.2 shows the part of the data sheet for the 74HC138 3-to-8 line decoder/demultiplexer.

The operation of the circuit is summarized by its function table. When the IC is used as a decoder its inputs, G_1, $\bar{G}_{2A}$ and $\bar{G}_{2B}$, function as enable inputs, but when the device is used as a demultiplexer the input data is applied to any one of these three inputs. G_1

Fig. 5.2 *Data sheet for the 74HC138 3-to-8 line decoder. (Courtesy of Texas Instruments)*

SCLS107B – DECEMBER 1982 – REVISED JULY 1996

- Designed Specifically for High-Speed Memory Decoders and Data Transmission Systems
- Incorporate Three Enable Inputs to Simplify Cascading and/or Data Reception
- Package Options Include Plastic Small-Outline (D), Thin Shrink Small-Outline (PW), and Ceramic Flat (W) Packages, Ceramic Chip Carriers (FK), and Standard Plastic (N) and Ceramic (J) 300-mil DIPs

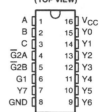

SN54HC138 . . . J OR W PACKAGE
SN74HC138 . . . D, N, OR PW PACKAGE
(TOP VIEW)

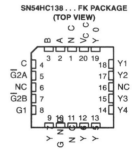

SN54HC138 . . . FK PACKAGE
(TOP VIEW)

NC – No internal connection

description

The 'HC138 are designed to be used in high-performance memory-decoding or data-routing applications requiring very short propagation delay times. In high-performance memory systems, these decoders can be used to minimize the effects of system decoding. When employed with high-speed memories utilizing a fast enable circuit, the delay times of these decoders and the enable time of the memory are usually less than the typical access time of the memory. This means that the effective system delay introduced by the decoders is negligible.

The conditions at the binary-select inputs at the three enable inputs select one of eight output lines. Two active-low and one active-high enable inputs reduce the need for external gates or inverters when expanding. A 24-line decoder can be implemented without external inverters and a 32-line decoder requires only one inverter. An enable input can be used as a data input for demultiplexing applications.

The SN54HC138 is characterized for operation over the full military temperature range of –55°C to 125°C. The SN74HC138 is characterized for operation from –40°C to 85°C.

FUNCTION TABLE

INPUTS						OUTPUTS							
ENABLE			SELECT										
G1	$\overline{G2A}$	$\overline{G2B}$	C	B	A	Y0	Y1	Y2	Y3	Y4	Y5	Y6	Y7
X	H	X	X	X	X	H	H	H	H	H	H	H	H
X	X	H	X	X	X	H	H	H	H	H	H	H	H
L	X	X	X	X	X	H	H	H	H	H	H	H	H
H	L	L	L	L	L	L	H	H	H	H	H	H	H
H	L	L	L	L	H	H	L	H	H	H	H	H	H
H	L	L	L	H	L	H	H	L	H	H	H	H	H
H	L	L	L	H	H	H	H	H	L	H	H	H	H
H	L	L	H	L	L	H	H	H	H	L	H	H	H
H	L	L	H	L	H	H	H	H	H	H	L	H	H
H	L	L	H	H	L	H	H	H	H	H	H	L	H
H	L	L	H	H	H	H	H	H	H	H	H	H	L

Fig. 5.2 *(Cont'd)*

logic symbols (alternatives)

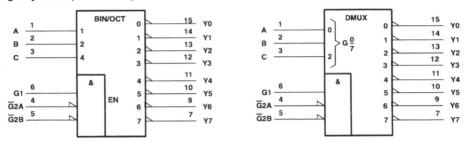

absolute maximum ratings over operating free-air temperature range[†]

Supply voltage range, V_{CC}	−0.5 V to 7 V
Input clamp current, I_{IK} ($V_I < 0$ or $V_I > V_{CC}$) (see Note 1)	±20 mA
Output clamp current, I_{OK} ($V_O < 0$ or $V_O > V_{CC}$) (see Note 1)	±20 mA
Continuous output current, I_O ($V_O = 0$ to V_{CC})	±25 mA
Continuous current through V_{CC} or GND	±50 mA
Maximum power dissipation at $T_A = 55°C$ (in still air) (see Note 2): D package	1.3 W
N package	1.1 W
PW package	0.5 W
Storage temperature range, T_{stg}	−65°C to 150°C

[†] Stresses beyond those listed under "absolute maximum ratings" may cause permanent damage to the device. These are stress ratings only, and functional operation of the device at these or any other conditions beyond those indicated under "recommended operating conditions" is not implied. Exposure to absolute-maximum-rated conditions for extended periods may affect device reliability.

NOTES: 1. The input and output voltage ratings may be exceeded if the input and output current ratings are observed.
2. The maximum package power dissipation is calculated using a junction temperature of 150°C and a board trace length of 750 mils, except for the N package, which has a trace length of zero.

recommended operating conditions

			SN54HC138			SN74HC138			UNIT
			MIN	NOM	MAX	MIN	NOM	MAX	
V_{CC}	Supply voltage		2	5	6	2	5	6	V
V_{IH}	High-level input voltage	$V_{CC} = 2$ V	1.5			1.5			V
		$V_{CC} = 4.5$ V	3.15			3.15			
		$V_{CC} = 6$ V	4.2			4.2			
V_{IL}	Low-level input voltage	$V_{CC} = 2$ V	0		0.5	0		0.5	V
		$V_{CC} = 4.5$ V	0		1.35	0		1.35	
		$V_{CC} = 6$ V	0		1.8	0		1.8	
V_I	Input voltage		0		V_{CC}	0		V_{CC}	V
V_O	Output voltage		0		V_{CC}	0		V_{CC}	V
t_t	Input transition (rise and fall) time	$V_{CC} = 2$ V	0		1000	0		1000	
		$V_{CC} = 4.5$ V	0		500	0		500	ns
		$V_{CC} = 6$ V	0		400	0		400	
T_A	Operating free-air temperature		−55		125	−40		85	°C

Fig. 5.2 *(Cont'd)*

electrical characteristics over recommended operating free-air temperature range (unless otherwise noted)

PARAMETER	TEST CONDITIONS		V_CC	T_A = 25°C			SN54HC138		SN74HC138		UNIT
				MIN	TYP	MAX	MIN	MAX	MIN	MAX	
V_{OH}	$V_I = V_{IH}$ or V_{IL}	$I_{OH} = -20\ \mu A$	2 V	1.9	1.998		1.9		1.9		V
			4.5 V	4.4	4.499		4.4		4.4		
			6 V	5.9	5.999		5.9		5.9		
		$I_{OH} = -4\ mA$	4.5 V	3.98	4.3		3.7		3.84		
		$I_{OH} = -5.2\ mA$	6 V	5.48	5.8		5.2		5.34		
V_{OL}	$V_I = V_{IH}$ or V_{IL}	$I_{OL} = 20\ \mu A$	2 V		0.002	0.1		0.1		0.1	V
			4.5 V		0.001	0.1		0.1		0.1	
			6 V		0.001	0.1		0.1		0.1	
		$I_{OL} = 4\ mA$	4.5 V		0.17	0.26		0.4		0.33	
		$I_{OL} = 5.2\ mA$	6 V		0.15	0.26		0.4		0.33	
I_I	$V_I = V_{CC}$ or 0		6 V		±0.1	±100		±1000		±1000	nA
I_{CC}	$V_I = V_{CC}$ or 0,	$I_O = 0$	6 V			8		160		80	μA
C_i			2 V to 6 V		3	10		10		10	pF

switching characteristics over recommended operating free-air temperature range, C_L = 50 pF (unless otherwise noted) (see Figure 1)

PARAMETER	FROM (INPUT)	TO (OUTPUT)	V_CC	T_A = 25°C			SN54HC138		SN74HC138		UNIT
				MIN	TYP	MAX	MIN	MAX	MIN	MAX	
t_{pd}	A, B, or C	Any Y	2 V		67	180		270		225	ns
			4.5 V		18	36		54		45	
			6 V		15	31		46		38	
	Enable	Any Y	2 V		66	155		235		195	
			4.5 V		18	31		47		39	
			6 V		15	26		40		33	
t_t		Any	2 V		38	75		110		95	ns
			4.5 V		8	15		22		19	
			6 V		6	13		19		16	

operating characteristics, T_A = 25°C

PARAMETER		TEST CONDITIONS	TYP	UNIT
C_{pd}	Power dissipation capacitance	No load	85	pF

is active-HIGH but the other two are active-LOW. The address of the output to which the data is to be directed is applied to the select inputs A, B and C (A = LSB). The Boolean equation that describes the operation of the circuit can be derived from the function table, and is

$$F = \bar{A}\bar{B}\bar{C}Y_0 + A\bar{B}\bar{C}Y_1 + \bar{A}B\bar{C}Y_2 + AB\bar{C}Y_3 + \bar{A}\bar{B}CY_4$$
$$+ A\bar{B}CY_5 + \bar{A}BCY_6 + ABCY_7 \qquad (5.1)$$

When the device is used as a demultiplexer its action is to route a single input data line to any one of the eight output lines Y_0 through to Y_7. Figure 5.3 shows how the 74HC138 is connected for this purpose. The input data is shown to be connected to the $\bar{G}_2$ inputs so G_1 must be held HIGH. The outputs are then active-LOW. If the input data were to be applied to the G_1 input instead then the $\bar{G}_2$ inputs should be connected to earth and the outputs will be active-HIGH.

The number of output lines can be increased by connecting two 138s together as shown in Fig. 5.4. The address lines A, B and C are connected to the A, B and C inputs

Fig. 5.3 *74HC138 generates a Boolean logic function*

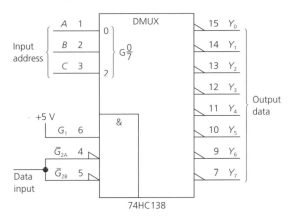

Fig. 5.4 *Connecting two 74HC138 ICs to increase the number of outputs*

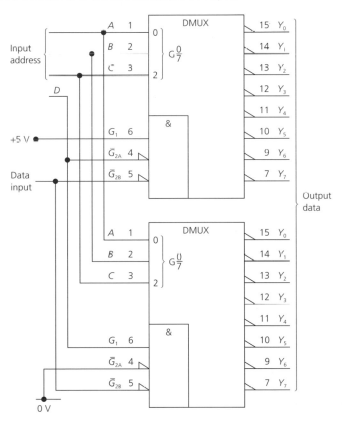

of both ICs. The address line D is connected to the $\overline{G}_{2A}$ input of the low-address IC and to the G_1 input of the high-address IC. The input data line is connected to the $\overline{G}_{2B}$ inputs of both ICs. Lastly, the unconnected G input of both ICs must be connected to its respective active level, i.e. G_1 to +5 V and $\overline{G}_{2A}$ to earth.

When the device is employed as a decoder its action is to detect a particular code and then produce an output to signal the presence of that code. For the IC to operate as a decoder the G_1 input must be held HIGH, and the two $\bar{G}_2$ inputs must be held LOW. The detected code is the address of the output terminal that becomes active. Decoders are often employed as *address decoders* to allow specific chips in a complex system to be addressed.

EXAMPLE 5.1

(a) From the data sheet of the 74HC138 3-to-8 line decoder state
    (i) which output goes LOW when the input word is 011
    (ii) what happens to all the outputs when input $\bar{G}_{2B}$ (labelled 'G2B' on the data sheet) is HIGH
    (iii) the meaning of 'output clamp current' = ±20 mA
    (iv) the propagation delay between any data input and any data output and between an enable input and any output
    (v) how this device differs from the 74HCT138
    (vi) the maximum low-level input voltage.

(b) Explain the meaning of the enable input labels in the logic symbol of the device.

Solution

(a) (i) Y_6 (*Ans.*)
    (ii) all HIGH (*Ans.*)
    (iii) the maximum current that can safely flow into, or out of, an output terminal at a voltage outside the normal operating range (*Ans.*)
    (iv) 54 ns (*Ans.*)
    (v) it has TTL level compatible inputs
    (vi) 0.8 V (*Ans.*)
(b) G_1 must be HIGH and both $\bar{G}_{2A}$ and $\bar{G}_{2B}$ must be LOW to enable the circuit (*Ans.*)

[Note that Y_6, G_1, $\bar{G}_{2A}$ and $\bar{G}_{2B}$ are designated as Y6, G1, G2A and G2B, respectively, on the data sheet.]

EXAMPLE 5.2

A 74HC138 3-to-8 line decoder is to be used to detect the occurrence of any of the codes (0, 3, 5, 7). Draw the required circuit.

Solution

The circuit is shown in Fig. 5.5. The required outputs of the decoder are connected to the inputs of four NOR gates. When an output goes LOW, signalling that a particular code has been detected at the input, the output of a NOR gate goes HIGH. When any of the OR gate's inputs goes HIGH its output also is HIGH (*Ans.*)

Fig. 5.5

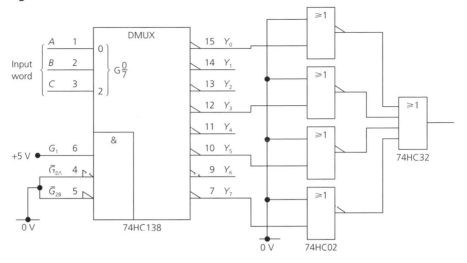

Table 5.2

Inputs							Multiplication				
Denary		Binary					Denary	Binary			
A	B	A_1	A_0	B_1	B_0	Output		P_4	P_3	P_2	P_1
0	0	0	0	0	0	Y_0	0	0	0	0	0
1	0	0	1	0	0	Y_1	0	0	0	0	0
2	0	1	0	0	0	Y_2	0	0	0	0	0
3	0	1	1	0	0	Y_3	0	0	0	0	0
0	1	0	0	0	1	Y_4	0	0	0	0	0
1	1	0	1	0	1	Y_5	1	0	0	0	1
2	1	1	0	0	1	Y_6	2	0	0	1	0
3	1	1	1	0	1	Y_7	3	0	0	1	1
0	2	0	0	1	0	Y_8	0	0	0	0	0
1	2	0	1	1	0	Y_9	2	0	0	1	0
2	2	1	0	1	0	Y_{10}	4	0	1	0	0
3	2	1	1	1	0	Y_{11}	6	0	1	1	0
0	3	0	0	1	1	Y_{12}	0	0	0	0	0
1	3	0	1	1	1	Y_{13}	3	0	0	1	1
2	3	1	0	1	1	Y_{14}	6	0	1	1	0
3	3	1	1	1	1	Y_{15}	9	1	0	0	1

EXAMPLE 5.3

Design a circuit using a 4-to-16 line decoder that will multiply two 2-bit numbers together.

Solution

The truth table of the required operation is given in Table 5.2. From the table,

Fig. 5.6

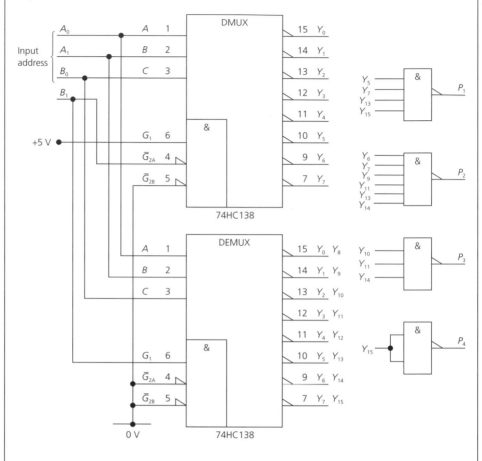

$$P_1 = \bar{5} + \bar{7} + \overline{13} + \overline{15} = \overline{5 \cdot 7 \cdot 13 \cdot 15}$$

$$P_2 = \bar{6} + \bar{7} + \bar{9} + \overline{11} + \overline{13} + \overline{14} = \overline{6 \cdot 7 \cdot 9 \cdot 11 \cdot 13 \cdot 14}$$

$$P_3 = \overline{10} + \overline{11} + \overline{14} = \overline{10 \cdot 11 \cdot 14}$$

$$P_4 = \overline{15}.$$

A 4-to-16 line decoder is not available in any of the logic families and hence two 74HC138 3-to-8 line decoders will have to be combined in the manner shown in Fig. 5.4. The required circuit is shown in Fig. 5.6 (*Ans.*)

One 5-input NAND gate, one 4-input NAND gate, one 3-input AND gate and one inverter are needed, but there are no NAND gates with more than four inputs in the 74HC logic family. A 5-input NAND gate can be obtained using the arrangement shown in Fig. 5.7 and this will allow the circuit to be built using two 4-input gates, one triple 3-input NAND gate, and two 2-input NAND gates.

Fig. 5.7 *Obtaining a 5-input NAND gate*

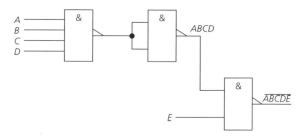

PRACTICAL EXERCISE 5.1

Aim: to build the circuit shown in Fig. 5.7 and confirm its operation.
Components and equipment: one 74HC20 dual 4-input NAND gate IC, one 74HC11 triple 3-input NAND gate IC, one 74HC00 quad 2-input NAND gate IC, one LED and one 270 Ω resistor. Breadboard. Power supply.
Procedure:

(a) Build the circuit shown in Fig. 5.7.
(b) Apply each of the possible binary numbers 00, 01, 10 and 11 to the inputs *AB* and *CD* of the circuit. Each time note the logical state of the LED and hence obtain the truth table for the circuit.
(c) Check the truth table to see if the circuit correctly multiplies the two 2-bit numbers together.

Use of a decoder to implement Boolean equations

A decoder can be employed in conjunction with two or more gates to generate a Boolean function which has more than one output. If the decoder outputs are active-LOW, NAND gates will be necessary. The 74HC138 is only able to generate logic functions with three variables but the 74LS154, which is a 4-to-16 line decoder, can generate four-variable functions.

EXAMPLE 5.4

Use a 74LS138 to implement the functions $F = \Sigma(1, 3)$, $G = \Sigma(0, 7)$, and $H = \Sigma(2, 4)$.

Solution

The outputs Y_0 through to Y_7 of the decoder go LOW when selected by the data word *CBA*. The output of a 2-input NAND gate goes LOW only when both its inputs are HIGH. Hence, the required circuit is shown in Fig. 5.8. Only the gate that has one input LOW will have a HIGH output (*Ans.*)

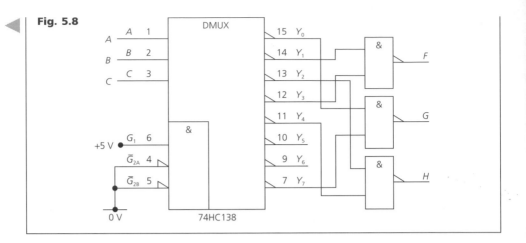

Fig. 5.8

Table 5.3 *74HC42 BCD-to-decimal decoder*

A_0	0	1	0	1	0	1	0	1	0	1
A_1	0	0	1	1	0	0	1	1	0	0
A_2	0	0	0	0	1	1	1	1	0	0
A_3	0	0	0	0	0	0	0	0	1	1
Decimal digit	0	1	2	3	4	5	6	7	8	9

BCD-to-decimal decoder

A BCD-to-decimal decoder converts an input BCD coded binary word into one of ten decimal digits. The circuit is often known as a 4-to-10 line decoder. The 74HC42 4-to-10 BCD-to-decimal decoder has the function table given in Table 5.3. The device cannot be used as a demultiplexer because it does not have an enable input terminal.

EXAMPLE 5.5

A 74HC42 4-to-10 line decoder has the input digital word applied to its terminals A_0, A_1, A_2 and A_3. Determine which output goes LOW.

Solution

From the function table in the data sheet, output 9 (*Ans.*)

Fig. 5.9 *Logic symbol of the 74LS47 BCD-to-seven-segment decoder/driver*

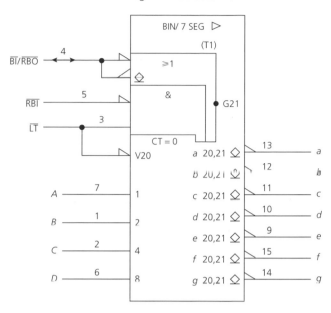

The 74LS47 BCD-to-seven-segment decoder/driver

The 74LS47 BCD-to-seven-segment decoder/driver accepts an input BCD digital word, decodes it and drives a seven-segment display. The logic symbol for such a circuit is given in Fig. 5.9. Most of the labels are self-evident but some are special features of the device. When $\overline{LT}$ is LOW and $\overline{BI/RBO}$ is HIGH all the segments in the seven-segment display should be turned ON. This is used as a test of the correct operation of the display. Inputs $\overline{RBI}$, the ripple blanking input, and $\overline{BI/RBO}$, the blanking input/ripple blanking output, are used to provide leading- and trailing-zero suppression when displays are cascaded to give a multi-digit display.

Comparators

A comparator is a logic circuit that indicates whether two input data words A and B are equal to one another, and, if not, which of them is the larger. The data sheet for the 74HC682 comparator is shown in Fig. 5.10. If $A_3 = 1$ and $B_3 = 0$ then $A > B$; if $A_3 = 0$ and $B_3 = 1$ then $B > A$. If $A_3 = B_3$ then the next lower bit position must be examined and so on. The function table shows that if A (P) = B (Q) the output $P = Q$ is LOW and the output $P > Q$ is HIGH; if $P > Q$ output $P = Q$ is HIGH and output $P > Q$ is LOW. If $P < Q$ both outputs are HIGH.

SN54HC682, SN74HC682
8-BIT MAGNITUDE COMPARATORS

SCLS018B – MARCH 1984 – REVISED JANUARY 1996

- **Compare Two 8-Bit Words**
- **100-kΩ Pullup Resistors Are on the Q Inputs**
- **Package Options Include Plastic Small-Outline (DW) and Ceramic Flat (W) Packages, Ceramic Chip Carriers (FK), and Standard Plastic (N) and Ceramic (J) 300-mil DIPs**

SN54HC682 . . . J OR W PACKAGE
SN74HC682 . . . DW OR N PACKAGE
(TOP VIEW)

```
$\overline{P > Q}$ [ 1    20 ] V_CC
      P0 [ 2    19 ] $\overline{P = Q}$
      Q0 [ 3    18 ] Q7
      P1 [ 4    17 ] P7
      Q1 [ 5    16 ] Q6
      P2 [ 6    15 ] P6
      Q2 [ 7    14 ] Q5
      P3 [ 8    13 ] P5
      Q3 [ 9    12 ] Q4
     GND [ 10   11 ] P4
```

description

These magnetude comparators perform comparisons of two 8-bit binary or BCD words. The 'HC682 feature 100-kΩ pullup termination resistors on the Q inputs for analog or switch data.

The SN54HC682 is characterized for operation over the full military temperature range of –55°C to 125°C. The SN74HC682 is characterized for operation from –40°C to 85°C.

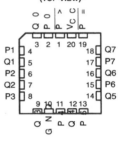

SN54HC682 . . . FK PACKAGE
(TOP VIEW)

```
        P1 [ 4    3 2 1 20 19   18 ] Q7
        Q1 [ 5              17 ] P7
        P2 [ 6              16 ] Q6
        Q2 [ 7              15 ] P6
        P3 [ 8              14 ] Q5
              9 10 11 12 13
```

FUNCTION TABLE

DATA INPUTS	OUTPUTS	
P, Q	$\overline{P = Q}$	$\overline{P > Q}$
P = Q	L	H
P > Q	H	L
P < Q	H	H

NOTE: The P < Q function can be generated by applying $\overline{P = Q}$ and $\overline{P > Q}$ to a 2-input NAND gate.

logic symbol

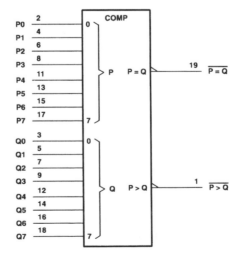

Fig. 5.10 *(Cont'd)*

recommended operating conditions

			SN54HC682			SN74HC682			UNIT
			MIN	NOM	MAX	MIN	NOM	MAX	
V_{CC}	Supply voltage		2	5	6	2	5	6	V
V_{IH}	High-level input voltage	$V_{CC} = 2$ V	1.5			1.5			V
		$V_{CC} = 4.5$ V	3.15			3.15			
		$V_{CC} = 6$ V	4.2			4.2			
V_{IL}	Low-level input voltage	$V_{CC} = 2$ V	0		0.5	0		0.5	V
		$V_{CC} = 4.5$ V	0		1.35	0		1.35	
		$V_{CC} = 6$ V	0		1.8	0		1.8	
V_I	Input voltage		0		V_{CC}	0		V_{CC}	V
V_O	Output voltage		0		V_{CC}	0		V_{CC}	V
t_t	Input transition (rise and fall) time	$V_{CC} = 2$ V	0		1000	0		1000	ns
		$V_{CC} = 4.5$ V	0		500	0		500	
		$V_{CC} = 6$ V	0		400	0		400	
T_A	Operating free-air temperature		-55		125	-40		85	°C

electrical characteristics over recommended operating free-air temperature range (unless otherwise noted)

PARAMETER	TEST CONDITIONS		V_{CC}	$T_A = 25$°C			SN54HC682		SN74HC682		UNIT
				MIN	TYP	MAX	MIN	MAX	MIN	MAX	
V_{OH}	$V_I = V_{IH}$ or V_{IL}	$I_{OH} = 20$ µA	2 V	1.9	1.998		1.9		1.9		V
			4.5 V	4.4	4.400		4.4		4.4		
			6 V	5.9	5.999		5.9		5.9		
		$I_{OH} = -4$ mA	4.5 V	3.98	4.3		3.7		3.84		
		$I_{OH} = -5.2$ mA	6 V	5.48	5.8		5.2		5.34		
V_{OL}	$V_I = V_{IH}$ or V_{IL}	$I_{OL} = 20$ µA	2 V		0.002	0.1		0.1		0.1	V
			4.5 V		0.001	0.1		0.1		0.1	
			6 V		0.001	0.1		0.1		0.1	
		$I_{OL} = 4$ mA	4.5 V		0.17	0.26		0.4		0.33	
		$I_{OL} = 5.2$ mA	6 V		0.15	0.26		0.4		0.33	
I_{IH}	$V_I = V_{CC}$		6 V		0.1	100		1000		1000	nA
I_{IL}	$V_I = 0$	Q inputs	6 V		-50	-90		-160		-140	µA
		All other inputs	6 V		-0.1	-100		-1000		-1000	nA
I_{CC}	$V_I = V_{CC}$ or 0,	$I_O = 0$	6 V		480	700		1300		1100	µA
C_i			2 V to 6 V		3	10		10		10	pF

switching characteristics over recommended operating free-air temperature range, $C_L = 50$ pF (unless otherwise noted) (see Figure 1)

PARAMETER	FROM (INPUT)	TO (OUTPUT)	V_{CC}	$T_A = 25$°C			SN54HC682		SN74HC682		UNIT
				MIN	TYP	MAX	MIN	MAX	MIN	MAX	
t_{pd}	P or Q	Any	2 V		130	275		413		344	ns
			4.5 V		26	55		88		69	
			6 V		22	47		70		58	
t_t		Any	2 V		38	75		110		95	ns
			4.5 V		8	15		22		19	
			6 V		6	13		19		16	

operating characteristics, $T_A = 25$°C

PARAMETER		TEST CONDITIONS	TYP	UNIT
C_{pd}	Power dissipation capacitance	No load	40	pF

EXAMPLE 5.6

Design a circuit using the 74HC682 that will indicate when an input digital word is not a valid BCD number.

Solution

Valid BCD numbers are 0 through to 9. Any number greater than 9 is not valid. Denary 9 = 1001 in binary. If the data word to be checked is connected to the inputs P_0 through to P_3, then the inputs P_4 through to P_7 must be connected to earth. The binary number 1001 must then be set up on the Q input. This means that Q_0 and Q_3 are connected to +5 V and all the other Q inputs are connected to earth (see Fig. 5.11). If the input word is equal to denary 10 or more the $P > Q$ output will go LOW to indicate that an invalid BCD word is present (*Ans.*)

Fig. 5.11

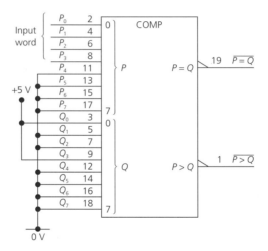

EXAMPLE 5.7

The 74HC682 has the binary numbers

(a) $P = 0111$ and $Q = 1000$.
(b) $P = 1011$ and $Q = 1010$.
(c) $P = 0110$ and $Q = 0110$ applied to its inputs.

Determine the logical states of the outputs in each case.

Solution

(a) $P < Q$, so both terminals $\overline{P = Q}$ and $\overline{P > Q}$ are HIGH (*Ans.*)
(b) $P > Q$ so $\overline{P = Q}$ is HIGH and $\overline{P > Q}$ is LOW (*Ans.*)
(c) $P = Q$ so $\overline{P = Q}$ is LOW and $\overline{P > Q}$ is HIGH (*Ans.*)

Encoders

An encoder performs the opposite function to that performed by a decoder. An 8-to-3 encoder, for example, encodes eight input data lines to three output lines and can be used as an octal-to-binary converter. Only one of its inputs is active at any one time and each of them produces a unique output codeword. A priority encoder always encodes the highest magnitude input and ignores all other inputs whose magnitude is lower. A part of the data sheet of the 74HC148 8-to-3 line priority encoder is shown in Fig. 5.12. The remaining parts give data on absolute maximum ratings, operating conditions, and electrical/switching characteristics similar to a gate data sheet. The operation of the circuit is summarized by its function table and by its logic symbol. Note that the outputs are active-LOW so that, for example, if 5_{10} is encoded the output will be 010. The input E_1 must be held LOW for encoding to occur. GS is the output strobe pin and E_0 is the output enable; it is used to indicate if any of the outputs are active. The top label indicates the function of the device, i.e. convert the active input with the highest priority to its BCD code. The label Z indicates interconnection dependency, indicating that the data inputs on the left are connected to the outputs on the right with the same numbered label. V indicates OR dependency.

Fig. 5.12 *Data sheet for the 74HC148 priority encoder. (Courtesy of Texas Instruments)*

SN54HC148, SN74HC148
8-LINE TO 3-LINE PRIORITY ENCODERS

SCLS109C – MARCH 1984 – REVISED JULY 1996

- **Encode Eight Data Lines to 3-Line Binary (Octal)**
- **Applications Include:**
  - n-Bit Encoding
  - Code Converters and Generators
- **Package Options Include Plastic Small-Outline (D) and Ceramic Flat (W) Packages, Ceramic Chip Carriers (FK), and Standard Plastic (N) and Ceramic (J) 300-mil DIPs**

description

The 'HC148 feature priority decoding of the inputs to ensure that only the highest-order data line is encoded. These devices encode eight data lines to 3-line (4-2-1) binary (octal). Cascading circuitry (enable input EI and enable output EO) has been provided to allow octal expansion without the need for external circuitry. Data inputs and outputs are active at the low logic level.

The SN54HC148 is characterized for operation over the full military temperature range of –55°C to 125°C. The SN74HC148 is characterized for operation from –40°C to 85°C.

SN54HC148 . . . J OR W PACKAGE
SN74HC148 . . . D OR N PACKAGE
(TOP VIEW)

4 [1	16] V$_{CC}$
5 [2	15] EO
6 [3	14] GS
7 [4	13] 3
EI [5	12] 2
A2 [6	11] 1
A1 [7	10] 0
GND [8	9] A0

SN54HC148 . . . FK PACKAGE
(TOP VIEW)

6 [4	18] GS
7 [5	17] 3
NC [6	16] NC
EI [7	15] 2
A2 [8	14] 1

NC – No internal connection

Fig. 5.12 *(Cont'd)*

FUNCTION TABLE

	INPUTS									OUTPUTS				
EI	0	1	2	3	4	5	6	7	A2	A1	A0	GS	EO
H	X	X	X	X	X	X	X	X	H	H	H	H	H
L	H	H	H	H	H	H	H	H	H	H	H	H	L
L	X	X	X	X	X	X	X	L	L	L	L	L	H
L	X	X	X	X	X	X	L	H	L	L	H	L	H
L	X	X	X	X	X	L	H	H	L	H	L	L	H
L	X	X	X	X	L	H	H	H	L	H	H	L	H
L	X	X	X	L	H	H	H	H	H	L	L	L	H
L	X	X	L	H	H	H	H	H	H	L	H	L	H
L	X	L	H	H	H	H	H	H	H	H	L	L	H
L	L	H	H	H	H	H	H	H	H	H	H	L	H

logic symbol

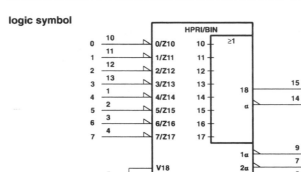

Two 74HC148 8-to-3 line priority encoders are to be connected to form a hexadecimal priority encoder. State how it may be achieved.

Solution

Connect the E_1 input of the lower-order encoder to the E_0 terminal of the higher-order encoder, and leave the E_0 and GS terminals of the lower-order encoder unconnected. Connect the E_1 terminal of the higher-order encoder to a HIGH voltage. The GS pin of the higher-order encoder will give the 2^3 bit of the output number. Connect the A_0 terminal of both encoders to the inputs of a 2-input AND gate; the output of this gate will give the 2^0 bit. Connect the A_1 terminal of both encoders to another AND gate; its output gives the 2^1 bit. Lastly, connect the two A_2 terminals to a third AND gate and the gate output will give the 2^2 bit of the result (*Ans.*)

Fig. 5.13 *Half-adder*

Table 5.4 *Half-adder truth table*				
A	0	1	0	1
B	0	0	1	1
Sum	0	1	1	0
Carry C_{out}	0	0	0	1

Binary arithmetic circuits

Very often in digital circuitry the need arises for the addition, subtraction, multiplication or division of two binary numbers. For all of these arithmetic processes the basic circuit employed is the *binary adder* since it can be modified, with additional logic circuits, to perform the other functions.

Half-adder

The simplest form of binary adder is known as the *half-adder*. A half-adder accepts two binary digits as inputs and produces their sum and perhaps a carry at its output terminals. A half-adder has two inputs A and B but it does not have a 'carry input' from some previous stage. The half-adder can be used to add together the two least significant bits A_0 and B_0 of two numbers where there is no input carry. The block diagram of a half-adder is shown in Fig. 5.13.

The truth table of a half-adder is given in Table 5.4. Two different Boolean equations can be obtained from Table 5.4 that express the sum S and the carry C_{out}. These equations are

$$S = A\bar{B} + \bar{A}B \qquad C_{out} = AB \tag{5.2}$$

$$S = (A + B)(\bar{A} + \bar{B}) \qquad C_{out} = AB \tag{5.3}$$

Full-adder

The full-adder is a circuit which adds two input bits A and B together with a possible carry input bit from a previous stage. The block diagram of a full-adder is shown in Fig. 5.14, and its truth table is given in Table 5.5. From it, Boolean expressions for the sum S and the output carry C_{out} can be obtained:

Fig. 5.14 *Full-adder*

Table 5.5 *Full-adder truth table*

A	0	1	0	1	0	1	0	1
B	0	0	1	1	0	0	1	1
C_{in}	0	0	0	0	1	1	1	1
Sum	0	1	1	0	1	0	0	1
C_{out}	0	0	0	1	0	1	1	1

Fig. 5.15 *Connection of two half-adders to give one full-adder*

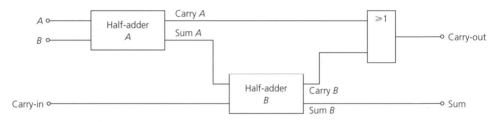

$$S = A\bar{B}\bar{C}_{in} + \bar{A}B\bar{C}_{in} + \bar{A}\bar{B}C_{in} + ABC_{in}$$
$$= (A\bar{B} + \bar{A}B)\bar{C}_{in} + (\bar{A}\bar{B} + AB)C_{in} \tag{5.4}$$
$$= (A \oplus B)\bar{C}_{in} + (\overline{A \oplus B})C_{in} \tag{5.5}$$

Also,

$$C_{out} = AB\bar{C}_{in} + A\bar{B}C_{in} + \bar{A}BC_{in} + ABC_{in}$$
$$= AB(\bar{C}_{in} + C_{in}) + C_{in}(A\bar{B} + \bar{A}B)$$
$$= AB + C_{in}(A \oplus B) \tag{5.6}$$

A full-adder can also be constructed by connecting two half-adders together in the manner shown in Fig. 5.15. The two outputs of the left-hand half-adder are

$$\text{Sum } A = A\bar{B} + \bar{A}B \quad \text{and} \quad \text{Carry } A = AB$$

Hence, for the right-hand half-adder,

$$\text{Sum } B = (A\bar{B} + \bar{A}B)\bar{C}_{in} + (\overline{A\bar{B} + \bar{A}B})C_{in}$$
$$= (A\bar{B} + \bar{A}B)\bar{C}_{in} + (AB + \bar{A}\bar{B})C_{in}$$

Fig. 5.16 *Four-bit parallel full-adder*

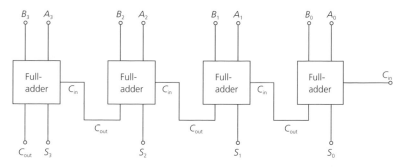

which is equation (5.5) and the sum output of the circuit. Also,

Carry $B = (A\bar{B} + \bar{A}B)C_{in}$

This means that the carry-out of the circuit is

$C_{out} = AB + (AB + \bar{A}B)C_{in}$

which is equation (5.6).

Adding multi-bit numbers

When two multi-bit numbers are to be added either parallel or serial addition can be used, with the former the more common. Parallel addition requires the use of n full-adders to add two n-bit binary numbers and it is very fast in its operation. The way in which four full-adders should be connected to form a 4-bit parallel adder is shown in Fig. 5.16.

Two 4-bit binary numbers $A_3A_2A_1A_0$ and $B_3B_2B_1B_0$ are applied to the A and B inputs of the four full-adders. A_0 and B_0 are the least significant bits. Any carry from a previous stage is applied to the C_{in} terminal of the right-hand (least significant) full-adder. The carry-out C_{out} terminal of each full-adder is connected directly to the C_{in} terminal of the next more significant full-adder. The C_{out} terminal of the left-hand (most significant) full-adder provides the carry-out of the complete circuit. The carry-out bits pass through the circuit stage by stage and an individual sum will be correct only when the carry (if any) from the preceding stage has been applied. This means that the output sum and carry of the circuit must not be read until enough time has elapsed, from the instant the input numbers were applied, for all of the carry bits to have propagated through the circuit. This means that ripple addition is slow if large numbers are to be added and a technique known as *look-ahead carry* is usually employed. This technique involves obtaining the final carry-out of the addition using separate circuitry.

Carry look-ahead

At each stage a carry-out will occur if both the bits that are being added are 1, or if either bit is a 1 and there is a carry-in from the previous stage. Therefore

$$C_{on} = A_n B_n + (A_n \oplus B_n)C_{in}$$

The term $A_n B_n$ is known as the *carry-generate* G_n and the term $(A_n \oplus B_n)$ is known as the *carry-propagate* P_n.

For each stage in the circuit,

$$C_{on} = G_n + P_n C_{in} = G_n + P_n C_{o(n-1)} \tag{5.7}$$

For the first (least significant) stage,

$$C_{o1} = G_1 + P_1 C_{i1} = C_{i2} \tag{5.8}$$

For the second stage,

$$C_{o2} = G_2 + P_2 C_{i2} = G_2 + P_2(G_1 + P_1 C_{i1})$$
$$= G_2 + P_2 G_1 + P_2 P_1 C_{i1} = C_{i3} \tag{5.9}$$

For the third stage,

$$C_{o3} = G_3 + P_3 C_{i3} = G_3 + P_3(G_2 + P_2 G_1 + P_2 P_1 C_{i1})$$
$$= G_3 + P_3 G_2 + P_3 P_2 G_1 + P_3 P_2 P_1 C_{i1} = C_{i4} \tag{5.10}$$

For the most significant stage,

$$C_{o4} = G_4 + P_4 C_{i4}$$
$$= G_4 + P_4(G_3 + P_3 G_2 + P_3 P_2 G_1 + P_3 P_2 P_1 C_{i1})$$
$$= G_4 + P_4 G_3 + P_4 P_3 G_2 + P_4 P_3 P_2 G_1 + P_4 P_3 P_2 P_1 C_{i1} \tag{5.11}$$

It can be seen that although the expressions get longer as the bit significance increases, the overall delay in producing the carry-out remains constant. This means that there are no cumulative effects and so the ripple effect is eliminated. The look-ahead circuitry is required to implement equations (5.7) through to (5.11).

Only one full-adder is still available in an IC package. This device is the 74LS283 4-bit full-adder which is a parallel adder and incorporates carry look-ahead. The two 4-bit numbers to be added are applied to pins A_1 through to A_4, and B_1 through to B_4, respectively, and any input carry is applied to pin C_0. The result of the addition then appears at pins Σ_1 through to Σ_4 and any output carry appears at pin C_4.

Two 74LS283s can be used to add two 8-bit numbers by connecting the C_{out} output pin of the low-order adder to the C_{in} pin of the high-order adder. The C_{in} pin of the low-order adder is connected to earth. Similarly, four 74LS283s can be cascaded to form a 16-bit adder.

Binary subtraction

The subtraction of binary numbers is usually carried out using *2s complement arithmetic* since binary adders can then be employed. Figure 5.17 shows how a full-adder is connected to act as either an adder or a subtractor. When the control signal is low, the carry-in C_{in} terminal will be LOW and the exclusive-OR gates will act to pass each input through to the full-adder itself. The inputs are not altered in any way and the circuit acts as an adder. When the control line is HIGH, the C_{in} terminal will also be HIGH and each of the exclusive-OR gates will invert its input signal. Thus the B input number is inverted *and* has 1 added to it and so it is changed into its 2s complement form.

Fig. 5.17 *Binary adder/subtractor*

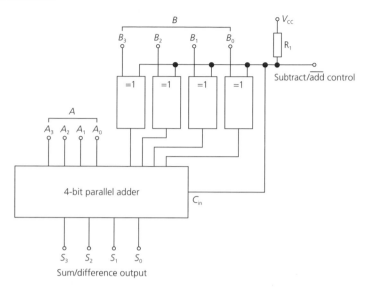

EXERCISES

5.1 A 74HC682 comparator has the following input signals
(a) $P = 1100$ and $Q = 1001$.
(b) $P = 1000$ and $Q = 1011$.
(c) $P = 0101$ and $Q = 0101$.
Determine the output in each case.

5.2 (a) Consult the data sheet of the 74LS283 full-adder to find out how to connect two such devices as an 8-bit adder.
(b) The 74LS283 incorporates carry look-ahead. Explain what this term means and how it works.

5.3 Show how four 74LS283 full-adders can be connected to give a 16-bit full-adder.

5.4 (a) Calculate the noise margin of the 74HC138 3-to-8 line decoder.
(b) Determine the output that goes LOW when the input digital word is
(i) 101
(ii) 111.
(c) The input pins G_1 and $\bar{G}_{2A}$, and inputs A, B and C, are all HIGH. What is the logical state of each of the eight outputs?

5.5 Implement the Boolean equations $F = ABC + \bar{A}\bar{B}\bar{C}$ and $G = \bar{B}C + \bar{A}B$ using the 74HC138 decoder.

5.6 The 74--157 quad 2-to-1 data selector/multiplexer is to be used in a system which will operate at 30 MHz. Minimum power dissipation is required so, if possible, the power supply voltage should be 3.3 V. List the various devices that are available and compare their important (for this purpose) characteristics. Choose a device. Availability and second-sourcing should be considered in the choice.

5.7   How can a 32-to-1 line multiplexer be obtained using the 74HC151 and 74HC153 devices?

5.8   (a)   Draw the logic symbol of the 74HC138 decoder and explain the meaning of each of the labels.

        (b)   The enable inputs have dual functions. Explain what they are.

5.9   Show how two 74HC138 decoders can be connected to act as a 1-to-12 demultiplexer.

5.10 Design a logic circuit to compare two 3-bit binary numbers A and B. The circuit should give outputs to indicate each of the conditions $A > B$, $A = B$ and $A < B$.

5.11 Design, with the aid of a truth table, a majority decision circuit that will give an output at logical 1 whenever two out of its three inputs are at 1.

5.12 Design a logic circuit to have the following characteristics: inputs A, B and C representing, in binary, the denary numbers 0 through to 7, where A is the least significant bit. There should be two outputs X and Y; X should be at 1 only when the input is an even number but not zero; Y should be at 1 only when the input is an odd number.

6 MSI flip-flops and counters

After reading this chapter you should be able to:

- Explain the difference between a latch and a flip-flop.
- Explain the difference between an S-R latch and a D latch.
- Explain the difference between D, J-K and T flip-flops.
- Use the data sheet of a device to obtain information such as the timing requirements and the switching characteristics.
- Use a data sheet to select the best device for a particular application.
- Explain the importance of parameters such as propagation delay, set-up time, hold time and maximum operating frequency.
- Use two, or more, flip-flops to obtain a counter with a straight counting sequence.
- Explain the difference between a non-synchronous and a synchronous counter.
- Analyse a counter to determine its counting sequence and its modulus.
- Understand the data sheet of an MSI counter.
- Modify the count of an MSI counter.
- Use two, or more. MSI counters connected in cascade to obtain a higher count.
- Use a book of data sheets to select a counter for a particular application.
- Understand the logic symbols for latches, flip-flops and counters.

The latch and the flip-flop are circuits that have two stable states and are able to stay in either state indefinitely. The circuit will change state only when switching is initiated by an input pulse. Once the circuit has switched it remains in the other stable state until another input pulse is received to force it to return to the original state. A latch or flip-flop has two output terminals, labelled as Q and $\bar{Q}$; the logical state of Q is always the complement of the logical state of $\bar{Q}$. When the circuit is in the state $Q = 1$ and $\bar{Q} = 0$ it is said to be set, and when in the state $Q = 0$ and $\bar{Q} = 1$ it is said to be reset.

Two types of latch are available, the S-R and D; and there are three kinds of flip-flop, the J-K, the D and the T. A latch is level triggered; this means that its operation is initiated by the voltage level applied to its input, provided the circuit is enabled by the clock waveform, and not by a change or transition in the clock waveform. A latch may be enabled by its clock, or enable, input being held HIGH (or LOW for some devices). Most flip-flops are edge-triggered devices that change state at either the leading edge, or

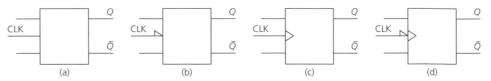

at the trailing edge of a clock pulse. The clock is a rectangular waveform of constant frequency. The symbol for a latch or flip-flop, which is shown in Fig. 6.1, indicates how the device is triggered. A circuit that operates when the clock is held HIGH is given the symbol shown in Fig. 6.1(a) while Fig. 6.1(b) gives the symbol for an active-LOW device. An edge-triggered circuit is indicated by a small wedge at the clock input; if there is no triangle at the clock input the device is leading-edge triggered, while the presence of a triangle indicates that the circuit is trailing-edge triggered. Leading-edge triggering is also known as positive-edge triggering, and trailing-edge triggering is known as negative-edge triggering.

A counter consists of a number n of flip-flops connected together in such a way that the count follows a predefined sequence. If the count moves from 0 through to 2^n in sequence the circuit is an n-bit binary counter and if the count goes from 0 to 9 in sequence it is a decade counter.

Latches

The S-R latch

The S-R latch has two input terminals, S (for set) and R (for reset), and two output terminals Q and $\bar{Q}$. Q is always the complement of $\bar{Q}$. The symbol for an S-R latch is shown in Fig. 6.2, and Table 6.1 gives the truth table for such a device. In the table Q represents the *present state* of the Q output terminal and Q^+ represents the *next state* of the Q terminal after the input terminals have been taken HIGH or LOW. Any change in the logical state of the output terminals will take place only when the clock waveform is HIGH (or LOW for some devices). When both S and R are held LOW the logical states of the output terminals do not change. This is shown by $Q = Q^+$. When $S = 1$ and $R = 0$ the logical state of the output Q will become 1 no matter what its value was before. This is known as the *set* operation. If $S = 0$ and $R = 1$ the next state of the Q output will be 0

Fig. 6.2 *S-R latch*

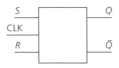

Table 6.1 S-R latch truth table

S	R	Q	Q⁺	
0	0	0	0	} No change in state
0	0	1	1	
1	0	0	1	} Set operation
1	0	1	1	
0	1	0	0	} Reset operation
0	1	1	0	
1	1	0	×	} Indeterminate operation
1	1	1	×	

whatever its value was before. This is the *reset* operation. If both S and R are at 1 the operation of the latch is unpredictable and this combination of input states should not be allowed to occur. There is one S-R latch IC only, in the 74LS logic family (the 74LS279A).

The D latch

A D latch has one data input, labelled D, one enable or clock input, and one output Q. The logical state of the D input is transferred to the output whenever the clock input is HIGH. The logical state of the Q output will follow any changes in the logical state of the D input as long as the clock input remains HIGH. When the clock input goes LOW the logical state of the D input at that moment will be retained at the Q output no matter what changes may occur at the D input. When the clock input goes HIGH again the Q output will once again follow any changes in the logical state of the D input. A D latch is said to be transparent when the clock is HIGH.

A part of the data sheet of the 74HC373 octal D latch is given in Fig. 6.3. The pinout, the function table and the logic symbol are shown and a description of the device is given in the first part of the data sheet. The remainder of the data sheet offers information similar to that given for gates (see Fig. 6.4 as a guide).

The J-K flip-flop

A J-K flip-flop has two data input terminals, labelled as J and K, a clock input and two output terminals Q and $\bar{Q}$. Non-synchronous preset PRE and clear CLR inputs may also be provided. A J-K flip-flop will change its output state at a clock transition, either at the leading edge, or at the trailing edge, of a clock pulse. The operation of a J-K flip-flop is summarized by the truth table given in Table 6.2. A J-K flip-flop always changes state when $J = K = 1$ and the device is said to *toggle*.

SN54HC373, SN74HC373
OCTAL TRANSPARENT D-TYPE LATCHES
WITH 3-STATE OUTPUTS
SCLS140A – DECEMBER 1982 – REVISED JANUARY 1996

Eight High-Current Latches in a Single Package

High-Current 3-State True Outputs Can Drive up to 15 LSTTL Loads

Full Parallel Access for Loading

Package Options Include Plastic Small-Outline (DW), Shrink Small-Outline (DB), Thin Shrink Small-Outline (PW), and Ceramic Flat (W) Packages, Ceramic Chip Carriers (FK), and Standard Plastic (N) and Ceramic (J) 300-mil DIPs

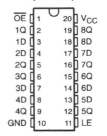

SN54HC373 . . . J OR W PACKAGE
SN74HC373 . . . DB, DW, N, OR PW PACKAGE
(TOP VIEW)

$\overline{OE}$	1	20	V_{CC}
1Q	2	19	8Q
1D	3	18	8D
2D	4	17	7D
2Q	5	16	7Q
3Q	6	15	6Q
3D	7	14	6D
4D	8	13	5D
4Q	9	12	5Q
GND	10	11	LE

SN54HC373 . . . FK PACKAGE
(TOP VIEW)

description

These 8-bit latches feature 3-state outputs designed specifically for driving highly capacitive or relatively low-impedance loads. They are particularly suitable for implementing buffer registers, I/O ports, bidirectional bus drivers, and working registers.

The eight latches of the 'HC373 are transparent D-type latches. While the latch-enable (LE) input is high, the Q outputs follow the data (D) inputs. When LE is taken low, the Q outputs are latched at the levels that were set up at the D inputs.

An output-enable ($\overline{OE}$) input places the eight outputs in either a normal logic state (high or low logic levels) or the high-impedance state. In the high-impedance state, the outputs neither load nor drive the bus lines significantly. The high-impedance state and increased drive provide the capability to drive bus lines without interface or pullup components.

$\overline{OE}$ does not affect the internal operations of the latches. Old data can be retained or new data can be entered while the outputs are off.

The SN54HC373 is characterized for operation over the full military temperature range of –55°C to 125°C. The SN74HC373 is characterized for operation from –40°C to 85°C.

logic symbol

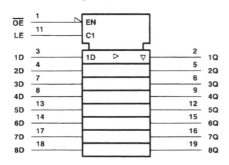

OE	1	EN
LE	11	C1
1D	3	1D ▷ ▽	2	1Q
2D	4		5	2Q
3D	7		6	3Q
4D	8		9	4Q
5D	13		12	5Q
6D	14		15	6Q
7D	17		16	7Q
8D	18		19	8Q

FUNCTION TABLE
(each latch)

INPUTS			OUTPUT
$\overline{OE}$	LE	D	Q
L	H	H	H
L	H	L	L
L	L	X	Q_0
H	X	X	Z

Table 6.2 *J-K flip-flop truth table*

J	K	Q	Q⁺	
0	0	0	0	} No change in state
0	0	1	1	
1	0	0	1	} Set operation
1	0	1	1	
0	1	0	0	} Reset operation
0	1	1	0	
1	1	0	1	} Toggle
1	1	1	0	

Data sheet

There are two J-K flip-flops in the 74CMOS logic families, the '109 and the '112, while the 74LS family includes these two devices plus the '73 and '107. The '109 is a leading-edge-triggered J-K flip-flop and the '112 is a trailing-edge-triggered J-K flip-flop. The data sheet for a J-K flip-flop gives values for the same parameters as for gates, i.e. I_{OH}, I_{IH}, V_{OL} etc. These parameters are all defined in the same way as for gates but there may be some variations depending upon which input is specified. There are figures quoted for the propagation delay between different inputs and the Q output. In the function table the symbol × indicates a 'don't care' state, as for the Karnaugh map. The symbol ↑ indicates that a positive-going clock transition initiates the read-in of data, and the symbol ↓ indicates that read-in of data occurs during a negative-going clock transition. $\overline{CLR}$ means that when the $\overline{CLR}$ pin is LOW, no matter what the logical states of the synchronous inputs are, the flip-flop will be cleared. When $\overline{CLR}$ is held HIGH it has no effect on the circuit. Toggle means that the state of the Q outputs changes at each clock pulse. A part of the data sheet for the 74HC112 dual J-K flip-flop is given in Fig. 6.4.

Non-synchronous preset and clear

Most clocked flip-flops have one, or two, non-synchronous inputs as well as the synchronous data and clock inputs. The non-synchronous inputs are labelled $\overline{PRE}$ and $\overline{CLR}$. When a non-synchronous input is HIGH it is non-active and the flip-flop's normal operation can take place. When the $\overline{PRE}$ input is held LOW the flip-flop will immediately set to have $Q = 1$ and $\overline{Q} = 0$, regardless of the logical states of any of the synchronous inputs. When the $\overline{CLR}$ input is taken LOW the flip-flop will clear to have $Q = 0$ and $\overline{Q} = 1$. In normal operation the condition $\overline{PRE} = \overline{CLR} = 0$ ought not to occur; if it does the operation of the flip-flop cannot be predicted. Some flip-flops may have active-HIGH non-synchronous inputs.

Fig. 6.4 *Data sheet for the 74HC112 dual J-K flip-flop. (Courtesy of Texas Instruments)*

SN54HC112, SN74HC112
DUAL J-K NEGATIVE-EDGE-TRIGGERED FLIP-FLOPS
WITH CLEAR AND PRESET
SCLS099A – DECEMBER 1982 – REVISED JANUARY 1996

Package Options Include Plastic Small-Outline (D) and Ceramic Flat (W) Packages, Ceramic Chip Carriers (FK), and Standard Plastic (N) and Ceramic (J) 300-mil DIPs

description

The 'HC112 contain two independent J-K negative-edge-triggered flip-flops. A low level at the preset ($\overline{\text{PRE}}$) or clear ($\overline{\text{CLR}}$) inputs sets or resets the outputs regardless of the levels of the other inputs. When $\overline{\text{PRE}}$ and $\overline{\text{CLR}}$ are inactive (high), data at the J and K inputs meeting the setup time requirements are transferred to the outputs on the negative-going edge of the clock (CLK) pulse. Clock triggering occurs at a voltage level and is not directly related to the rise time of CLK. Following the hold-time interval, data at the J and K inputs may be changed without affecting the levels at the outputs. These versatile flip-flops perform as toggle flip-flops by tying J and K high.

The SN54HC112 is characterized for operation over the full military temperature range of –55°C to 125°C. The SN74HC112 is characterized for operation from –40°C to 85°C.

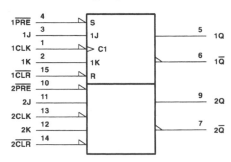

SN54HC112 . . . J OR W PACKAGE
SN74HC112 . . . D OR N PACKAGE
(TOP VIEW)

SN54HC112 . . . FK PACKAGE
(TOP VIEW)

NC – No internal connection

FUNCTION TABLE

INPUTS					OUTPUTS	
$\overline{\text{PRE}}$	$\overline{\text{CLR}}$	CLK	J	K	Q	$\overline{\text{Q}}$
L	H	X	X	X	H	L
H	L	X	X	X	L	H
L	L	X	X	X	H†	H†
H	H	↓	L	L	Q_0	$\overline{Q}_0$
H	H	↓	H	L	H	L
H	H	↓	L	H	L	H
H	H	↓	H	H	Toggle	
H	H	H	X	X	Q_0	$\overline{Q}_0$

† This configuration is nonstable; that is, it will not persist when either $\overline{\text{PRE}}$ or $\overline{\text{CLR}}$ returns to its inactive (high) level.

logic symbol

1PRE 4 S
1J 3 1J 5 1Q
1CLK 1 C1
1K 2 1K 6 1$\overline{\text{Q}}$
1$\overline{\text{CLR}}$ 15 R
2$\overline{\text{PRE}}$ 10
2J 11 9 2Q
2CLK 13
2K 12 7 2$\overline{\text{Q}}$
2$\overline{\text{CLR}}$ 14

absolute maximum ratings over operating free-air temperature range†

Supply voltage range, V_{CC} . . . . –0.5 V to 7 V
Input clamp current, I_{IK} ($V_I < 0$ or $V_I > V_{CC}$) (see Note 1) . . . . ±20 mA
Output clamp current, I_{OK} ($V_O < 0$ or $V_O > V_{CC}$) (see Note 1) . . . . ±20 mA
Continuous output current, I_O ($V_O = 0$ to V_{CC}) . . . . ±25 mA
Continuous current through V_{CC} or GND . . . . ±50 mA
Maximum power dissipation at $T_A = 55°C$ (in still air) (see Note 2): D package . . . . 1.3 W
N package . . . . 1.1 W
Storage temperature range, T_{stg} . . . . –65°C to 150°C

† Stresses beyond those listed under "absolute maximum ratings" may cause permanent damage to the device. These are stress ratings only, and functional operation of the device at these or any other conditions beyond those indicated under "recommended operating conditions" is not implied. Exposure to absolute-maximum-rated conditions for extended periods may affect device reliability.

NOTES: 1. The input and output voltage ratings may be exceeded if the input and output current ratings are observed.
2. The maximum package power dissipation is calculated using a junction temperature of 150°C and a board trace length of 750 mils, except for the N package, which has a trace length of zero.

Fig. 6.4 *(Cont'd)*

recommended operating conditions

			SN54HC112			SN74HC112			UNIT
			MIN	NOM	MAX	MIN	NOM	MAX	
V_{CC}	Supply voltage		2	5	6	2	5	6	V
V_{IH}	High-level input voltage	$V_{CC} = 2$ V	1.5			1.5			V
		$V_{CC} = 4.5$ V	3.15			3.15			
		$V_{CC} = 6$ V	4.2			4.2			
V_{IL}	Low-level input voltage	$V_{CC} = 2$ V	0		0.5	0		0.5	V
		$V_{CC} = 4.5$ V	0		1.35	0		1.35	
		$V_{CC} = 6$ V	0		1.8	0		1.8	
V_I	Input voltage		0		V_{CC}	0		V_{CC}	V
V_O	Output voltage		0		V_{CC}	0		V_{CC}	V
t_t‡	Input transition (rise and fall) time	$V_{CC} = 2$ V	0		1000	0		1000	ns
		$V_{CC} = 4.5$ V	0		500	0		500	
		$V_{CC} = 6$ V	0		400	0		400	
T_A	Operating free-air temperature		−55		125	−40		85	°C

‡ If this device is used in the threshold region (from V_{IL}max = 0.5 V to V_{IH}min = 1.5 V), there is a potential to go into the wrong state from induced grounding, causing double clocking. Operating with the inputs at t_t = 1000 ns and V_{CC} = 2 V will not damage the device; however, functionally, the CLK inputs are not ensured while in the shift, count, or toggle operating modes.

electrical characteristics over recommended operating free-air temperature range (unless otherwise noted)

PARAMETER	TEST CONDITIONS		V_{CC}	$T_A = 25°C$			SN54HC112		SN74HC112		UNIT
				MIN	TYP	MAX	MIN	MAX	MIN	MAX	
V_{OH}	$V_I = V_{IH}$ or V_{IL}	$I_{OH} = -20$ μA	2 V	1.9	1.998		1.9		1.9		V
			4.5 V	4.4	4.499		4.4		4.4		
			6 V	5.9	5.999		5.9		5.9		
		$I_{OH} = -4$ mA	4.5 V	3.98	4.3		3.7		3.84		
		$I_{OH} = -5.2$ mA	6 V	5.48	5.8		5.2		5.34		
V_{OL}	$V_I = V_{IH}$ or V_{IL}	$I_{OL} = 20$ μA	2 V		0.002	0.1		0.1		0.1	V
			4.5 V		0.001	0.1		0.1		0.1	
			6 V		0.001	0.1		0.1		0.1	
		$I_{OL} = 4$ mA	4.5 V		0.17	0.26		0.4		0.33	
		$I_{OL} = 5.2$ mA	6 V		0.15	0.26		0.4		0.33	
I_I	$V_I = V_{CC}$ or 0		6 V		±0.1	±100		±1000		±1000	nA
I_{CC}	$V_I = V_{CC}$ or 0,	$I_O = 0$	6 V			4		80		40	μA
C_i			2 V to 6 V		3	10		10		10	pF

timing requirements over recommended operating free-air temperature range (unless otherwise noted)

			V_{CC}	$T_A = 25°C$		SN54HC112		SN74HC112		UNIT
				MIN	MAX	MIN	MAX	MIN	MAX	
f_{clock}	Clock frequency		2 V	0	5	0	3.4	0	4	MHz
			4.5 V	0	25	0	17	0	20	
			6 V	0	29	0	20	0	24	
t_w	Pulse duration	PRE or CLR low	2 V	100		150		125		ns
			4.5 V	20		30		25		
			6 V	17		25		21		
		CLK high or low	2 V	100		150		125		
			4.5 V	20		30		25		
			6 V	17		25		21		
t_{su}	Setup time before CLK↓	Data (J, K)	2 V	100		150		125		ns
			4.5 V	20		30		25		
			6 V	17		25		21		
		PRE or CLR inactive	2 V	100		150		125		
			4.5 V	20		30		25		
			6 V	17		25		21		
t_h	Hold time, data after CLK↓		2 V	0		0		0		ns
			4.5 V	0		0		0		
			6 V	0		0		0		

Fig. 6.4 *(Cont'd)*

switching characteristics over recommended operating free-air temperature range, $C_L = 50$ pF (unless otherwise noted) (see Figure 1)

PARAMETER	FROM (INPUT)	TO (OUTPUT)	V_{CC}	$T_A = 25°C$			SN54HC112		SN74HC112		UNIT
				MIN	TYP	MAX	MIN	MAX	MIN	MAX	
f_{max}			2 V	5	10		3.4		4		MHz
			4.5 V	25	50		17		20		
			6 V	29	60		20		24		
t_{pd}	$\overline{PRE}$ or $\overline{CLR}$	Q or $\overline{Q}$	2 V		54	165		245		205	ns
			4.5 V		16	33		49		41	
			6 V		13	28		42		35	
	CLK	Q or $\overline{Q}$	2 V		56	125		185		155	
			4.5 V		16	25		37		31	
			6 V		13	21		31		26	
t_t		Q or $\overline{Q}$	2 V		29	75		110		95	ns
			4.5 V		9	15		22		19	
			6 V		8	13		19		16	

operating characteristics, $T_A = 25°C$

PARAMETER		TEST CONDITIONS	TYP	UNIT
C_{pd}	Power dissipation capacitance per flip-flop	No load	35	pF

Timing parameters of a J-K flip-flop

The timing parameters of a J-K flip-flop are the propagation delay parameters, set-up time, hold time, maximum clock frequency, propagation delay, and minimum pulse width for both the clock and the synchronous inputs. Ideally a flip-flop switches at either the leading edge, or the trailing edge of a clock pulse but, in practice, it will take up to 30 ns to switch. The input data must be held steady for some time before and after the clock transition to ensure that the data is transferred to the Q output terminal. This time is known as the *set-up time*. The data sheet of a device gives the minimum allowable set-up time $t_{s(min)}$. The *hold time* t_h of an edge-triggered flip-flop is the time interval after the clock transition that changes the logical state of the output for which the input data pulse must be maintained constant. If this requirement is not satisfied the flip-flop may not operate correctly. For reliable operation the data input(s) must be kept constant for a time equal to at least $t_{s(min)} + t_{h(min)}$. If this time requirement is not met the flip-flop may not respond reliably when a clock pulse is applied.

There are propagation delays from the synchronous inputs to the output(s) and also from the non-synchronous inputs to the outputs. The delays are measured between the 50 per cent amplitude points on the input and output waveforms. Propagation delay is quoted in the data sheet for all the possible combinations of inputs and output. Values for the set-up time, the hold time and the propagation delays for the various combinations of inputs and output are given in the data sheet of a device. The maximum clock frequency is also of importance and it is also quoted in the data sheet. The maximum clock frequency is limited by the propagation delay through the circuit. The minimum clock pulse width, measured at the 50 per cent amplitude points, is also

Fig. 6.5 *(a) Propagation delay, (b) pulse width, (c) set-up time and (d) hold time*

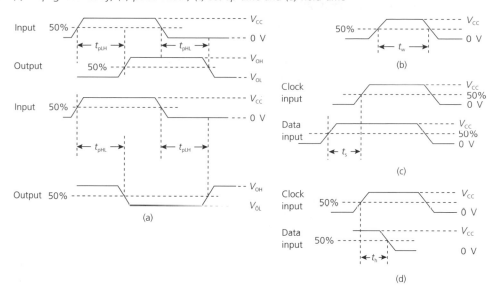

quoted: in some cases different figures are quoted for the HIGH and LOW pulse widths. The meanings of set-up and hold times, propagation delays and pulse widths are shown in Fig. 6.5.

From the data sheet of the 74HC112, when $V_{CC} = 4.5$ V, t_{pd} CLK $\rightarrow$ Q is 31 ns maximum, and t_{pd} PRE or CLR to Q is 41 ns maximum.

The frequency of operation cannot be so large that a clock pulse arrives before the circuit has finished its response to the preceding pulse. Operation at any higher frequency will result in unpredictable behaviour. The timing parameters become of increasing importance as the frequency of operation is increased towards the maximum frequency quoted in the data sheet. As long as the propagation delay of each driving flip-flop is longer than the hold time of the driven flip-flop the circuit ought to work in the manner predicted by the function table given in the data sheet.

EXAMPLE 6.1

Determine the propagation delays and the maximum clock frequency of the 74HC112 J-K flip-flop, $V_{CC} = 4.5$ V.

Solution

From the data sheet, t_{pd} $\overline{\text{PRE}}$ to Q or $\overline{Q} = 41$ ns (*Ans.*)
t_{pd} CLK to Q or $\overline{Q} = 31$ ns (*Ans.*)
Maximum clock frequency = 20 MHz (*Ans.*)

EXAMPLE 6.2

(a) Will the 74HC112 J-K flip-flop, whose clock, J, K and $\overline{CLR}$ waveforms are shown in Fig. 6.6, set?

(b) If the set-up and hold times are 25 and 0 ns, respectively, draw the J waveform required for the circuit to set.

Solution

(a) The flip-flop will not set because input J must be HIGH for a time at least equal to the set-up time before the leading edge of a clock pulse occurs (*Ans.*)

(b) Since the minimum set-up time is 25 ns at $V_{CC} = 4.5$ V the logical states of the J and K inputs may be changed at any time more than 25 ns before the trailing edge of a clock pulse. At 25 ns before the clock transition the voltage of the J input must be held at the required voltage level. The logical state of the data input must be held for at least the hold time after the clock transition. Since, for this flip-flop, the hold time is 0 ns there is no problem. This means that the J and K inputs may be changed immediately after the clock transition. The required waveforms are given in Fig. 6.7 (*Ans.*)

Fig. 6.6

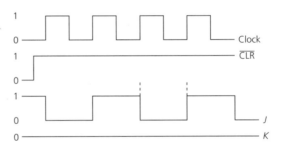

Fig. 6.7

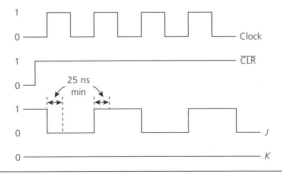

Aim: to measure the propagation delay of a J-K flip-flop. To observe the effect of operating the device at a frequency:

(a) Near the maximum frequency.
(b) At the maximum frequency.
(c) Above the maximum frequency.

Components and equipment: one 74HC112 dual J-K flip-flop. Pulse generator. Breadboard. CRO. Power supply.
Procedure:

(a) Connect pins 2, 3, 4, 15 and 16 to +5 V, and connect pin 8 to earth.
(b) Apply the pulse generator to the 1CLK terminal at pin 1. Connect the CRO, channel A to pin 4 and channel B to pin 5 (*1Q*).
(c) The flip-flop is connected to operate as a divide-by-2 circuit. Set, in turn, the frequency of the pulse generator to
 (i) 10 MHz
 (ii) 25 MHz
 (iii) 30 MHz
 (iv) 50 MHz.
 Each time use the CRO to observe the input and output waveforms and measure the propagation delay and the division ratio.
(d) Comment on the results obtained.

EXAMPLE 6.3

The *J-K* and clock inputs and the *Q* outputs of a flip-flop are shown in Fig. 6.8. From the data given in Table 6.3 determine which device ought to be used.

Fig. 6.8

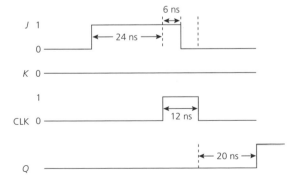

Table 6.3

	74HC112	74LVC112	74LS112	74ALS112	74S112	74F112
Max clock frequency	24	150	30	30	45	100 MHz
Max propagation delay	26/21	5.5/5.5	23/23	15/18	23/23	7.5/7.5 ns
Min set-up time	21	23	20	22	7	3.5 ns
Min hold time	0	0.7	0	0	0	0 ns
Min clock pulse width	21	3.3	20	16.5	8	5 ns

Solution

From the waveforms $t_s = 24$ ns, $t_h = 6$ ns, $t_w = 12$ ns and $t_{pd} = 20$ ns. The HC and LVC devices fail to satisfy the set-up time requirement t_s; all the devices pass for the hold time t_h; the 74ALS112 has too large a t_w figure so the choice is between the S and F devices. The S series is virtually obsolete so choose the 74F112 (*Ans.*)

PRACTICAL EXERCISE 6.2

Aim: to investigate the need for the *J* and *K* inputs of a J-K flip-flop to be kept steady during the set-up time.
Components and equipment: one 74HC112 dual J-K flip-flop IC, one 74HC164 shift register, one LED and one 270 Ω resistor. Pulse generator. Breadboard. CRO. Power supply.
Procedure:

(a) Build the circuit shown in Fig. 6.9.

Fig. 6.9

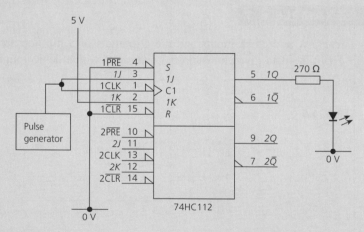

(b) Connect the pulse generator to both the *1J* and 1CLK inputs of the flip-flop. Set the pulse generator to 10 Hz, 5 V and 50 per cent duty cycle. Connect the CRO to both the output of the pulse generator and the *Q* output terminal of the flip-flop.

(c) Does the LED turn ON and OFF at regular intervals? If it does note the waveform displayed on the CRO. Repeat at frequencies of
  (i) 1 kHz
  (ii) 1 MHz
  (iii) 10 MHz
      30 MHz.
  Explain the reason for the results obtained.
(d) The minimum set-up time for flip-flop is quoted in the data sheet as 25 ns. The shift register can be used to provide a delay in the application of the clock waveform to the CLK input. Remove the connections between the pulse generator and the CLK input terminal but leave the connection between the 1J terminal and the generator intact. Connect the pulse generator to the B input of the shift register (pinout on p. 191) and restore its frequency to 1 kHz. Connect terminal A to +5 V and Q_H to the CLK input of the flip-flop. The circuit is shown in Fig. 6.10. The set-up time is now equal to the total delay of the shift register. Note that the circuit now works so that this delay is adequate.

Fig. 6.10

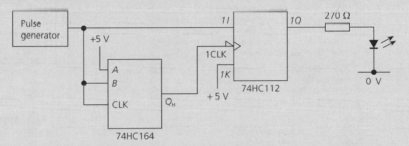

(e) Note the waveforms displayed by the CRO at each of the frequencies used in (c).
(f) Look up (p. 193) the propagation delay of one stage of the shift register and hence determine the set-up time for the flip-flop.
(g) Is this time greater than the minimum set-up time? Does it depend upon the clock frequency? Determine how many shift register stages are needed and modify the circuit to use that number.

The D flip-flop

A D flip-flop has one data (D) input, one clock input, and two outputs Q and $\bar{Q}$. The logical state of the D input will be transferred to the Q output at a clock transition. The transition is either the leading edge, or the trailing edge, of a clock pulse depending on the type of flip-flop. A D flip-flop is not transparent since the Q output will only take up the logical state of the D input at either the leading edge or the trailing edge of a clock

SN54HC174, SN74HC174
HEX D-TYPE FLIP-FLOPS
WITH CLEAR
SCLS119A – DECEMBER 1982 – REVISED JANUARY 1996

Contain Six Flip-Flops With Single-Rail Outputs

Applications Include:
- Buffer/Storage Registers
- Shift Registers
- Pattern Generators

Package Options Include Plastic Small-Outline (D) and Ceramic Flat (W) Packages, Ceramic Chip Carriers (FK), and Standard Plastic (N) and Ceramic (J) 300-mil DIPs

SN54HC174 . . . J OR W PACKAGE
SN74HC174 . . . D OR N PACKAGE
(TOP VIEW)

$\overline{CLR}$	1	16	V_{CC}
1Q	2	15	6Q
1D	3	14	6D
2D	4	13	5D
2Q	5	12	5Q
3D	6	11	4D
3Q	7	10	4Q
GND	8	9	CLK

description

These monolithic positive-edge-triggered D-type flip-flops have a direct clear ($\overline{CLR}$) input.

Information at the data (D) inputs meeting the setup time requirements is transferred to the outputs on the positive-going edge of the clock (CLK) pulse. Clock triggering occurs at a particular voltage level and is not directly related to the transition time of the positive-going edge of CLK. When CLK is at either the high or low level, the D input has no effect at the output.

The SN54HC174 is characterized for operation over the full military temperature range of –55°C to 125°C. The SN74HC174 is characterized for operation from –40°C to 85°C.

SN54HC174 . . . FK PACKAGE
(TOP VIEW)

1D	4	18 6D
2D	5	17 5D
NC	6	16 NC
2Q	7	15 5Q
3D	8	14 4D

NC – No internal connection

FUNCTION TABLE
(each flip-flop)

INPUTS			OUTPUT
$\overline{CLR}$	CLK	D	Q
L	X	X	L
H	↑	H	H
H	↑	L	L
H	L	X	Q_0

logic symbol

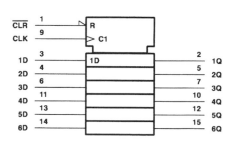

pulse. A D flip-flop can be used as a toggle by connecting its input terminal to its *Q* output. Figure 6.11 gives a part of the data sheet for the 74HC174 hex D flip-flop IC. All the flip-flops in the 74HC174 D hex flip-flop have a common clock, no $\overline{PRE}$ pins and their *Q* outputs are not brought out. This is done so that six devices can be fitted into the one package. The 74HC74 dual D flip-flop is similar but having fewer flip-flops allows *Q* terminals to be accessible, and there is a $\overline{PRE}$ terminal.

EXAMPLE 6.4

If the 74HC174 hex D flip-flop is used with a supply voltage V_{CC} of 4.5 V determine:

(a) The minimum possible clock pulse width.
(b) The minimum time for which a change in the logical state of the D input from LOW to HIGH must be held before the arrival of a clock pulse.
(c) The minimum time for which the logical state of the D input must be held after the leading edge of a clock pulse.

Solution

(a) The minimum clock pulse width is quoted as 16 ns, but since the maximum clock frequency is 30 MHz the minimum clock pulse width is $0.5 \times [1/(30 \times 10^{-6})] = 16.16$ ns (*Ans.*)
(b) $t_s = 20$ ns (*Ans.*)
(c) $t_h = 5$ ns (*Ans.*)

EXAMPLE 6.5

The data sheet of a D flip-flop gives the set-up time as 20 ns, the hold time as 5 ns and the minimum clock pulse width as 16 ns. Calculate the highest frequency at which the device can be clocked.

Solution

Maximum frequency $= 1/(t_s + t_h + t_w) = 1/(41 \times 10^{-9}) = 24.39$ MHz (*Ans.*)

PRACTICAL EXERCISE 6.3

Aim: to show how a D flip-flop can be employed to prevent pulses in a simple system being shortened.
Components and equipment: one 74HC08 quad 2-input AND gate IC, one 74HC174 hex D flip-flop IC. Pulse generator. Breadboard. CRO. Power supply.
Procedure:

(a) Connect up the circuit given in Fig. 6.12(a).
(b) Set the pulse generator to 1 kHz, 5 V and 50 per cent duty cycle and observe the waveform displayed on the CRO as the switch is repeatedly operated between its two positions.
(c) Note that the displayed waveform has the width of some pulses shortened. Explain why shortening occurs and say why it may be undesirable in a digital system. Repeat with the frequency of the pulse generator increased to:
  (i) 1 MHz
  (ii) 30 MHz.

(d) Use the 74HC174 IC to remove this pulse shortening effect. Retain the circuit in Fig. 6.12(a) and build another circuit with the clock waveform connected to both the AND gate (as in Fig. 6.12(a)), and the clock input of a D flip-flop. Connect the same switch as in Fig. 6.12(a) to the *D* input of the flip-flop, and connect the *Q* output to the other AND gate input. Connect the other CRO channel to the output of this AND gate (see Fig. 6.12(b)).

Fig. 6.12

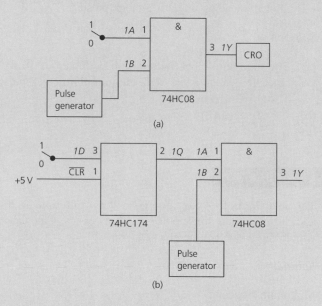

(a)

(b)

(e) With the frequency of the pulse generator set, in turn, to
 (i) 1 kHz
 (ii) 1 MHz
 (iii) 30 MHz
 compare the two displayed waveforms. Has the pulse shortening effect been overcome? Is any contact bounce observable? If it is state how it might be overcome.

(f) Determine a method of using the 74HC174 to build a circuit whose output is HIGH when both inputs *A* and *B* are HIGH and input *A* went HIGH before *B*. The output of the circuit should not go HIGH if *B* goes HIGH before *A*. Construct the circuit and check its operation.

Merits of J-K and D flip-flops

Each J-K flip-flop has two input terminals compared with just the one for a D flip-flop. In a VLSI circuit the number of literals and the total length of conductor is of greater importance than the number of devices used in a circuit. For this reason D flip-flops are almost always employed and not J-K flip-flops.

Fig. 6.13 *Use of D flip-flops as registers*

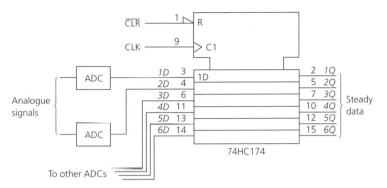

Registers

D flip-flops are often employed as registers to hold data ready for application to another circuit. An example of this is shown in Fig. 6.13. The data on each input when the leading (or trailing) edge of a clock pulse occurs is held steady at the Q output terminal. If the magnitude of the input data should vary the Q output will remain constant until the next clock transition takes place.

Metastability

If the set-up and hold time requirements of a flip-flop are not satisfied the flip-flop will have to make a random decision whether to set or reset its Q output to 0 or to 1. A flip-flop may become *metastable* if its data input(s) is/are changed as the circuit is clocked. The effects of metastability are that either the flip-flop will oscillate for a while before it settles in one stable state or the other, or its clock-to-Q delay will be considerably increased.

Non-synchronous counters

A single J-K flip-flop can be connected to operate as a divide-by-2 circuit by holding the J and K inputs at the logical 1 voltage level, and applying the signal to be divided to the clock input terminal. A D flip-flop can be operated as a divide-by-2 circuit by connecting its Q output to its D input. The two divide-by-2 counters are shown in Fig. 6.14. For a count in excess of 2, a number n of flip-flops must be connected in cascade to give a maximum count of $2^n - 1$. Since the first state is 0 this means that n flip-flops can give a count of 2^n binary states. The concept is illustrated in Fig. 6.15(a) and (b). Figure 6.15(a) shows how two D flip-flops can be connected to give a maximum count of 3 (divide-by-4), and Fig. 6.15(b) shows three J-K flip-flops connected as a divide-by-8 circuit.

Each stage is clocked at one-half of the clocking frequency of the preceding stage; as a result the logical states of the Q outputs seem to ripple through the counter. Because of this, non-synchronous counters are often known as ripple counters.

Fig. 6.14 *Divide-by-2 counters: (a) J-K flip-flop and (b) D flip-flop*

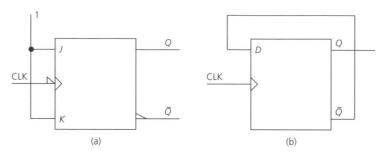

(a)   (b)

Fig. 6.15 *(a) Divide-by-4 counter using D flip-flops and (b) divide-by-8 counter using J-K flip-flops*

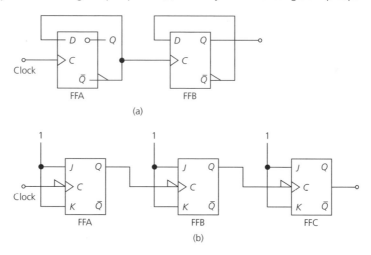

The main disadvantages of this type of circuit are that the propagation delays of the individual stages are additive and the total delay may severely limit the maximum possible frequency of operation. Also, the count can follow only a straight binary code sequence.

EXAMPLE 6.6

The flip-flops in a 4-bit counter each introduce a maximum delay of 40 ns. Calculate the maximum clock frequency.

Solution

Total propagation delay = $4 \times 40 = 160$ ns.
Maximum clock frequency = $1/(160 \times 10^{-9}) = 6.25 \times 10^{6}$ Hz (*Ans.*)

Synchronous counters

For many applications of counters the time taken for a clock pulse to ripple through the circuit and/or the glitches that often arise are not acceptable. The speed of operation can be considerably increased if all the flip-flops are clocked simultaneously. This is known as synchronous operation.

The clock input of each flip-flop is connected directly to the clock line. The design of a synchronous counter, when the count is required to be 2^n, where n is the number of stages, is determined as described below.

Suppose that a divide-by-8 synchronous counter is to be designed. Three stages are necessary since $2^3 = 8$. Table 6.4 shows the required states of the Q outputs of each stage at each clock pulse.

Table 6.4

Clock pulse	Q_C	Q_B	Q_A
0	0	0	0
1	0	0	1
2	0	1	0
3	0	1	1
4	1	0	0
5	1	0	1
6	1	1	0
7	1	1	1
8	0	0	0

J_A/K_A: Q_A is required to change state, or toggle, at each clock pulse transition. Hence, connect both J_A and K_A to 1.

J_B/K_B: Q_B is required to change state at each clock transition whenever Q_A is 1. Hence, connect Q_A to both J_B and K_B.

J_C/K_C: Q_C is required to change state at each clock transition whenever both Q_A and Q_B are 1. Hence, connect Q_A and Q_B to the inputs of a 2-input AND gate and connect the gate output to both J_C and K_C. The required circuit is shown in Fig. 6.16.

This principle is easily extended to obtain a 4-bit (divide-by-16) counter.

Fig. 6.16 *Divide-by-8 synchronous counter*

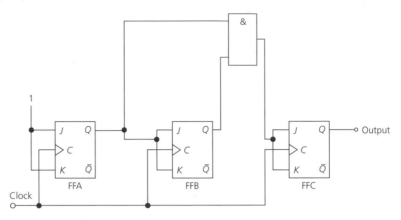

Clock

EXAMPLE 6.7

Design a BCD (decade) synchronous counter using J-K flip-flops.

Solution

A four-stage counter will be necessary since three stages have a maximum count of 8. The state table for the required counter is given in Table 6.5 (*Ans.*)

Table 6.5

Clock pulse	Q_D	Q_C	Q_B	Q_A
0	0	0	0	0
1	0	0	0	1
2	0	0	1	0
3	0	0	1	1
4	0	1	0	0
5	0	1	0	1
6	0	1	1	0
7	0	1	1	1
8	1	0	0	0
9	1	0	0	1
10	0	0	0	0

From the state table: J_A/K_A: Q_A is to toggle at each clock transition, so connect both J_A and K_A to 1.

J_B/K_B: Q_B changes state at the next clock transition when $Q_A = 1$ and $Q_D = 0$. Therefore, $J_B = K_B = Q_A\bar{Q}_D$. Connect Q_A and $\bar{Q}_D$ to the inputs of a 2-input AND gate and connect the output of the gate to J_B and K_B.

J_C/K_C: Q_C changes state at a clock transition whenever $Q_A = Q_B = 1$. Hence, connect Q_A and Q_B to the inputs of a 2-input AND gate and the gate output to J_C and K_C.

J_D/K_D: Q_D changes state at a clock transition whenever $Q_A = Q_B = Q_C = 1$ OR $Q_A = Q_D = 1$. Connect the output of the AND gate already connected to J_C and K_C to one input of a 2-input AND gate. Connect the other input of this gate to Q_C. Connect the two inputs of another AND gate to Q_A and Q_D. The outputs of the two AND gates must be connected to the inputs of a 2-input OR gate whose output is connected to both J_D and K_D. The required circuit is shown in Fig. 6.17.

Fig. 6.17 *Synchronous decade counter*

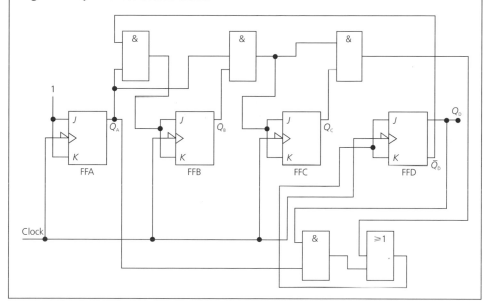

MSI non-synchronous counters

The non-synchronous counters in the 74HC logic family are the 74HC393 dual 4-bit ripple counter, the 74HC4020 14-bit ripple counter, the 74HC4040 12-bit ripple counter, and the 74HC6060 14-bit ripple counter. The first page of the data sheet for the 74HC4040 12-bit non-synchronous binary counter is given in Fig. 6.18. The package includes 12 flip-flops with all 12 Q outputs being accessible. No $\bar{Q}$ outputs are available. The non-synchronous clear terminal is active-HIGH. Each of the 14 flip-flops, except the first, is triggered by a negative-going transition at the preceding flip-flop's Q output terminal. Because of the inherent propagation delay (t_{pd}) of each flip-flop this means that the second flip-flop will not operate until t_{pd} seconds after the trailing-edge clock transition at the first flip-flop. The third flip-flop will not operate until $2t_{pd}$ seconds after that clock transition and so on for the later stages. This means that the propagation delays of the individual stages add up and hence the twelfth flip-flop will not change state until a time of $12t_{pd}$ after the clock transition. The 74HC4040 has a maximum clock frequency of 22 MHz at $V_{CC} = 4.5$ V and 25 MHz at $V_{CC} = 6$ V, while $t_{pd} = 38$ ns maximum at 4.5 V. The remainder of the data sheet gives similar information to other data sheets, such as the absolute maximum ratings, recommended operating conditions, electrical and switching characteristics, and timing requirements.

Fig. 6.18 *Data sheet for the 74HC4040 12-bit non-synchronous binary counter. (Courtesy of Texas Instruments)*

SN54HC4040, SN74HC4040
12-BIT ASYNCHRONOUS BINARY COUNTERS

SCLS160B – DECEMBER 1982 – REVISED MAY 1997

Package Options Include Plastic Small-Outline (D), Shrink Small-Outline (DB), Thin Shrink Small-Outline (PW), and Ceramic Flat (W) Packages, Ceramic Chip Carriers (FK), and Standard Plastic (N) and Ceramic (J) 300-mil DIPs

description

The 'HC4040 are 12-stage asynchronous binary counters with the outputs of all stages available externally. A high level at the clear (CLR) input asynchronously clears the counter and resets all outputs low. The count is advanced on a high-to-low transition at the clock (CLK) input. Applications include time-delay circuits, counter controls, and frequency-dividing circuits.

The SN54HC4040 is characterized for operation over the full military temperature range of –55°C to 125°C. The SN74HC4040 is characterized for operation from –40°C to 85°C.

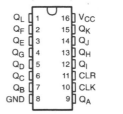

SN54HC4040 . . . J OR W PACKAGE
SN74HC4040 . . . D, DB, N, OR PW PACKAGE
(TOP VIEW)

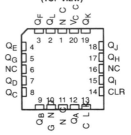

SN54HC4040 . . . FK PACKAGE
(TOP VIEW)

NC – No internal connection

logic symbol

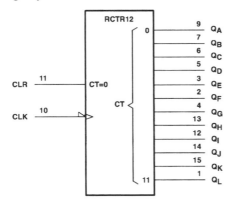

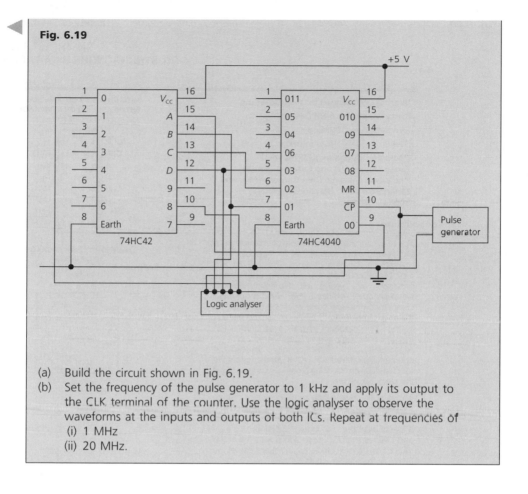

Fig. 6.19

(a) Build the circuit shown in Fig. 6.19.
(b) Set the frequency of the pulse generator to 1 kHz and apply its output to the CLK terminal of the counter. Use the logic analyser to observe the waveforms at the inputs and outputs of both ICs. Repeat at frequencies of
 (i) 1 MHz
 (ii) 20 MHz.

MSI synchronous counters

The four examples of synchronous counters in the 74HC logic family are:

(a) The 74HC162/163 4-bit binary counters which are also in the 74ALS, 74AS and 74LS logic families.
(b) The 74HC191/193 4-bit up-down counters which are in the 74ALS and 74LS families.

Part of the data sheet for the 74HC161/163 4-bit counters is given in Fig. 6.20. The difference between the two counters lies in their clear arrangements: the 161 has non-synchronous clear and the 163 has synchronous clear.

The switching characteristics in a data sheet give information on how quickly the counter operates and the conditions under which the test was conducted. f_{max} gives

SN54HC161, SN74HC161
4-BIT SYNCHRONOUS BINARY COUNTERS

SCLS297 – JANUARY 1996

Internal Look-Ahead for Fast Counting

Carry Output for n-Bit Cascading

Synchronous Counting

Synchronously Programmable

**Package Options Include Plastic
Small-Outline (D) and Ceramic Flat (W)
Packages, Ceramic Chip Carriers (FK), and
Standard Plastic (N) and Ceramic (J)
300-mil DIPs**

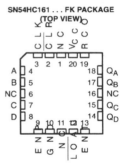

SN54HC161 . . . J OR W PACKAGE
SN74HC161 . . . D OR N PACKAGE
(TOP VIEW)

$\overline{CLR}$	1	16 V_{CC}
CLK	2	15 RCO
A	3	14 Q_A
B	4	13 Q_B
C	5	12 Q_C
D	6	11 Q_D
ENP	7	10 ENT
GND	8	9 $\overline{LOAD}$

SN54HC161 . . . FK PACKAGE
(TOP VIEW)

NC – No internal connection

description

These synchronous, presettable counters feature an internal carry look-ahead for application in high-speed counting designs. The 'HC161 are 4-bit binary counters. Synchronous operation is provided by having all flip-flops clocked simultaneously so that the outputs change coincident with each other when so instructed by the count-enable (ENP, ENT) inputs and internal gating. This mode of operation eliminates the output counting spikes that are normally associated with synchronous (ripple-clock) counters. A buffered clock (CLK) input triggers the four flip-flops on the rising (positive-going) edge of the clock waveform.

These counters are fully programmable; that is, they can be preset to any number between 0 and 9 or 15. As presetting is synchronous, setting up a low level at the load input disables the counter and causes the outputs to agree with the setup data after the next clock pulse, regardless of the levels of the enable inputs.

The clear function for the 'HC161 is asynchronous. A low level at the clear ($\overline{CLR}$) input sets all four of the flip-flop outputs low, regardless of the levels of the CLK, load ($\overline{LOAD}$), or enable inputs.

The carry look-ahead circuitry provides for cascading counters for n-bit synchronous applications without additional gating. Instrumental in accomplishing this function are ENP, ENT, and a ripple-carry (RCO) output. Both ENP and ENT must be high to count, and ENT is fed forward to enable RCO. Enabling RCO produces a high-level pulse while the count is maximum (9 or 15 with Q_A high). This high-level overflow ripple-carry pulse can be used to enable successive cascaded stages. Transitions at ENP or ENT are allowed, regardless of the level of CLK.

These counters feature a fully independent clock circuit. Changes at control inputs (ENP, ENT, or $\overline{LOAD}$) that modify the operating mode have no effect on the contents of the counter until clocking occurs. The function of the counter (whether enabled, disabled, loading, or counting) is dictated solely by the conditions meeting the stable setup and hold times.

The SN54HC161 is characterized for operation over the full military temperature range of –55°C to 125°C. The SN74HC161 is characterized for operation from –40°C to 85°C.

logic symbol

CTRDIV16

$\overline{CLR}$	1	CT=0
$\overline{LOAD}$	9	M1
		M2
ENT	10	G3
ENP	7	G4
CLK	2	C5/2,3,4+

3CT=15 — 15 — RCO

A	3	1,5D [1] — 14 — Q_A
B	4	[2] — 13 — Q_B
C	5	[4] — 12 — Q_C
D	6	[8] — 11 — Q_D

Fig. 6.20 *(Cont'd)*

typical clear, preset, count, and inhibit sequence

The following sequence is illustrated below:

1. Clear outputs to zero (asynchronous)
2. Preset to binary 12
3. Count to 13, 14, 15, 0, 1, and 2
4. Inhibit

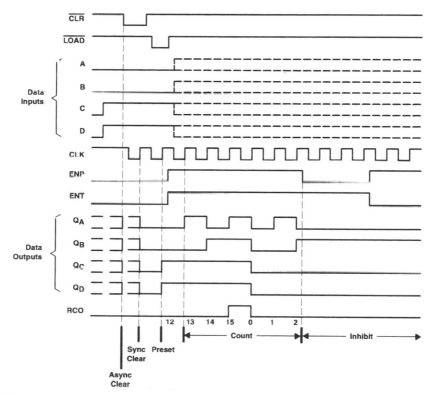

electrical characteristics over recommended operating free-air temperature range (unless otherwise noted)

PARAMETER	TEST CONDITIONS		V_{CC}	$T_A = 25°C$			SN54HC161		SN74HC161		UNIT
				MIN	TYP	MAX	MIN	MAX	MIN	MAX	
V_{OH}	$V_I = V_{IH}$ or V_{IL}	$I_{OH} = -20$ μA	2 V	1.9	1.998		1.9		1.9		V
			4.5 V	4.4	4.499		4.4		4.4		
			6 V	5.9	5.999		5.9		5.9		
		$I_{OH} = -4$ mA	4.5 V	3.98	4.3		3.7		3.84		
		$I_{OH} = -5.2$ mA	6 V	5.48	5.8		5.2		5.34		
V_{OL}	$V_I = V_{IH}$ or V_{IL}	$I_{OL} = 20$ μA	2 V		0.002	0.1		0.1		0.1	V
			4.5 V		0.001	0.1		0.1		0.1	
			6 V		0.001	0.1		0.1		0.1	
		$I_{OL} = 4$ mA	4.5 V		0.17	0.26		0.4		0.33	
		$I_{OL} = 5.2$ mA	6 V		0.15	0.26		0.4		0.33	
I_I	$V_I = V_{CC}$ or 0		6 V		+0.1	+100		±1000		±1000	nA
I_{CC}	$V_I = V_{CC}$ or 0, $I_O = 0$		6 V			8		160		80	μA
C_i			2 V to 6 V		3	10		10		10	pF

Fig. 6.20 *(Cont'd)*

timing requirements over recommended operating free-air temperature range (unless otherwise noted)

		V_{CC}	$T_A = 25°C$		SN54HC161		SN74HC161		UNIT
			MIN	MAX	MIN	MAX	MIN	MAX	
f_{clock}	Clock frequency	2 V	0	6	0	4.2	0	5	
		4.5 V	0	31	0	21	0	25	MHz
		6 V	0	36	0	25	0	29	
t_w	Pulse duration	CLK high or low	2 V	80		120		100	
			4.5 V	16		24		20	
			6 V	14		20		17	ns
		$\overline{CLR}$ low	2 V	80		120		100	
			4.5 V	16		24		20	
			6 V	14		20		17	
t_{su}	Setup time before CLK↑	A, B, C, or D	2 V	150		225		190	
			4.5 V	30		45		38	
			6 V	26		38		32	
		$\overline{LOAD}$ low	2 V	135		205		170	
			4.5 V	27		41		34	
			6 V	23		35		29	ns
		ENP, ENT	2 V	170		255		215	
			4.5 V	34		51		43	
			6 V	29		43		37	
		$\overline{CLR}$ inactive	2 V	125		190		155	
			4.5 V	25		38		31	
			6 V	21		32		26	
t_h	Hold time, all synchronous inputs after CLK↑		2 V	0		0		0	
			4.5 V	0		0		0	ns
			6 V	0		0		0	

switching characteristics over recommended operating free-air temperature range, C_L = 50 pF (unless otherwise noted)

PARAMETER	FROM (INPUT)	TO (OUTPUT)	V_{CC}	$T_A = 25°C$			SN54HC161		SN74HC161		UNIT
				MIN	TYP	MAX	MIN	MAX	MIN	MAX	
f_{max}			2 V	6	14		4.2		5		
			4.5 V	31	40		21		25		MHz
			6 V	36	44		25		29		
t_{pd}	CLK	RCO	2 V		83	215		325		270	
			4.5 V		24	43		65		54	
			6 V		20	37		55		46	
		Any Q	2 V		80	205		310		255	
			4.5 V		25	41		62		51	ns
			6 V		21	35		53		43	
	ENT	RCO	2 V		62	195		295		245	
			4.5 V		17	39		59		49	
			6 V		14	33		50		42	
t_{PHL}	$\overline{CLR}$	Any Q	2 V		105	210		315		265	
			4.5 V		21	42		63		53	
			6 V		18	36		54		45	ns
		RCO	2 V		110	220		330		275	
			4.5 V		22	44		66		55	
			6 V		19	37		56		47	
t_t		Any	2 V		38	75		110		95	
			4.5 V		8	15		22		19	ns
			6 V		6	13		19		16	

Fig. 6.20 *(Cont'd)*

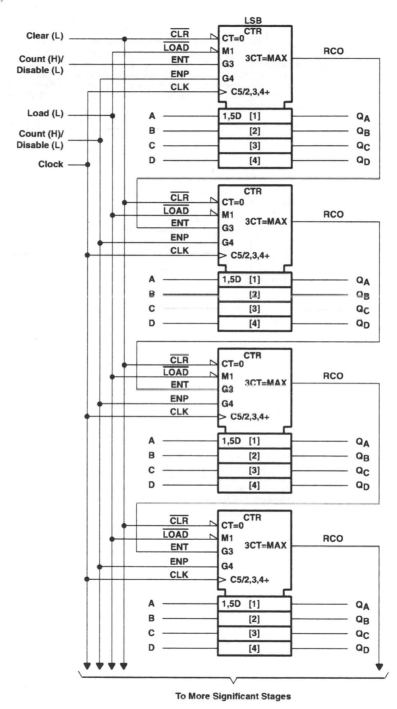

Figure 2

Fig. 6.20 *(Cont'd)*

The glitch on RCO is caused because the propagation delay of the rising edge of Q_A of the second stage is shorter than the propagation delay of the falling edge of ENT. RCO is the product of ENT, Q_A, Q_B, Q_C, and Q_D (ENT $\times Q_A \times Q_B \times Q_C \times Q_D$). The resulting glitch is about 7–12 ns in duration. Figure 3 shows the condition in which the glitch occurs. For simplicity, only two stages are being considered, but the results can be applied to other stages. Q_B, Q_C, and Q_D of the first and second stage are at logic one, and Q_A of both stages are at logic zero (1110 1110) after the first clock pulse. On the rising edge of the second clock pulse, Q_A and RCO of the first stage go high. On the rising edge of the third clock pulse, Q_A and RCO of the first stage return to a low level, and Q_A of the second stage goes to a high level. At this time, the glitch on RCO of the second stage appears because of the race condition inside the chip.

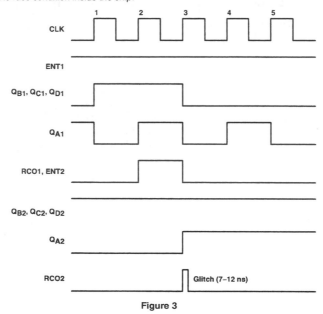

Figure 3

The glitch causes a problem in the next stage (stage three) if the glitch is still present when the next rising clock edge appears (clock pulse 4). To ensure that this does not happen, the clock frequency must be less than the inverse of the sum of the clock-to-RCO propagation delay and the glitch duration (t_g). In other words, $f_{max} = 1/(t_{pd} \text{ CLK-to-RCO} + t_g)$. For example, at 25°C at 4.5-V V_{CC}, the clock-to-RCO propagation delay is 43 ns and the maximum duration of the glitch is 12 ns. Therefore, the maximum clock frequency that the cascaded counters can use is 18 MHz. The following tables contain the f_{clock}, t_w, and f_{max} specifications for applications that use more than two 'HC161 devices cascaded together.

		V_{CC}	$T_A = 25°C$		SN54HC161		SN74HC161		UNIT
			MIN	MAX	MIN	MAX	MIN	MAX	
f_{clock}	Clock frequency	2 V	0	3.6	0	2.5	0	2.9	MHz
		4.5 V	0	18	0	12	0	14	
		6 V	0	21	0	14	0	17	
t_w	Pulse duration, CLK high or low	2 V	140		200		170		ns
		4.5 V	28		40		36		
		6 V	24		36		30		

switching characteristics over recommended operating free-air temperature range, C_L = 50 pF (unless otherwise noted) (see Note 3)

PARAMETER	FROM (INPUT)	TO (OUTPUT)	V_{CC}	$T_A = 25°C$		SN54HC161		SN74HC161		UNIT
				MIN	MAX	MIN	MAX	MIN	MAX	
f_{max}			2 V	3.6		2.5		2.9		MHz
			4.5 V	18		12		14		
			6 V	21		14		17		

NOTE 3:   These limits apply only to applications that use more than two 'HC161 devices cascaded together.

minimum and typical figures for the clock frequency; t_{pLH} and t_{pHL} give the time in nanoseconds that it takes for a change in the output from LOW to HIGH, or from HIGH to LOW, to take place when $\overline{PRE}$ is active. The times are measured from the instant $\overline{PRE}$ is taken LOW. Similar figures are then given for the $\overline{CLR}$ input.

In the logic symbol the label CTRDIV16 indicates a divide-by-16 counter. CT = 0 indicates the clear function; clear occurs when $\overline{CLR}$ is taken LOW and a clock pulse arrives – this is shown by the figure 5 in front of CT. A 5 before CT indicates that the input clears the counter to 0 but under the control of the clock (C5). The label 3CT = 15 means that when the input is active the circuit will count up to 15.

Cascaded counters

The data sheet shows how two, or more, counters may be connected in cascade to provide a greater count. If the 'HC161 is used as a single unit, or only two cascaded together, then the maximum clock frequency that the device can use is not limited because of the glitch. In these situations, the device can be operated at the maximum specifications. The maximum count of two counters in cascade is 16^2 or 256, of three counters in cascade is $16^3 = 4096$ and so on. Suppose a divide-by-153 counter is to be designed. There are two ways of achieving this:

(a) $153_{10} = 99H = 1001\ 1001$. If the count starts from 0, the circuit must be cleared when the count reaches 153_{10} or $1001\ 1001$. Hence, the Q_A and Q_D terminals of both counters must be connected to the four inputs of a 4-input NAND gate. The output of the NAND gate is then connected to the $\overline{CLR}$ input of each IC. When the count reaches 153_{10} the counter will move to the all 0s state and then resume the count.

(b) Subtract 153_{10} from 256_{10} (i.e. 16^2). This gives $103_{10} = 67H = 0110\ 0111$. Load 0110 into the most significant stage and load 0111 into the least significant stage. The count then starts from $0110\ 0110$ and progresses to $1111\ 1111$; at the next clock pulse the circuit should go back to its initial state of 67H to start another count sequence. Hence, connect all the Q outputs of the most significant counter to the four inputs of the 4-input NAND gate, and connect the output of the gate to the LOAD terminal of each counter.

PRACTICAL EXERCISE 6.5

Aim: to build and test a divide-by-153 counter.
Components and equipment: two 74HC161 or 74HC163 ICs, ten LEDs and ten 270 Ω resistors. Pulse generator. Breadboard. Power supply.
Procedure:

(a) Build the circuit shown in Fig. 6.21.

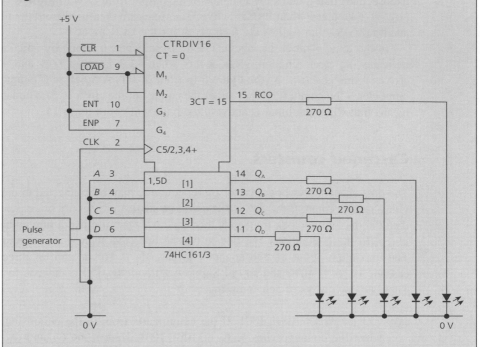

Fig. 6.21

(b) Set the pulse generator to 1 kHz, 5 V and 50 per cent duty cycle and connect it to the CLK input. Observe the logical state of the LEDs. Check that the circuit is counting correctly from 0 through to 15 using binary arithmetic. Check that the RCO output goes HIGH when the count reaches 15, and then goes LOW again as soon as the count reverts to 0. Check the operation of the circuit at a frequency near to the maximum frequency for the IC.

(c) Build the circuit shown in Fig. 6.22 which is the first circuit cascaded with a second counter.

(d) Confirm that the circuit counts correctly from 0 to 255. The frequency of the pulse generator should be chosen to make it easier to note the output frequency. A digital frequency meter would assist in this if available.

(e) Modify the circuit so that it operates as a divide-by-153 counter using method (a). The $\overline{CLR}$ pins of the two counters will be held HIGH until all four NAND gate inputs go HIGH and then they will go LOW to clear the circuit.

(f) Check that the circuit operates as a divide-by-153 counter. Set the frequency of the pulse generator to a suitable value to make the count of the higher-order IC easier to see, or use a frequency counter.

(g) Now modify the circuit again (or build it anew) so that it divides by 153 using method (b).

(h) Confirm that this circuit also counts from 0 to 153_{10}.

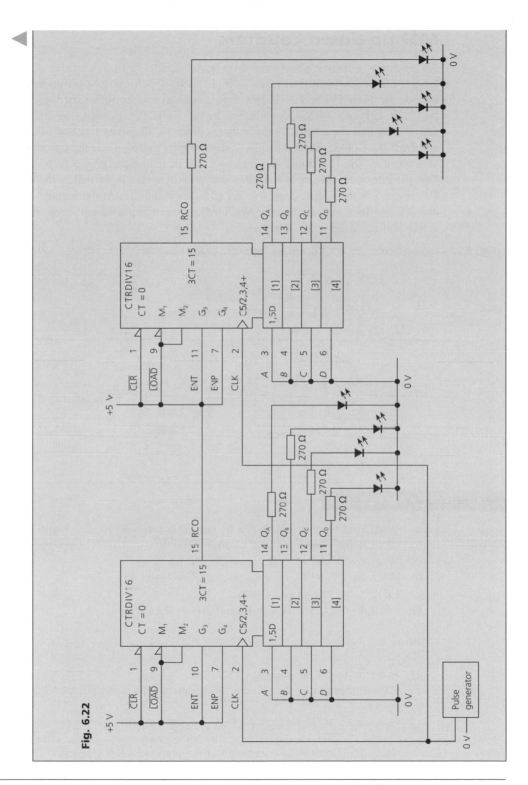

Fig. 6.22

MSI up-down counters

An up-down counter is able to count up from 0 to $2^n - 1$, or down from $2^n - 1$ to 0, i.e. for a 4-bit circuit from 0 to 15 or from 15 to 0. Synchronous 4-bit up-down counters include the 74HC191 and 74HC193. Figure 6.23(a) and (b) shows the logic symbols for the 74HC191 and 74HC193 counters. In the 74HC191 symbol, the $\overline{LOAD}$ input is active-LOW and the label C5 means that it controls all inputs labelled 5D, i.e. A, B, C and D. If the $\overline{LOAD}$ pin is HIGH the load facility is disabled and the circuit can then act as a counter. The $D/\overline{U}$ input creates two mode dependencies M_2(DOWN) and M_3(UP). The triangle at the M_3 input indicates that a LOW at this input makes the circuit count UP; if pin 5 is HIGH the circuit will act as a down-counter. At the outputs $2(CT = 0)Z6$ and $3(CT = 0)Z6$ indicate that the MAX/MIN output at pin 12 will count down if M_2 is active and count up if M_3 is active.

Fig. 6.23 *Logic symbols: (a) 74HC191 and (b) 74HC193 up-down counters*

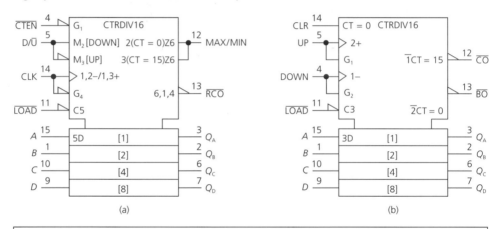

EXAMPLE 6.8

(a) What kind of circuit is shown by the logic symbol given in Fig. 6.23(b)?
(b) What is the meaning of the symbol $CT = 0$?
(c) What do 2+ and G_1 indicate on the UP pin 5?
(d) What do 1– and G_2 indicate on the DOWN pin 4?
(e) Explain the meaning of the label C3.
(f) What do the labels $\overline{1}CT = 15$ and $\overline{2}CT = 0$ indicate?

Solution

(a) A divide-by-16 counter (*Ans.*)
(b) $CT = 0$ means that the counter will reset to zero when pin 14 (CLR) is HIGH (*Ans.*)
(c) G_1 means that the input is ANDed with the output(s) having the same number (3), i.e. $1CT = 15$. The + sign indicates an up-count. Hence, $\overline{CO}$ goes LOW when the circuit is up-counting and the count reaches 15 (*Ans.*)

(d) G_2 means that the input is ANDed with any output(s) having the same number, i.e. 2CT = 0. The minus sign indicates a down-count. $\overline{CO}$ goes LOW when the count reaches 0 (*Ans.*)

(e) C3 indicates that the input controls the inputs with a 3 label, i.e. *A*, *B*, *C* and *D* (*Ans.*)

(f) Count up to 15 and count down to 0 (*Ans.*)

EXAMPLE 6.9

Figure 6.24 shows a typical load, count and inhibit sequence for a 74HC191 4-bit synchronous up-down binary counter.

Fig. 6.24

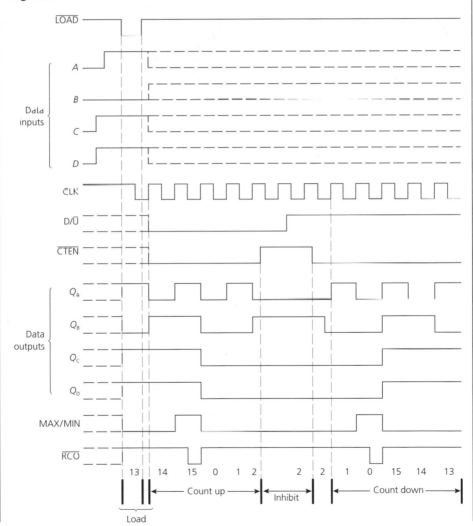

(a) What binary number has been loaded?
(b) Determine the counting sequence of the circuit.
(c) What happens when $\overline{\text{CTEN}}$ is HIGH?
(d) When does MAX/MIN go HIGH?
(e) What happens if $\overline{\text{LOAD}}$ is held LOW?

Solution

(a) 13_{10} (Ans.)
(b) 13, 14, 15, 0, 1, 2, stop, 1, 0, 15, 14, 13 (Ans.)
(c) The counter is disabled (Ans.)
(d) When the count is 15, or 0 depending on the count direction (Ans.)
(e) The circuit is disabled (Ans.)

PRACTICAL EXERCISE 6.6

Aim: to employ the 74HC191 as a down-counter with a count starting from 7_{10} and ending at 0_{10}. To discover the effect of changing the count direction.
Components and equipment: one 74HC191 up-down counter IC, five LEDs and five 270 Ω resistors. Pulse generator. Breadboard. Power supply.
Procedure:

(a) Build the circuit shown in Fig. 6.25. Set the pulse generator to 1 kHz, 5 V and 50 per cent duty cycle. Connect the pulse generator to the CLK input.

Fig. 6.25

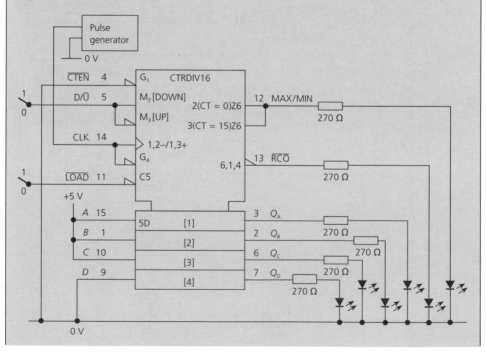

(b) Set the initial count to 7, i.e. load 0111, and connect 5 V to the D/U̅ pin. Load the input data. Remove the LOW from the LOAD̅ pin and the circuit will start to count down from 7_{10}. Note the logical states of the LEDs and find out if the operation of the circuit is correct.

(c) Modify the circuit so that it restores to 7_{10} after 0_{10} instead of going to 15_{10}.

(d) Now connect the D/U̅ pin to 5 V instead of to 0 V and again determine the action of the circuit. Comment on the result.

(e) If a logic analyser is available use it to display the input and output waveforms of the circuit and check the operation of the counter in this way.

(f) Try running the circuit first at 1 MHz and then at 30 MHz and each time note the action of the circuit.

EXERCISES

6.1   (a) Each stage in a 14-bit binary counter has a propagation delay of 20 ns. Calculate the maximum clock frequency.

(b) Figure 6.26 shows the logic symbol of a counter
   (i)   what kind of counter is it?
   (ii) is it leading edge or trailing-edge triggered?
   (iii) can the circuit be programmed to have some other count?
   (iv) is the counter cleared synchronously or non-synchronously?

6.2   The logic symbol of a device is shown in Fig. 6.27.
(a) State the kind of circuit it represents.
(b) When are the outputs in their high-impedance state?
(c) When is the input data latched?
(d) What does the triangle point to the right indicate?

Fig. 6.26

Fig. 6.27

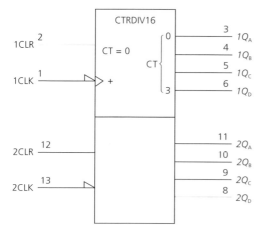

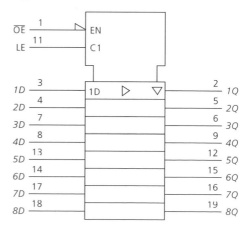

Fig. 6.28

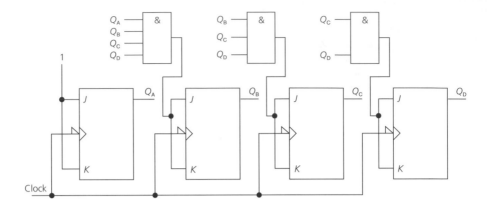

Fig. 6.29

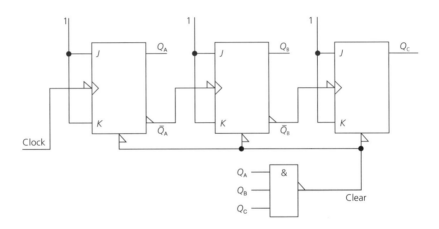

6.3 Show how three J-K flip-flops can be connected to give a divide-by-5 ripple counter. State two disadvantages of the circuit.

6.4 Show how 74HC112 J-K flip-flops can be connected to form a divide-by-6 ripple counter.

6.5 Connect four flip-flops to give a count of 16. The circuit should clear all stages to 0 when the count reaches 13. Design the circuit. State its disadvantages.

6.6 Determine the count sequence of the circuit shown in Fig. 6.28.

6.7 A 20 MHz clock is applied to a 4-bit ripple counter. Each flip-flop has $t_{pd} = 20$ ns. Determine if any of the counter states will not occur because of the propagation delays.

6.8 Determine the counting sequence of the circuit shown in Fig. 6.29.

6.9 Determine the count sequence of the circuit given in Fig. 6.30.

6.10 Design a circuit that will repeatedly first count up from 0 to 15 and then count down from 15 to 0.

Fig. 6.30

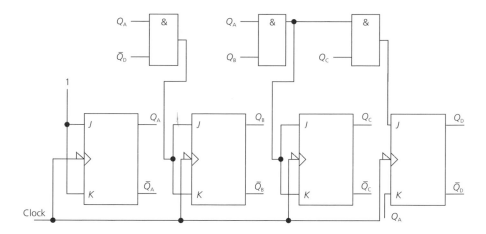

6.11 The 74HC174 hex D flip-flop IC is employed in a circuit that uses all six flip-flops. The flip-flops are clocked at 1.5 MHz and $V_{CC} = 5$ V. Calculate the power dissipation.

6.12 The 74HC174 hex D flip-flop is connected as a binary counter using all six flip-flops.
(a) What is the modulus of the counter?
(b) What is the maximum clock frequency?

6.13 (a) How many 74HC163 synchronous 4-bit binary counters are required to give a count of
(i) 200
(ii) 4002?
(b) What connections are required for each circuit?

6.14 (a) The data sheet of a flip-flop specifies that the clock pulses must have a minimum HIGH width of 28 ns and minimum LOW width of 33 ns. Calculate the maximum operating frequency.
(b) Determine the d.c. current taken from a 5 V power supply and the static power dissipated.
(c) If the minimum set-up time is 20 ns and the propagation delay from CLK to Q is 50 ns, calculate the maximum clock frequency.

6.15 Design a non-synchronous counter to have a modulus of 12 that follows the normal binary sequence.

6.16 Design a synchronous divide-by-7 counter.

6.17 A 74HC174 hex D flip-flop IC is used with $V_{CC} = 4.5$ V.
(a) Determine from the data sheet the minimum clock pulse width for a 50 per cent duty cycle.
(b) When the D input changes from LOW to HIGH, for how long must it stay HIGH before the next clock transition occurs?
(c) For how long must the D input stay HIGH after the clock transition has occurred?

6.18 For the 74HC109 dual J-K flip-flop determine from the data sheet:

    (a)   The maximum time between the 50 per cent points on the leading edge of a clock pulse and the Q output pulse when Q changes from LOW to HIGH.

    (b)   The minimum time between the 50 per cent points of the $\overline{\text{PRE}}$ pulse input and the Q output.

    (c)   The maximum number of $J\text{-}K$ inputs of other 74HC109 devices that can be driven.

6.19 Two octal D flip-flop or latch ICs are to be employed to store a byte of data at the leading edge of a clock pulse. The D ICs available are the 74HC273, 74HC373, the 74HC374, the 74HC534, the 74HC563, the 74HC573 and the 74HC564.

    (a)   Draw the basic circuit.

    (b)   List the brief data on each device and suggest which might best be used.

6.20 Design a divide-by-365 counter using 74HC193 synchronous counter ICs.

6.21 Four modulus 16 synchronous counters are cascaded.

    (a)   What is the maximum count?

    (b)   The counters are loaded with the hex number 77A1H. Determine the actual count.

    (c)   If the clock frequency is 15 MHz calculate the frequency of the output of the most significant counter.

6.22 Determine the count of the circuit given in Fig. 6.31.

Fig. 6.31

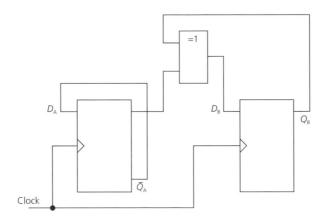

7 Shift registers and shift register counters

After reading this chapter you should be able to:

- Connect J-K or D flip-flops to produce a shift register.
- Explain how data is entered into, taken out of, and moved through a shift register.
- Understand the differences between SISO, SIPO, PIPO and PISO shift registers.
- Understand the data sheet of a shift register.
- Draw the logic symbol of a shift register and explain the meaning of each label used.
- State some applications for shift registers, such as a time delay circuit, short-term memory, and parallel-to-serial conversion.
- Connect a shift register to operate as a ring counter.
- Connect a shift register to operate as a Johnson counter.
- Compare the differences between ring and Johnson counters.
- Determine the count sequences of both ring and Johnson counters.
- Know that the count of an n-bit ring counter is n, and of an n-bit Johnson counter is $2n$.
- Understand the operation of a feedback shift register, both linear and non-linear.

Shift registers

A *shift register* consists of a number of J-K flip-flops connected in cascade as shown in Fig. 7.1(a). The connection of Q_A to J_B and of $\bar{Q}_A$ to K_B and so on ensures that each flip-flop will take up *either* the state $Q = 0$ *or* $Q = 1$ of the preceding flip-flop at the trailing edge of each clock pulse. Each flip-flop transfers its *bit* of information to the following flip-flop whenever a clock pulse occurs. Alternatively, D flip-flops can be used with the Q output of each stage connected to the D input of the next stage; this is shown in Fig. 7.1(b). The flip-flops are usually provided with a clear, or reset, terminal so that the register can be cleared or reset to 0. Often preset terminals are also provided.

A shift register can be converted into a *ring counter* by applying feedback between the output and the input terminals; see Fig. 7.2(a). A ring counter does not count in binary but follows a repetitive sequence of Q output states. A ring counter is often employed to control a sequence of events that occurs in some process. A twisted-ring or

Fig. 7.1 *Shift register using (a) J-K flip-flops and (b) D flip-flops*

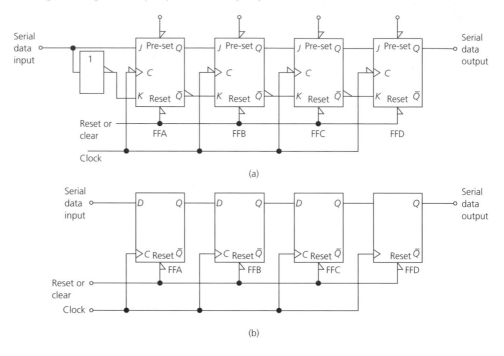

(a)

(b)

Fig. 7.2 *(a) Ring counter and (b) twisted-ring counter*

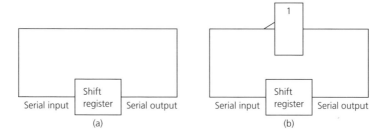

Johnson counter is similar to a ring counter but it has the logical state fed back from output to input inverted as shown in Fig. 7.2(b).

Operation of a shift register

The shift register shown in Fig. 7.1 is known as a *serial-in serial-out* (SISO) shift register. A SIPO shift register accepts input data serially, 1 bit at a time, from a single source. The stored data is also taken out of the register serially.

Suppose that, initially, all four flip-flops are cleared, i.e. $Q_A = Q_B = Q_C = Q_D = 0$. If the data to be stored by the register is applied serially to the input terminal, the input

Fig. 7.3 *Four-bit shift register: timing diagram*

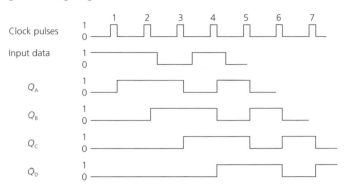

data will shift one stage to the right at the trailing edge of each clock pulse. This means that if the stored data is a binary number the most significant bit is on the right. If the data applied to the input terminal is 1101, the bits 11 are loaded first and the action of the circuit is as follows:

- At the end of the first clock pulse $Q_A = 1$.
- After the second clock pulse, $Q_A = 1$ and $Q_B = 1$.
- The third clock pulse will give $Q_B = Q_C = 1$ but the first flip-flop will have reset to give $Q_A = 0$.
- The fourth data bit is a 1 and this will be stored by flip-flop A at the end of the fourth clock pulse, when the bits stored by the other flip-flops all move one stage to the right. At this point the stored data is 1101.
- At the trailing edge of the fifth clock pulse flip-flop A clears and all the other flip-flops take up the logical state of the preceding stage. The most significant bit of the data word is shifted out of the register and is lost.
- After the sixth clock pulse, the two most significant bits have been lost.
- After eight clock pulses all the data has been shifted out of the register and lost.

The timing diagram illustrating the operation of the shift register is shown in Fig. 7.3. The Q_A waveform is delayed behind the input data waveform by a time that may not be equal (in the figure it isn't) to the periodic time of the clock waveform. However, the Q waveforms of all the other flip-flops are delayed behind the Q waveform of the preceding flip-flop by *exactly* the clock period.

The shift register can be used to multiply or divide a binary number by a factor of 2 by shifting the number one place to the right, or one place to the left, respectively. In either case, any part of the number must not be shifted right out of the register. Often a strobe, or enable, terminal is provided to prevent this from happening; the enable terminal is used to stop the right-shifting process after all 4 bits have been entered into the register. A SIPO shift register can be employed as a delay circuit or as a short-term memory.

A shift register can also be operated in any one of three different ways. The alternative methods of operation are:

(a) Serial-in parallel-out (SIPO).
(b) Parallel-in serial-out (PISO).
(c) Parallel-in parallel-out (PIPO).

In a SIPO shift register the Q output of each stage is accessible externally and once the data word has been stored each bit appears at a different output terminal. A data word is fed into the circuit in the same way as for a SISO register and, when the complete word has been entered, the data can be read out from the parallel output terminals simultaneously. The PISO shift register operates in the opposite way. The data to be stored is set up by first clearing all stages and then entering the data to be stored at the parallel input terminals simultaneously instead of sequentially. A LOAD command will then cause the data to be entered into the register. The stored data can then be read out serially under the control of the clock in the same way as for a SISO shift register. With a PIPO shift register the data is loaded into the register in the same way as for a PISO device; when the complete digital word has been stored all the bits in the word can be read out simultaneously from the Q output terminals.

Some shift registers are able to operate in more than one of the four modes of operation and/or have the capability to shift data either to the left or to the right. A *bi-directional* register has two data input terminals, one of which is used for shifting serial data to the right and the other for left-shifting. The direction in which the data is shifted is determined by the logic levels applied to shift left/shift right control terminals.

EXAMPLE 7.1

A 74HC174 hex D flip-flop is connected to provide a 6-stage SISO shift register.

(a) Determine the propagation delay through the circuit.
(b) What is the maximum clocking frequency?

Solution

(a) From the data sheet (p. 162) t_{pd} = 40 ns maximum. Therefore propagation delay = 240 ns (*Ans.*)
(b) Maximum clocking frequency = $1/(240 \times 10^{-9})$ = 4.17 MHz (*Ans.*)

MSI shift registers

Several different shift registers are offered in the various 74TTL logic families but the 74CMOS logic families include only the following 8-bit devices: 74HC164 SIPO, 74HC165 PISO, 74HC166 PISO and 74HC594/5 SIPO with output latches and 3-state outputs. The '164 SIPO shift register is also available in the 74LVC logic family. A part of the data sheet for the 74HC164 8-bit shift register is given in Fig. 7.4.

The logic symbol for the 74HC164 consists of the common-control block and a block that represents the eight flip-flops in the register. The label SRG8 indicates an 8-bit shift register. The CLK input terminal has an internal label C1/→. The / merely separates C1 and →: C1 indicates that the clock input controls the entry of data into the least significant flip-flop A, since only flip-flop A has the label 1D. The data entry into the circuit is shown as the AND combination of two inputs *A* and *B*; both inputs must be HIGH for data entry to take place. There are no external data inputs to flip-flops B through to H and the clock does not control any of these stages (no D label). The → label indicates right-shifting of data at the leading edge of a clock pulse (because of the wedge at the clock input and no small triangle). The CLR input provides non-synchronous clearing of all the flip-flops when it is taken LOW.

Fig. 7.4 *Data sheet for the 74HC164 8-bit shift register. (Courtesy of Texas Instruments)*

SN54HC164, SN74HC164
8-BIT PARALLEL-OUT SERIAL SHIFT REGISTERS

SCLS115A – DECEMBER 1982 – REVISED JANUARY 1996

AND-Gated (Enable/Disable) Serial Inputs
Fully Buffered Clock and Serial Inputs
Direct Clear
Package Options Include Plastic
Small-Outline (D) and Ceramic Flat (W)
Packages, Ceramic Chip Carriers (FK), and
Standard Plastic (N) and Ceramic (J)
300-mil DIPs

description

These 8-bit shift registers feature AND-gated serial inputs and an asynchronous clear ($\overline{CLR}$) input. The gated serial (A and B) inputs permit complete control over incoming data; a low at either input inhibits entry of the new data and resets the first flip-flop to the low level at the next clock (CLK) pulse. A high-level input enables the other input, which then determines the state of the first flip-flop. Data at the serial inputs can be changed while CLK is high or low, provided the minimum setup time requirements are met. Clocking occurs on the low-to-high-level transition of CLK.

The SN54HC164 is characterized for operation over the full military temperature range of –55°C to 125°C. The SN74HC164 is characterized for operation from –40°C to 85°C.

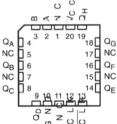

SN54HC164 . . . J OR W PACKAGE
SN74HC164 . . . D OR N PACKAGE
(TOP VIEW)

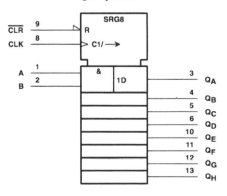

SN54HC164 . . . FK PACKAGE
(TOP VIEW)

NC – No internal connection

logic symbol

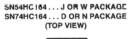

FUNCTION TABLE

INPUTS				OUTPUTS		
$\overline{CLR}$	CLK	A	B	Q_A	Q_B	. . . Q_H
L	X	X	X	L	L	L
H	L	X	X	Q_{A0}	Q_{B0}	Q_{H0}
H	↑	H	H	H	Q_{An}	Q_{Gn}
H	↑	L	X	L	Q_{An}	Q_{Gn}
H	↑	X	L	L	Q_{An}	Q_{Gn}

Q_{A0}, Q_{B0}, Q_{H0} = the level of Q_A, Q_B, or Q_H, respectively, before the indicated steady-state input conditions were established
Q_{An}, Q_{Gn} = the level of Q_A or Q_G before the most recent ↑ transition of CLK: indicates a 1-bit shift

Fig. 7.4 *(Cont'd)*

typical clear, shift, and clear sequence

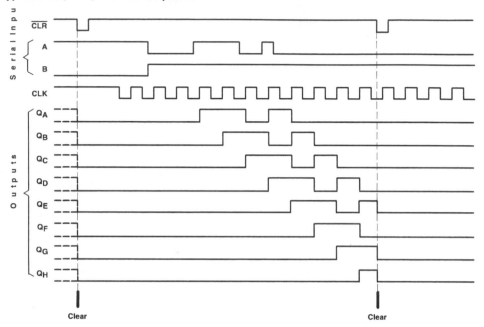

timing requirements over recommended operating free-air temperature range (unless otherwise noted)

			V_{CC}	$T_A = 25°C$		SN54HC164		SN74HC164		UNIT
				MIN	MAX	MIN	MAX	MIN	MAX	
f_{clock}	Clock frequency		2 V	0	6	0	4.2	0	5	MHz
			4.5 V	0	31	0	21	0	25	
			6 V	0	36	0	25	0	28	
t_w	Pulse duration	$\overline{CLR}$ low	2 V	100		150		125		ns
			4.5 V	20		30		25		
			6 V	17		25		21		
		CLK high or low	2 V	80		120		100		
			4.5 V	16		24		20		
			6 V	14		20		18		
t_{su}	Setup time before CLK↑	Data	2 V	100		150		125		ns
			4.5 V	20		30		25		
			6 V	17		25		21		
		$\overline{CLR}$ inactive	2 V	100		150		125		
			4.5 V	20		30		25		
			6 V	17		25		21		
t_h	Hold time, data after CLK↑		2 V	5		5		5		ns
			4.5 V	5		5		5		
			6 V	5		5		5		

Fig. 7.4 *(Cont'd)*

switching characteristics over recommended operating free-air temperature range, C_L = 50 pF (unless otherwise noted) (see Figure 1)

PARAMETER	FROM (INPUT)	TO (OUTPUT)	V_{CC}	T_A = 25°C			SN54HC164		SN74HC164		UNIT
				MIN	TYP	MAX	MIN	MAX	MIN	MAX	
f_{max}			2 V	6	10		4.2		5		MHz
			4.5 V	31	54		21		25		
			6 V	36	62		25		28		
t_{PHL}	$\overline{CLR}$	Any Q	2 V		140	205		295		255	ns
			4.5 V		28	41		59		51	
			6 V		24	35		51		46	
t_{pd}	CLK	Any Q	2 V		115	175		265		220	
			4.5 V		23	35		53		44	
			6 V		20	30		45		38	
t_t			2 V		38	75		110		95	ns
			4.5 V		8	15		22		19	
			6 V		6	13		19		16	

operating characteristics, T_A = 25°C

PARAMETER		TEST CONDITIONS	TYP	UNIT
C_{pd}	Power dissipation capacitance	No load	135	pF

PRACTICAL EXERCISE 7.1

Aim: to investigate the performance of the 74HC164 shift register.
Components and equipment: one 74HC164 8-bit shift register, eight LEDs and eight 270 Ω resistors. Breadboard. Power supply.
Procedure:

(a) Build the circuit shown in Fig. 7.5.

(b) Connect $\overline{CLR}$ to the 0 voltage level for a moment to ensure that all the stages are cleared. Then switch $\overline{CLR}$ to the logic 1 level. Set the pulse generator to 10 Hz, 5 V and 50 per cent duty cycle and with the B input connected to 1 note the logical states of the eight LEDs after eight clock pulses. Now connect the B input to logic 0 and repeat. Switch $\overline{CLR}$ to 0 and see if all the LEDs dim. Return $\overline{CLR}$ to 1.

(c) Operate the B switch to input the 8-bit data word 10101010. Immediately the eighth clock pulse has entered the register switch it off. Note the logical states of the LEDs; has the input data word been stored? Note that if the pulse generator is not turned OFF quickly enough one or more of the 'stored' bits will be clocked out of the circuit. With the digital word stored in the register see what happens when the A input is taken LOW.

(d) Try operating the B switch first rapidly and then slowly and note what happens to the data stored in the circuit. What conclusion do you draw from this?

(e) Connect another pulse generator to input B, set to 5 V at 1 kHz, and increase the frequency of the clock generator to 10 kHz. Connect a logic analyser to the CLK input and to all eight Q output terminals and observe the waveforms in the circuit.

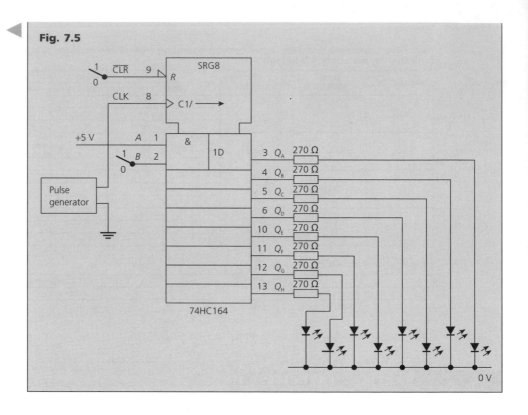

Fig. 7.5

Shift register counters

A shift register counter uses a SIPO device with the serial output terminal connected back to the serial input terminal. The feedback connection may be a direct link which gives a ring counter, it may include an inverter (or a register using J-K flip-flops will have the J-K to Q-$\bar{Q}$ connections reversed) giving a twisted-ring or Johnson counter, or it may include logic gates which results in a feedback shift register.

Ring counter

The circuit of a ring counter built using J-K flip-flops is shown in Fig. 7.6. The Q output of the highest-order flip-flop is connected to the J input terminal of the lowest-order flip-flop, and the $\bar{Q}$ output terminal is connected to the K input terminal. If D flip-flops are used the Q output of the highest-order flip-flop is connected to the D input of the lowest-order flip-flop as shown in Fig. 7.7. If an MSI shift register is employed the serial output terminal is connected to the serial input terminal.

To start a ring counter one of its stages must be preset to have $Q = 1$. The internal states of the shift register will now be circulated around the loop and, for an n-stage register, the binary pattern will be repeated at the output after every n clock pulses. The ring counter can hence be employed as a *delay circuit*; the data is stored and circulated internally in a multiple bit pattern and any particular bits can be read out as and when required.

Fig. 7.6 *Ring counter using J-K flip-flops*

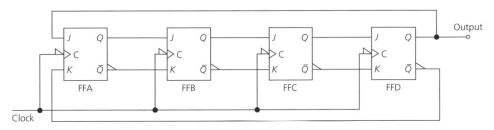

Fig. 7.7 *Ring counter using D flip-flops*

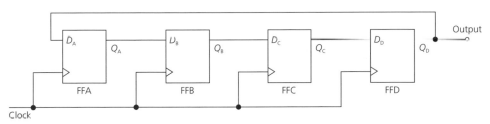

Table 7.1					
Clock pulse	Q_D	Q_C	Q_B	Q_A	Decimal number
0	0	0	0	1	1
1	0	0	1	0	2
2	0	1	0	0	4
3	1	0	0	0	8
4	0	0	0	1	1

If only a single bit is circulated around the counter the output of each flip-flop will give a uniquely timed *1-out-n* count. The count can be read by finding the flip-flop that is set and no decoding is required. A 4-stage shift register, for example, has a starting state of $Q_A = 1$, $Q_B = Q_C = Q_D = 0$. The bit pattern held in the circuit is given in Table 7.1 and the timing diagram is shown in Fig. 7.8(a). If both flip-flops A and B are initially set a double-width pulse will circulate around the register as shown in Fig. 7.8(b).

The four-stage ring counter goes through four different states before the count sequence is repeated. Hence, a modulus N counter requires N stages. This is more stages than a binary counter would need for the same modulus.

If a fifth stage is added to the register clock pulse 4 will give the state $Q_E = 1$, $Q_D = Q_C = Q_B = Q_A = 0$, and clock pulse 5 will result in the state $Q_E = Q_D = Q_C = Q_B = 0$ and $Q_A = 1$. The timing diagram for a 5-stage ring counter is shown in Fig. 7.9(a). If both flip-flops A and B are preset a double-width pulse will circulate around the circuit; see Fig. 7.9(b).

Fig. 7.8 *Timing diagram for a 4-bit ring counter*

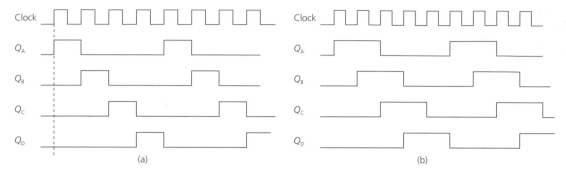

(a)

(b)

Fig. 7.9 *Timing diagram for a 5-bit ring counter*

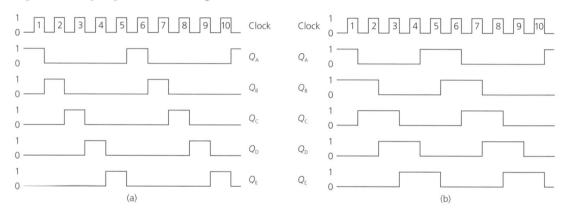

(a)

(b)

EXAMPLE 7.2

(a) How many stages would
   (i) a binary counter
   (ii) a ring counter
   need to give a count of 8?
(b) Determine the count sequence of
   (i) a four-stage ring counter
   (ii) a five-stage ring counter.

Solution

(a) (i) Number of stages = 2^n, so $n = 3$ (*Ans.*)
   (ii) Number of stages = modulus = 8 (*Ans.*)
(b) (i) From Table 7.1, the count sequence is 1, 2, 4, 8, 1, etc. (*Ans.*)
   (ii) Extending Table 7.1 gives the count sequence as 1, 2, 4, 8, 16, 1, etc. (*Ans.*)

Aim: to investigate the performance of a ring counter.

Components and equipment: one 74HC164 8-bit shift register, eight LEDs and eight 270 Ω resistors. Pulse generator. Breadboard. Power supply.

Procedure:

(a) Build the circuit given in Fig. 7.10.
(b) It will be found that the circuit will not start. Connect Q_H to the B switch in place of the earth connection so that input B may be switched between 5 V and Q_H.
(c) Clear the circuit by switching the $\overline{CLR}$ input to 0 V for a short time. When the $\overline{CLR}$ pin is taken HIGH the circuit still will not start. Now operate switch B to 5 V for a time short enough to allow just one pulse into the register. Note that now the single 1 bit will circulate around the register as expected
(d) Now that it has been confirmed that the circuit operates correctly once it has been started devise a means of self-starting the counter and then modify the circuit. Check that the modified circuit works correctly.

Fig. 7.10

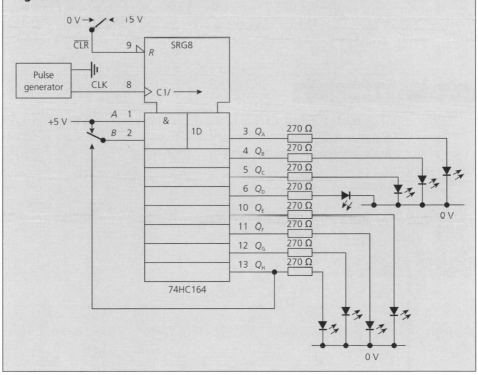

Self-starting a ring counter

For a ring counter to work it is necessary for one stage to be preset to store a 1 bit. This means that manual operation using a key or a switch, or automatic starting circuitry is

Fig. 7.11 *Self-starting ring counter*

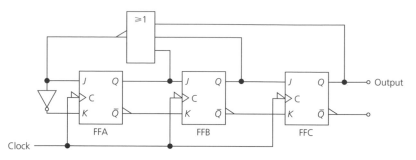

required. An example of a self-starting ring counter is shown in Fig. 7.11. The Q outputs of each stage are connected to the inputs of a NOR gate whose output is connected to the serial input of the register. When the power supply is first switched on the initial states of the stages are unpredictable but, no matter what these initial states happen to be, within a few clock pulses the output of the NOR gate will become 1 and then the counter will circulate just a single 1 bit.

Suppose that the initial state of the circuit is 1111. After one clock pulse $Q_A = 0$ and $Q_B = Q_C = Q_D = 1$; after clock pulse 2, $Q_A = Q_B = 0$ and $Q_C = Q_D = 1$; after clock pulse 3, only $Q_D = 1$ and now the single 1 bit will circulate around the circuit.

EXAMPLE 7.3

When the D flip-flop version of the ring counter shown in Fig. 7.11 is switched on its initial state is 0011. Determine how many clock pulses are necessary before a single 1 bit is circulated around the circuit.

Solution

The count sequence is given in Table 7.2.

Table 7.2

Clock pulse	Q_D	Q_C	Q_B	Q_A	D_D	D_C	D_B	D_A
0	0	0	1	1	0	1	1	0
1	0	1	1	0	1	1	0	0
2	1	1	0	0	1	0	0	1
3	1	0	0	1	0	0	1	0
4	0	0	1	0	0	1	0	0

Hence, after four clock pulses (*Ans.*)

Fig. 7.12 *Decade counter*

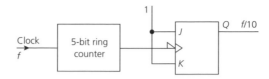

Decade counter

A ten-stage ring counter can be used as a decade counter that requires no feedback logic or decoding, but for some applications the lack of a binary output may be a disadvantage. A 5-bit ring counter can be easily converted to act as a decade counter by connecting the Q output of its most significant stage to the clock input of a J-K flip-flop. If the J and K input terminals are both held HIGH the output of the flip-flop will be at a frequency of $(f_{(clock)}/10)$. The circuit is shown in Fig. 7.12.

PRACTICAL EXERCISE 7.3

Aim: to build and test a ring decade counter.
Components and equipment: one 74HC174 hex D flip-flop IC, one 74HC112 J-K flip-flop IC, six LEDs and six 270 Ω resistors. Pulse generator. CRO. Breadboard. Power supply.
Procedure:

(a)   Build the circuit shown in Fig. 7.13.

Fig. 7.13

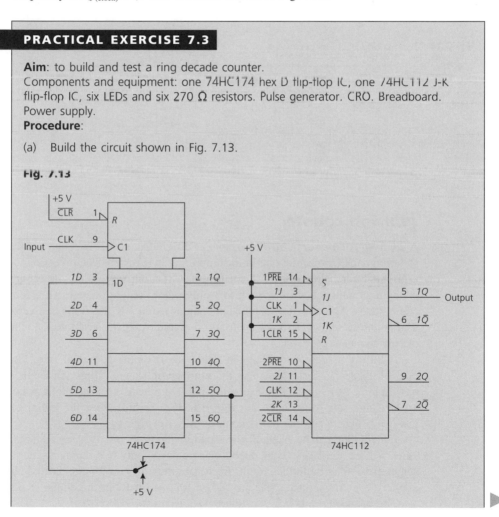

Fig. 7.14 *Twisted-ring or Johnson counter*

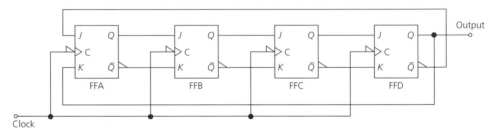

Johnson counter

A *Johnson counter* or *twisted-ring counter* differs from the ring counter in that the logical state fed back from serial output to serial input is inverted. The inversion is obtained for a circuit built using J-K flip-flops by interchanging or *twisting* the connections. The generated count sequence is such that only one Q output changes between consecutive count states. The circuit of a Johnson counter using J-K flip-flops is given in Fig. 7.14. The counting sequence of a 4-bit Johnson counter is given in Table 7.3. The D flip-flop circuit connects $\bar{Q}_D$ to D_A.

If all the stages are initially in either the 0 or the 1 logical state, the number of possible states is $2n$, where n is the number of stages. Thus, while the ring counter has a count of n, the Johnson counter has a count of $2n$. The eight possible states of a four-stage Johnson counter are shown in Fig. 7.15(a) when the initial count is 0 and in Fig. 7.15(b) when the initial count is 2.

If an output is to be taken off corresponding to each, or any, decimal number, decoding gates will be necessary. Each decoding gate needs only two inputs no matter how many stages there may be in the circuit. Each gate must have a unique pair of input variables applied to its inputs and these are shown for a 4-bit Johnson counter in Table 7.4.

The Johnson counter posses the advantages of:

Table 7.3

Clock pulse	Q_D	Q_C	Q_B	Q_A	Decimal number
0	0	0	0	0	0
1	0	0	0	1	1
2	0	0	1	1	3
3	0	1	1	1	7
4	1	1	1	1	15
5	1	1	1	0	14
6	1	1	0	0	12
7	1	0	0	0	8
8	0	0	0	0	0

Fig. 7.15 *Possible states for a four-stage Johnson counter*

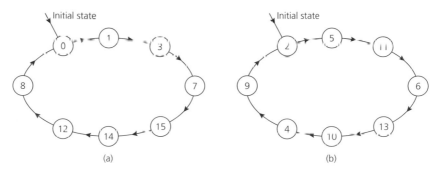

(a) (b)

Table 7.4

Decimal count	Inputs to AND gate for decoded output
0	$\bar{Q}_A\bar{Q}_D$
1	$Q_A\bar{Q}_B$
3	$Q_B\bar{Q}_C$
7	$Q_C\bar{Q}_D$
15	Q_AQ_D
14	$\bar{Q}_AQ_B$
12	$\bar{Q}_BQ_C$
8	$\bar{Q}_CQ_D$

(a) A high speed of operation, because it does not follow the true 8421 binary code sequence.

(b) The decoded outputs are free of glitches because any output is subject to the delay of only one stage.

Fig. 7.16 *Four cycles of possible states for a five-stage Johnson counter*

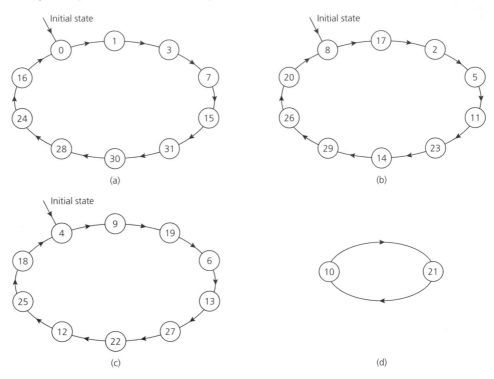

(a)　　　　　　　　　　　　　　(b)

(c)　　　　　　　　　　　　　　(d)

Table 7.5

Decimal number	Count	Q_E	Q_D	Q_C	Q_B	Q_A	Inputs to AND gate for decoded count
0	0	0	0	0	0	0	$\bar{Q}_A\bar{Q}_E$
1	1	0	0	0	0	1	$Q_A\bar{Q}_B$
2	3	0	0	0	1	1	$Q_B\bar{Q}_C$
3	7	0	0	1	1	1	$Q_C\bar{Q}_D$
4	15	0	1	1	1	1	$Q_D\bar{Q}_E$
5	31	1	1	1	1	1	Q_AQ_E
6	30	1	1	1	1	0	$\bar{Q}_AQ_B$
7	28	1	1	1	0	0	$\bar{Q}_BQ_C$
8	24	1	1	0	0	0	$\bar{Q}_CQ_D$
9	16	1	0	0	0	0	$\bar{Q}_DQ_E$

The disadvantage of the circuit is that the possibility exists that the counter may lock in an unwanted state.

The five-stage Johnson counter can operate in any one of four different groups of states and these are shown in Fig. 7.16. The total number of states is $2^5 = 32$. For the first cycle only (see Fig. 7.16(a)), the required decoding is given in Table 7.5.

Self-starting a Johnson counter

The Johnson counter is not self-starting unless some extra feedback circuitry is provided. To make the counter self-starting, logic representing one state from each of the possible unwanted count sequences should be used. In the case of a four-stage Johnson counter there is only one other count sequence. If, for example, the decimal number 5 is chosen the required logic is $F = Q_A\bar{Q}_BQ_C\bar{Q}_D$. If the output of the AND gate with these inputs is used to clear the counter the circuit will self-start within eight clock pulses.

EXAMPLE 7.4

Determine:

(a) The modulus of a three-stage Johnson counter.
(b) The number of decoding gates necessary.
(c) The necessary inputs to those gates.

Solution

(a) $M = 2n = 2 \times 3 = 6$ (*Ans.*)
(b) Number of gates $= M = 6$ (*Ans.*)
(c) The sequence followed is shown in Table 7.6. Each pair of gate inputs must be unique, hence from Table 7.6 the gate inputs are:

$$0, \bar{Q}_A\bar{Q}_C; \; 1, Q_A\bar{Q}_B; \; 3, Q_B\bar{Q}_C; \; 7, Q_AQ_C; \; 6, \bar{Q}_AQ_B; \text{ and } 4, \bar{Q}_BQ_C \; (Ans.)$$

Table 7.6

Q_C	Q_B	Q_A	Clock pulse	Decimal number
0	0	0	0	0
0	0	1	1	1
0	1	1	2	3
1	1	1	3	7
1	1	0	4	6
1	0	0	5	4
0	0	0	6	0

Unwanted count sequences

If the counter should start in an unwanted sequence it will remain in it unless logic is employed to switch it back into the required count sequence. The Karnaugh map showing the decimal equivalents of the states in the required count sequence of a four-stage Johnson counter is

Q_CQ_D \ Q_AQ_B	00	01	11	10
00	0	U	3	1
01	8	U	U	U
11	12	14	15	U
10	U	U	7	U

The cells that represent unwanted states have been marked U (refer to Fig. 7.16). The Boolean expression that represents these states is $F = Q_A\bar{Q}_BQ_C + \bar{Q}_AQ_C\bar{Q}_D + Q_A\bar{Q}_CQ_D + \bar{Q}_AQ_B\bar{Q}_C$. The logic of these states can be employed to clear the counter if the wrong state should be entered.

PRACTICAL EXERCISE 7.4

Aim:

(a) To build a six-stage Johnson counter and to determine its count sequence.
(b) To build and test a self-starting Johnson counter.

Components and equipment: one 74HC164 shift register IC, one 74HC04 hex inverter IC, eight LEDs and eight 270 Ω resistors. Pulse generator. Switches. Breadboard. Power supply.

Procedure:

(a) Build the Johnson counter circuit shown in Fig. 7.17 using the 74HC164 shift register and one of the inverters in the 74HC04.
(b) Note that the circuit does not self-start. Add the switch marked 'Start' and use it to get the circuit started in a count sequence.
(c) Note the counting sequence obtained. Switch the circuit OFF and then ON again. There is more than one possible counting sequence that the circuit may go into; determine what sequence the circuit has entered into this time. If it is a different sequence say which one it is.
(d) Try turning the circuit OFF and then ON again several times. See if it still goes through the same counting sequence. Choose one of the counting sequences that the circuit has not entered and suggest a way in which the circuit could be made to always follow that sequence. Implement the modification you have suggested and see if it works.

PRACTICAL EXERCISE 7.5

Aim: to build a self-starting Johnson counter using a D flip-flop IC.
Components and equipment: one 74HC175 quad D flip-flop IC, four LEDs and four 270 Ω resistors. Pulse generator. Breadboard. Power supply.

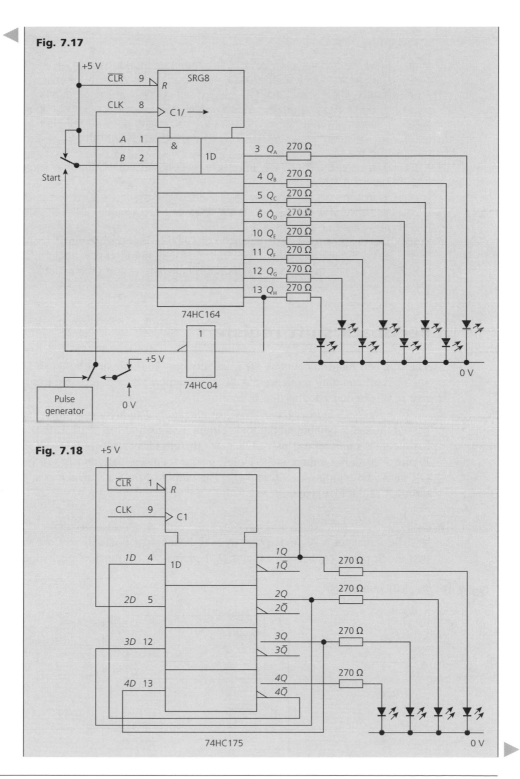

Fig. 7.17

Fig. 7.18

Feedback shift registers

The ring and Johnson counters are relatively easy to implement but have the disadvantage of short counting sequences. A decade counter, for example, requires ten stages, or five if an external J-K flip-flop is employed.

In a *feedback shift register* (FSR) or *maximum length ring counter* the feedback is passed through a combinational logic circuit as shown in Fig. 7.19. If the combinational logic consists *only* of exclusive-OR gates the circuit is a *linear feedback shift register*. Figure 7.20 shows a three-stage circuit whose feedback path consists of one exclusive-OR gate whose inputs are Q_B and Q_C. The output of the gate, which is applied to the J terminal of the first stage, is

$$J_A = Q_B \bar{Q}_C + \bar{Q}_B Q_C$$

The initial set is $Q_A = 1$, $Q_B = Q_C = 0$, hence $J_A = 0$ and $J_B = 1$. At the end of the first clock pulse, Q_A becomes 0 and Q_B becomes 1, Q_C remains unchanged at 0. Now

Fig. 7.19 *Feedback shift register*

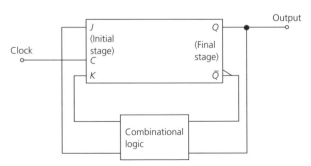

Fig. 7.20 *Linear feedback shift register*

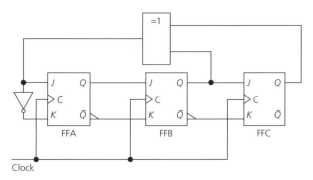

Table 7.7 State table of linear feedback register

Q_C	$Q_B = J_C$	$Q_A = J_A$	$J_A = Q_B\bar{Q}_C + \bar{Q}_B Q_C$
0	0	1	0
0	1	0	1
1	0	1	1
0	1	1	1
1	1	1	0
1	1	0	0
1	0	0	1
0	0	1	0

$J_A = 1.1 + 0.0 = 1$, $J_B = 1$, and $J_C = 1$. At the end of the next clock pulse, Q_A becomes 1 again, $Q_B = 0$ and $Q_C = 1$. Now $J_A = 0.0 + 1.1 = 1$, $J_B = 1$ and $J_C = 0$, and so the next state is $Q_A = Q_B = 1$ and $Q_C = 0$. The operation of the circuit is summarized in Table 7.7.

The 'all Qs zeros' state must be avoided at switch-on because it would lock the J_A input at 0. This requirement can be achieved by ANDing all the $\bar{Q}$ outputs and applying the output of the AND gate to the J_A input terminal. This means that the maximum length cycle is $2^n - 1$. The stages from which the feedback must be taken to give a maximum cycle length are shown in Table 7.8. Other feedback combinations will produce a shorter cycle length.

The output of a linear FSR forms a pseudo-random binary sequence which, if there is a large enough number of stages, may approximate to the white noise power density spectrum. For example, from Table 7.7 it can be seen that the output of a three-stage circuit is 100110111011101010001. This sequence contains all the possible 3-bit combinations except 000.

The maximum length counting sequence of a linear feedback shift register occurs when the exclusive-OR feedback is taken from the two most significant stages n and $n - 1$. The maximum length sequence (MLS) is equal to $2^n - 1$.

Table 7.8

Number of stages	Linear feedback	Cycle length $(2^n - 1)$
2	$1 \oplus 2$	3
3	$2 \oplus 3$	7
4	$3 \oplus 4$	15
5	$3 \oplus 5$	31
6	$5 \oplus 6$	63
7	$6 \oplus 7$	127
8	$2 \oplus 3 \oplus 4 \oplus 8$	255
9	$5 \oplus 9$	511
10	$7 \oplus 10$	1023

Note: $\oplus$ indicates exclusive-OR function.

EXAMPLE 7.5

A four-stage feedback shift register has a feedback path consisting of one exclusive-OR gate with inputs Q_A and Q_B. The initial state of the register is 0010.

(a) Write down a table to show the external states and the feedback signal and hence obtain the counting sequence.
(b) Determine the cycle length.
(c) What is the maximum cycle length?

Solution

(a) Table 7.9 gives the states of the circuit.
From the table the counting sequence is 2, 5, 10, 4, 8, 1, 2, etc. (*Ans.*)
(b) The cycle length is 6 (*Ans.*)
(c) The maximum length sequence is MLS = $2^4 - 1 = 15$ (*Ans.*)

Table 7.9

State	Q_D	Q_C	Q_B	Q_A	$Q_D \oplus Q_B$
S2	0	0	1	0	1
S5	0	1	0	1	0
S10	1	0	1	0	0
S4	0	1	0	0	0
S8	1	0	0	0	1
S1	0	0	0	1	0
S2	0	0	1	0	1

Fig. 7.21 *Determination of the possible states in a non-linear feedback shift register*

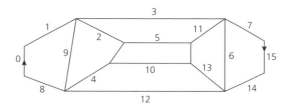

Table 7.10 *State table of non-linear feedback register*

State	Q_D	Q_C	Q_B	Q_A	Q_D^+	Q_C^+	Q_B^+	Q_A^+
0	0	0	0	0	0	0	0	1
1	0	0	0	1	0	0	1	0
2	0	0	1	0	0	1	0	1
5	0	1	0	1	1	0	1	0
10	1	0	1	0	0	1	0	0
4	0	1	0	0	1	0	0	0
8	1	0	0	0	0	0	0	0

Non-linear feedback shift registers

If the feedback from the serial output of the shift register to the serial input does *not* consist of exclusive-OR gates only then a *non-linear feedback shift register* is obtained. The two possible states of a four-stage Johnson counter are shown in Fig. 7.15(a) and (b). It is possible for the circuit to move out of one cycle and into the other at certain points in a cycle. Figure 7.21 shows the two cycles combined, with connections linking 2 and 4, and 10 and 11. There are 16 different states in all and so any count up to 15 can be designed.

If a count of 7 is required using the sequence 1, 2, 5, 10, 4, 8, 0 then the state table given in Table 7.10 applies. The feedback required to be applied to the J_A terminal, the same as Q_A^+, is taken from the appropriate outputs of the four flip-flops. The mapping for the feedback is

$Q_C Q_D$ \ $Q_A Q_B$	00	01	11	10
00	1	1	×	0
01	0	0	×	×
11	×	×	×	×
10	0	×	×	0

and, from this, $J_A = \bar{Q}_A \bar{Q}_C \bar{Q}_D$. The required circuit is shown in Fig. 7.22.

Fig. 7.22 *Non-linear feedback shift register*

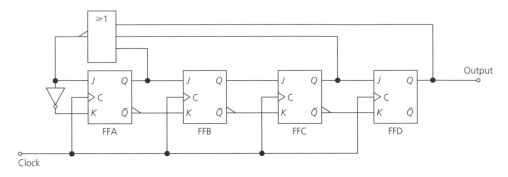

Design a non-linear feedback shift register to count from 0 to 15 using the counting sequence 1, 2, 4, 8, 15, 3, 5, 9, 14, 7, 10, 13, 6, 11, 12.

Table 7.11

State	Q_D	Q_C	Q_B	Q_A	Q_D^+	Q_C^+	Q_B^+	Q_A^+
1	0	0	0	1	0	0	1	0
2	0	0	1	0	0	1	0	0
3	0	1	0	0	1	0	0	0
4	1	0	0	0	1	1	1	1
5	1	1	1	1	0	0	1	1
6	0	0	1	1	0	1	0	1
7	0	1	0	1	1	0	0	1
8	1	0	0	1	1	1	1	0
9	1	1	1	0	0	1	1	1
10	0	1	1	1	1	0	1	0
11	1	0	1	0	1	1	0	1
12	1	1	0	1	0	1	1	0
13	0	1	1	0	1	0	1	1
14	1	0	1	1	1	1	0	0
15	1	1	0	0	0	0	0	1

Solution

The state table for the required counter is given in Table 7.11. All that matters is the logical state of the Q_D output. Mapping for each combination of $Q_A Q_B Q_C Q_D$ the logical state of Q_D^+ gives

Q_CQ_D \ Q_AQ_B	00	01	11	10
00	0	0	0	0
01	1	1	1	1
11	0	0	0	0
10	1	1	1	1

From the map $J(D) = Q_C\bar{Q}_D + \bar{Q}_CQ_D$ (*Ans.*)

EXERCISES

7.1 A 74HC164 shift register has its inputs A and B held HIGH and its Q_H output connected to the $\overline{CLR}$ pin via an inverter. Starting from the all stages cleared state, determine the counting sequence followed when a clock waveform is connected to the CLK input.

7.2 Refer to the data sheet of the 74HC164 shift register on p. 191.
   (a) Can the device be made to operate as a down-counter?
   (b) Is the clear synchronous?
   (c) What happens when the input data is applied to terminal A, and terminal B is
     (i) held HIGH
     (ii) held LOW?
   (d) When can data at the serial input terminal be changed?
   (e) The device is required to work at a frequency of 27 MHz. What supply voltage must be used?
   (f) What then are the values of the set-up time and the hold time for the non-synchronous inputs?

7.3 An 8-bit binary counter, an 8-bit ring counter and an 8-bit Johnson counter are connected in cascade. A clock of frequency 10 MHz is applied to the input of the first counter. Calculate the frequency of the output waveform.

7.4 (a) Show how five J-K flip-flops could be connected to form a twisted-ring counter.
   (b) What would be the count of this circuit?
   (c) With the aid of a timing diagram explain the operation of the circuit.

7.5 Draw the diagram, and explain the operation, of a shift register. Assume the circuit to be storing the digital word 0010.

7.6 Draw diagrams to show how a 4-bit shift register can be converted into
   (a) A 4-bit ring counter.
   (b) An 8-bit ring counter.

Fig. 7.23

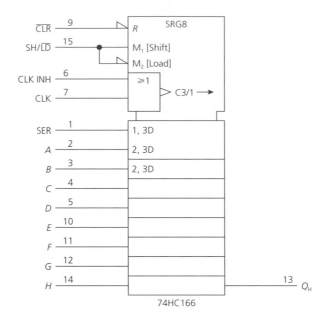

74HC166

7.7 A linear feedback shift register has four stages *A*, *B*, *C* and *D*. Feedback paths exist between the outputs of stages *A* and *D* to the input of stage *A*. Determine the code cycle that is generated.

7.8 Design a five-stage Johnson counter with decoding for alternate stages starting with $Q_A = Q_B = Q_C = Q_D = 0$.

7.9 Use a 74HC166 shift register to perform a parallel-to-serial conversion. The logic symbol for the device is given in Fig. 7.23.

7.10 (a) Johnson counters are to be used to obtain each of the following division ratios:
  (i) 6
  (ii) 10
  (iii) 20
  (iv) 30.
  How many stages are needed for each ratio?
  (b) What count would be obtained in each case if the circuit were connected as a ring counter?
  (c) State the relative merits of ring and twisted-ring counters.

7.11 An 8-bit digital word is applied to a 74HC164 8-bit SIPO shift register. The word is to be applied to three different sub-systems after time delays of:
  (a) 5 μs.
  (b) 7 μs.
  (c) 8 μs.
  State how these requirements can be satisfied.

7.12 (a) A 74HC164 shift register is to be used to convert serial data into parallel data. Draw the required circuit.

(b) The binary word 10110110 is to be converted. Which bit ought to be entered first and why?

(c) List the sequence of events in a conversion.

7.13 An 8-bit SISO shift register is loaded with the data word 10110110. The data word 180_{10} is then entered serially from right to left. After eight clock pulses the circuit is disabled. Write down a table that shows the bits stored after eight pulses.

7.14 A 74HC164 shift register is used to convert 8-bit serial input data words into parallel output data words. The output data is transferred to a memory whose capacity is 64k × 8. After each 8-bit word has been entered into the shift register the next clock pulse transfers the stored data to the memory. At the next clock pulse the next data word starts to be inputted. Calculate the time taken for the memory to be fully loaded if the clock frequency is 10 MHz.

8 Design of synchronous counters

After reading this chapter you should be able to:

- Understand what is meant by a state diagram and a state table.
- Draw the state diagram of a counter that is to be designed.
- Derive a state table from a state diagram.
- Obtain state maps from a state table and use these maps to design the logic required for a required counter.
- Design synchronous counters that follow a straight binary count sequence.
- Design a synchronous counter to follow a specified count sequence.
- Design a counter to be self-starting.
- Design an up-down counter.

Synchronous counters can be designed to follow any desired counting sequence and not just the binary sequence that an MSI device follows. The design of a synchronous counter involves the use of the *state diagram*, and the *state table*. The aim of the design method is to produce the minimal Boolean expression for the gating logic required to direct signals to the input(s) of the flip-flops employed. The flip-flops employed in a non-binary counter may be of the J-K type, or of the D type, or T flip-flops can be used. When a counter is to be implemented using a *programmable logic device* (PLD) only D flip-flops are available.

State tables and state diagrams

State diagrams

A *state diagram* is a graphical representation of the sequence of states that a sequential digital system follows. It consists of a number of circles, or nodes, each of which represents a state that might occur in the circuit. The circles are linked together by lines that represent the transition from one state to the next. Arrowheads indicate the direction in which a line is traversed from one circle to another, i.e. the order in which the circuit changes from one state to the next. Each transition line is labelled with the signal conditions that cause the state transition and the resulting output (0 or 1). When the next

Fig. 8.1 *Basic state diagram*

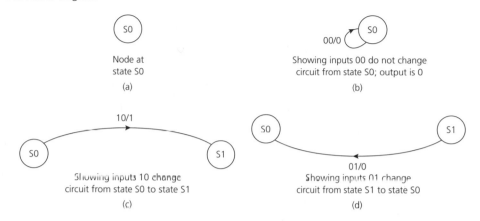

SO

Node at
state S0

(a)

00/0 ⟳ SO

Showing inputs 00 do not change
circuit from state S0; output is 0

(b)

10/1

SO ⟶ S1

Showing inputs 10 change
circuit from state S0 to state S1

(c)

SO ⟵ S1
01/0

Showing inputs 01 change
circuit from state S1 to state S0

(d)

Fig. 8.2 *Simple logic circuit*

Fig. 8.3 *State diagram for Fig. 8.2*

$AB/1, \bar{A}\bar{B}/1$

SO ⟶ S1

$\bar{A}B/0, A\bar{B}/0$

state is the same as the present state the transition line will start and finish at the same circle. Three examples of simple state diagrams are shown in Fig. 8.1.

Figure 8.2 shows a simple circuit for switching a light ON or OFF using either one of two switches. If the upward position of a switch is denoted by A or B, and the downward position by $\bar{A}$ or $\bar{B}$, then the light will be ON when both A and B are UP OR when both A and B are DOWN. Thus the Boolean equation describing the operation of the circuit is

$$L = AB + \bar{A}\bar{B} \tag{8.1}$$

The state diagram for this circuit is given in Fig. 8.3. The state S0 represents the light being turned OFF and the state S1 represents the light being turned ON.

A state table provides a way of representing the operation of a sequential circuit; it lists the present states and the next states for all the possible combinations of the inputs. The rows in the table represent the output states of the circuit and the columns represent the possible input states. The state table of a circuit is obtained from the state diagram of that circuit.

S-R latch

The state diagram for an S-R latch is shown in Fig. 8.4. The latch has two states S0 and S1; S0 is $Q = 0$ and $\bar{Q} = 1$, while S1 is $Q = 1$ and $\bar{Q} = 0$. The S0 circle has one line labelled 00/0 and 01/0 that leaves the circle and then re-enters it; this indicates that

Fig. 8.4 *State diagram for an S-R latch*

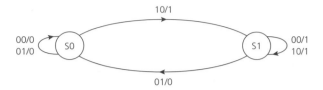

Table 8.1

Present state Q	Next state Q^+	Required inputs S	R
0	0	0	×
0	1	1	0
1	0	0	1
1	1	×	0

when the latch is in state S0, input $S = R = 0$, or $S = 0$ $R = 1$, does not change the state of the circuit. The top line linking circle S0 to circle S1 is labelled 10/1; this indicates that when the latch is in state S0 input $S = 1$ $R = 0$ makes the circuit switch to state S1. The line looping in and out of the S1 circle indicates that when the latch is in state S1 either $S = R = 0$ or $S = 1$ $R = 0$ will leave the state of the circuit unchanged. Lastly, the bottom line linking the S0 and S1 circles indicates that when the latch is set (state S1) inputs $S = 0$ $R = 1$ will switch the circuit to state S0.

The entries in the truth table of an S-R latch can be used to derive the state table. An entry in a particular position in the state table represents the *next state* Q^+ of the output of the latch for the circuit change, or *transition*, caused by applying the input shown by that column when the circuit is in the state represented by that row. The state table for the S-R latch is given in Table 8.1. A don't care condition is indicated by ×.

When the state map (a form of Karnaugh map) is drawn each cell represents one of the present states of the flip-flop. The values inserted into the cells are the required *next states* taken from Table 8.1. When $Q = Q^+ = 0$, S must be 0 and R can be either 1 or 0 (shown as × in the table). $Q = S = R = 0$ is the address of the top left-hand cell in the map and hence 0 is inserted into this cell. When $Q = Q^+ = 1$ $R = S = 0$ or $R = 0$ $S = 1$, so 1 is inserted into the cell QSR. The state map is

$^{SR}\!\diagdown$ Q	00	01	11	10
0	0	0	×	1
1	1	0	×	1

From the map,

$$Q^+ = S + Q\bar{R} \tag{8.2}$$

Fig. 8.5 *State diagram for a J-K flip-flop*

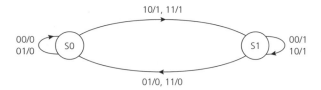

10/1, 11/1

00/0
01/0
S0

S1
00/1
10/1

01/0, 11/0

Table 8.2			
Present state Q	Next state Q^+	Required inputs J	K
0	0	0	×
0	1	1	×
1	0	×	1
1	1	×	0

and

$$\bar{Q}^+ = R + \bar{Q}\bar{S} \tag{8.3}$$

Equations (8.2) and (8.3) are known as the characteristic equations of the S-R latch.

J-K flip-flop

The state diagram of a J-K flip-flop is shown in Fig. 8.5. It is similar to the state diagram of the S-R latch shown in Fig. 8.4 differing only in that the two lines joining states S0 and S1 are labelled 11/1 and 11/0, respectively, in addition to the S-R labelling. The additional 11 labels indicate the toggling action of the J-K flip-flop when $J = K = 1$. The state table of a J-K flip-flop is given in Table 8.2.

The state map is (again each cell represents a present state and the entries into the cells are the required next states)

JK / Q	00	01	11	10
0	0	0	1	1
1	1	0	0	1

From the state map, the characteristic equations of the J-K flip-flop are:

$$Q^+ = J\bar{Q} + \bar{K}Q \tag{8.4}$$

$$Q^+ = \bar{J}\bar{Q} + KQ \tag{8.5}$$

D flip-flop

The state diagram for a D flip-flop is given in Fig. 8.6 and the state table is given in Table 8.3. From the state table the state map of a D flip-flop is

Fig. 8.6 *State diagram for a D flip-flop*

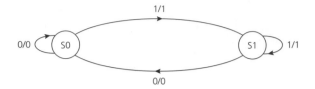

Table 8.3

Present state Q	Next state Q^+	Required input D
0	0	0
0	1	1
1	0	0
1	1	1

Fig. 8.7 *State diagram for a T flip-flop*

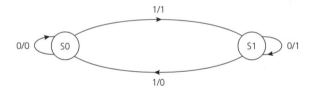

From the map, the characteristic equations of a D flip-flop are:

$$Q^+ = D \tag{8.6}$$

$$\bar{Q}^+ = \bar{D} \tag{8.7}$$

T flip-flop

The T flip-flop changes state whenever the signal applied to its T input terminal is at the logical 1 level and hence its state diagram is as shown in Fig. 8.7. From the state diagram the state table is obtained and is given in Table 8.4. The state map is

	T	
Q	0	1
0	0	1
1	1	0

Table 8.4

Present state Q	Next state Q^+	Required input T
0	0	0
0	1	1
1	0	1
1	1	0

and from this the characteristic equations for the T flip-flop are:

$$Q^+ = T\bar{Q} + \bar{T}Q \qquad (8.8)$$

$$\bar{Q}^+ = \bar{T}\bar{Q} + TQ \qquad (8.9)$$

Design of synchronous counters

The operation of a synchronous counter may be represented by a state diagram and a state table. The state diagram shows the present and next states of each flip-flop and includes all possible transitions. Each of the possible states that the counter can enter must be represented by a circle labelled as S0, S1, S2, etc. Possible transitions between these states are indicated by lines drawn between the circles. Each line is labelled with two binary numbers separated by a slash (/). The left-hand number indicates the input that determines the next state, and the right-hand number indicates the state that is then entered.

The *modulus* of a counter is the number of different states into which the counter may enter. A binary counter has a modulus of 2^n, where n is the number of stages. This means that a four-stage counter has a modulus of $2^4 = 16$ and hence its count is from 0000 to 1111, i.e. from 0_{10} to 15_{10}. A decade counter has a modulus of 10 and a count of from 0_{10} to 9_{10}.

A 2-bit counter will have four possible state: these are 00, 01, 10 and 11 and they are labelled S0, S1, S2 and S3, respectively, in the state diagram shown in Fig. 8.8(a). If a clear facility exists so that the counter can be cleared to state S0 whenever the $\overline{CLR}$ line is taken LOW the state diagram is modified to that shown in Fig. 8.8(b). State diagrams for straight counting sequence 4-bit binary and decade counters are shown in Fig. 8.9(a) and (b), respectively. A counter that follows a straight binary numbering sequence can be obtained using an MSI device and would be unlikely to be designed using state diagrams and tables. However, some designs of such counters are included as examples.

Fig. 8.8 *State diagrams for (a) a 2-bit counter and (b) a 2-bit counter with clear*

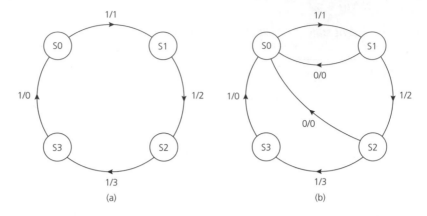

Fig. 8.9 *State diagrams for straight binary sequence counters: (a) 4-bit binary and (b) decade*

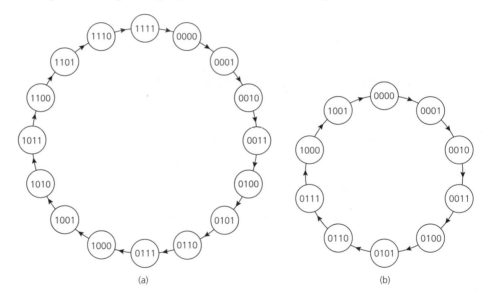

Use of D flip-flops

The design of a synchronous counter is easier using D flip-flops rather than one using J-K flip-flops because there are fewer inputs to consider. On the other hand a design that employs J-K flip-flops needs simpler combinational logic circuitry to obtain the necessary signals to the inputs. This is because of the extra flexibility offered by two inputs as opposed to one. The sequence of states in a counter may not include all of the 2^n possible states. Then the initial state entered when the circuit is first switched

Fig. 8.10 *State diagram for a divide-by-5 counter*

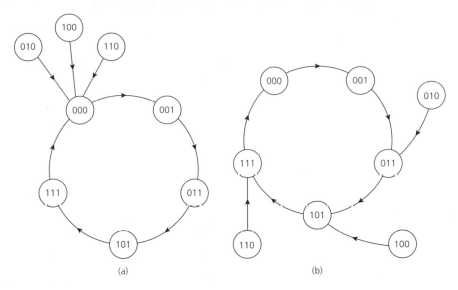

(a)   (b)

Table 8.5

	Present state		Next state		Required input	
Count	Q_B	Q_A	Q_B^+	Q_A^+	D_B	D_A
0	0	0	0	1	0	1
1	0	1	1	0	1	0
2	1	0	0	0	0	0

on may not be one of the states in the required sequence. A *self-starting* counter will eventually enter the count sequence for which it has been designed no matter what the initial state should happen to be. The state diagram of a self-starting counter shows every possible state, including those that are not wanted. Unwanted states should link to a required state. Two examples of this are given in Fig. 8.10(a) and (b) which shows the state diagram of a self-starting divide-by-5 counter that follows the count sequence 0, 1, 3, 5, 7, 0, etc. States 2, 4, and 6 are not wanted so their circles are linked to the S0 circle (Fig. 8.10(a)) if the circuit is to go to that state after entering an unwanted state. If (Fig. 8.10(b)) entry into an unwanted state is to be followed by entry into the next higher required state then S2 is linked to S3, S4 to S5 and S6 to S7.

Divide-by-3 counter

The state table for the required circuit must first be written down; this is given in Table 8.5, supposing that the counting sequence is to be 0, 1, 2, 0, etc. The state maps are

	Present state		Next state		Required inputs	
Count	Q_B	Q_A	Q_B^+	Q_A^+	D_B	D_A
0	0	0	1	0	1	0
2	1	0	0	1	0	1
1	0	1	0	1	0	0

Table 8.6 *State table of a divide-by-3 counter*

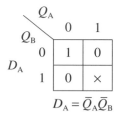

$$D_A = \bar{Q}_A\bar{Q}_B$$

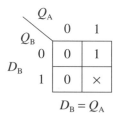

$$D_B = Q_A$$

If the count is required to follow the sequence 0, 2, 1, 0 then the state table would be slightly different and this is given in Table 8.6.

The state maps are

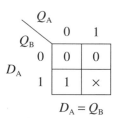

$$D_A = Q_B$$

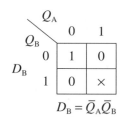

$$D_B = \bar{Q}_A\bar{Q}_B$$

PRACTICAL EXERCISE 8.1

Aim: to build and test a divide-by-3 counter to follow the counting sequences
(i)   0, 1, 2, 0, etc.
(ii)   0, 2, 1, 0, etc.
Components and equipment: one 74HC74 dual D flip-flop, one 74HC08 quad 2-input AND gate, two LEDs and two 270 Ω resistors. Pulse generator. CRO. Breadboard. Power supply.
Procedure:

(a)   Build the circuit shown in Fig. 8.11.
(b)   Observe the logical states of the LEDs and hence confirm that the circuit follows the required counting sequence. Use the CRO to observe the input and output waveforms. If a logic analyser is available it could be used for that purpose.
(c)   Modify the circuit so that its counting sequence is (ii) above. Observe the logical states of the LEDs and confirm that the modified circuit works correctly.

Fig. 8.11

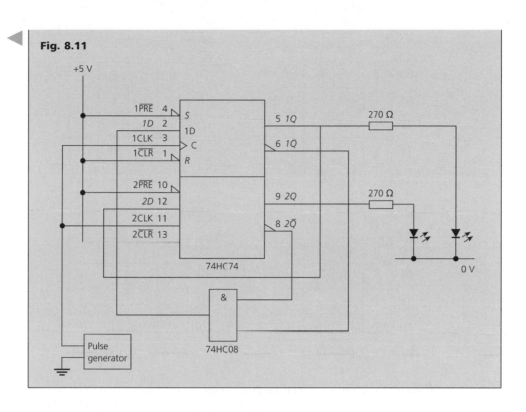

Table 8.7 State table of a decade counter

Count	Present state				Next state				Required inputs			
	Q_D	Q_C	Q_B	Q_A	Q_D^+	Q_C^+	Q_B^+	Q_A^+	D_D	D_C	D_B	D_A
0	0	0	0	0	0	0	0	1	0	0	0	1
1	0	0	0	1	0	0	1	0	0	0	1	0
2	0	0	1	0	0	0	1	1	0	0	1	1
3	0	0	1	1	0	1	0	0	0	1	0	0
4	0	1	0	0	0	1	0	1	0	1	0	1
5	0	1	0	1	0	1	1	0	0	1	1	0
6	0	1	1	0	0	1	1	1	0	1	1	1
7	0	1	1	1	1	0	0	0	1	0	0	0
8	1	0	0	0	1	0	0	1	1	0	0	1
9	1	0	0	1	0	0	0	0	0	0	0	0

Decade counter

The design of a decade counter that follows the straight binary sequence (state diagram in Fig. 8.9) starts with the state table. This is given in Table 8.7.

All the remaining states are don't cares. When a state is a don't care the D input is also a don't care. The state maps for each of the D input terminals are

$$D_A = \bar{Q}_A$$

$$D_B = Q_A\bar{Q}_B\bar{Q}_D + \bar{Q}_A Q_B$$

$$D_C = Q_A Q_B \bar{Q}_C + \bar{Q}_B Q_C + \bar{Q}_A Q_C$$

$$D_D = \bar{Q}_A Q_D + Q_A Q_B \bar{Q}_C$$

Clearly, the required circuit is quite complex. It requires two 74HC74 dual D flip-flop ICs, one 74HC11 triple 3-input AND gate IC, and one 74HC08 quad 2-input AND gate IC. If the 74HC174 hex D flip-flop IC is used instead of the '74 then a 74HC04 hex inverter IC will also be required to get Q_A, Q_B, etc. The complexity of the associated gating logic is a disadvantage of using D flip-flops.

Table 8.8 State table of a divide-by-3 counter

	Present state		Next state		Required inputs			
Count	Q_B	Q_A	Q_B^+	Q_A^+	J_B	K_B	J_A	K_A
0	0	0	0	1	0	×	1	×
1	0	1	1	0	1	×	×	1
2	1	0	0	0	×	1	0	×

Use of J-K flip-flops

Divide-by-3 counter

The state table for a divide-by-3 counter that follows the count sequence 0, 1, 2, 0, etc. is given in Table 8.8.

From the table it is clear that $K_A = K_B = 1$. The state maps for the other two inputs are

Table 8.9 Decade counter state table (J-K flip-flops)

	Present state				Next state					Required J and K inputs							
	Q_D	Q_C	Q_B	Q_A		Q_D^+	Q_C^+	Q_B^+	Q_A^+	J_D	K_D	J_C	K_C	J_B	K_B	J_A	K_A
0	0	0	0	0	1	0	0	0	1	0	×	0	×	0	×	1	×
1	0	0	0	1	2	0	0	1	0	0	×	0	×	1	×	×	1
2	0	0	1	0	3	0	0	1	1	0	×	0	×	×	0	1	×
3	0	0	1	1	4	0	1	0	0	0	×	1	×	×	1	×	1
4	0	1	0	0	5	0	1	0	1	0	×	×	0	0	×	1	×
5	0	1	0	1	6	0	1	1	0	0	×	×	0	1	×	×	1
6	0	1	1	0	7	0	1	1	1	0	×	×	0	×	0	1	×
7	0	1	1	1	8	1	0	0	0	1	×	×	1	×	1	×	1
8	1	0	0	0	9	1	0	0	1	×	0	0	×	0	×	1	×
9	1	0	0	1	10	0	0	0	0	×	1	0	×	0	×	×	1
10	×	×	×	×		×	×	×	×	×	×	×	×	×	×	×	×
↓																	
15	×	×	×	×		×	×	×	×	×	×	×	×	×	×	×	×

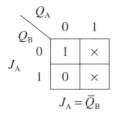

J_A	Q_A 0	1
Q_B 0	1	×
1	0	×

$J_A = \bar{Q}_B$

J_B	Q_A 0	1
Q_B 0	0	1
1	×	×

$J_B = Q_A$

Decade counter

The state table is given in Table 8.9. From the state table the state maps for each of the J and K inputs are drawn. When the next state is a don't care all the J and K inputs are also don't cares.

The first step is to draw a number of state maps, one for each J input and each K input. These maps plot Q_A, Q_B, Q_C and Q_D for each J and K value. $J_A = K_A = 1$.

J_B Q_CQ_D \ Q_AQ_B	00	01	11	10
00	0	×	×	1
01	0	×	×	0
11	×	×	×	×
10	0	×	×	1

$J_B = Q_A\bar{Q}_D$

K_B Q_CQ_D \ Q_AQ_B	00	01	11	10
00	×	0	1	×
01	×	×	×	×
11	×	×	×	×
10	×	0	1	×

$K_B = Q_A$

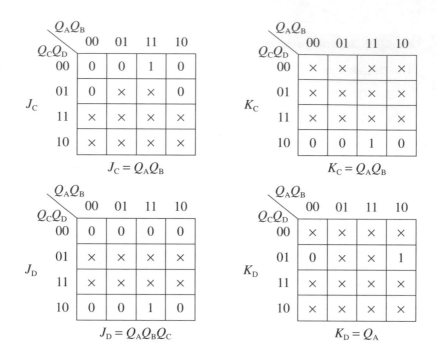

$$J_C = Q_A Q_B$$

$$K_C = Q_A Q_B$$

$$J_D = Q_A Q_B Q_C$$

$$K_D = Q_A$$

The designed circuit is given in Fig. 8.12.

Fig. 8.12 *Decade counter*

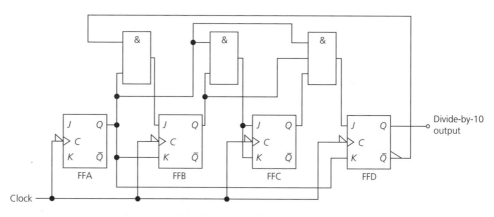

PRACTICAL EXERCISE 8.2

Aim: to build and test a decade counter using J-K flip-flops to follow the counting sequence 0, 2, 4, 6, 8, 1, 3, 5, 7, 9, 0, etc.
Components and equipment: two 74HC112 dual J-K flip-flop ICs, one 74HC11 triple 3-input AND gate IC, four LEDs and four 270 Ω resistors. Pulse generator. CRO. Breadboard. Power supply.

EXAMPLE 8.1

Design a divide-by-5 counter using D flip-flops that follows the count sequence 2, 4, 6, 1, 3, 2, etc.

Solution

The state diagram for the required circuit is shown in Fig. 8.13. From the state diagram the state table is obtained (see Table 8.10). The state maps are

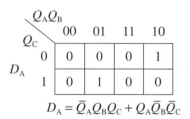

$$Q_A Q_B$$

Q_C \	00	01	11	10
D_A 0	0	0	0	1
1	0	1	0	0

$$D_A = \bar{Q}_A Q_B Q_C + Q_A \bar{Q}_B \bar{Q}_C$$

$$Q_A Q_B$$

Q_C \	00	01	11	10
D_B 0	1	0	1	1
1	1	0	1	1

$$D_B = Q_A + \bar{Q}_B$$

$$Q_A Q_B$$

Q_C \	00	01	11	10
D_C 0	0	1	0	0
1	1	0	0	0

$$D_C = \bar{Q}_A Q_B \bar{Q}_C + \bar{Q}_A \bar{Q}_B Q_C$$

Fig. 8.13 *State diagram for a divide-by-5 counter*

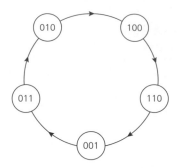

Table 8.10 *State table of a divide-by-5 counter*

Count	Present state Q_C	Q_B	Q_A	Next state Q_C^+	Q_B^+	Q_A^+	Required inputs D_C	D_B	D_A
2	0	1	0	1	0	0	1	0	0
4	1	0	0	1	1	0	1	1	0
6	1	1	0	0	0	1	0	0	1
1	0	0	1	0	1	1	0	1	1
3	0	1	1	0	1	0	0	1	0
2	0	1	0						

If this circuit is built it will be found that it probably won't start because the 'all zeros' state is not included in the counting sequence. There are various ways in which the counter can be started (see Example 8.2) but one way is to ensure that entry into any unwanted counting state is immediately followed by entry into a required state (*Ans.*)

EXAMPLE 8.2

Modify the circuit designed in Example 8.1 to be self-starting with entry on switch-on into any unwanted state leading to state 2.

Solution

The state diagram is shown in Fig. 8.14. The only difference between this diagram and the previous one is that the unwanted states 0, 5 and 7 are now included and each one has a line to state 010. The new state table is given in Table 8.11. From the state table the state maps are

Fig. 8.14 *State diagram for a self-starting divide-by-5 counter*

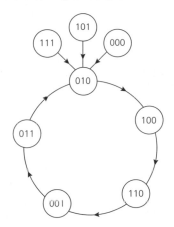

Table 8.11 *State table for self-starting divide-by-5 counter*

	Present state			Next state			Required inputs		
Count	Q_C	Q_B	Q_A	Q_C^+	Q_B^+	Q_A^+	D_C	D_B	D_A
0	0	0	0	0	1	0	0	1	0
1	0	0	1	0	1	1	0	1	1
2	0	1	0	1	0	0	1	0	0
3	0	1	1	0	1	0	0	1	0
4	1	0	0	1	1	0	1	1	0
5	1	0	1	0	1	0	0	1	0
6	1	1	0	0	0	1	0	0	1
7	1	1	1	0	1	0	0	1	0

$Q_A Q_B$

Q_C	00	01	11	10
0	0	0	0	1
1	0	1	0	0

D_A

$$D_A = \bar{Q}_A Q_B Q_C + Q_A \bar{Q}_B \bar{Q}_C$$

$Q_A Q_B$

Q_C	00	01	11	10
0	1	0	1	1
1	1	0	1	1

D_B

$$D_B = Q_A + \bar{Q}_B$$

$Q_A Q_B$

Q_C	00	01	11	10
0	0	1	0	0
1	1	0	0	0

D_C

$$D_C = \bar{Q}_A Q_B \bar{Q}_C + \bar{Q}_A \bar{Q}_B Q_C$$

Note the differences in the logic circuitry used for this circuit and the non-self-starting counter in Example 8.2. A counter can also be made self-starting by treating the unwanted states as don't cares. See exercise 8.8 (*Ans.*)

EXAMPLE 8.3

Design the counter in Example 8.2 using J-K flip-flops.

Table 8.12 *State table of a J-K divide-by-5 counter*

Count	Present state			Next state			Required inputs					
	Q_C	Q_B	Q_A	Q_C^+	Q_B^+	Q_A^+	J_C	K_C	J_B	K_B	J_A	K_A
0	0	0	0	0	1	0	0	×	1	×	0	×
1	0	0	1	0	1	1	0	×	1	×	×	0
2	0	1	0	1	0	0	1	×	×	1	0	×
3	0	1	1	0	1	0	0	×	×	0	×	1
4	1	0	0	1	1	0	×	0	1	×	0	×
5	1	0	1	0	1	0	×	1	1	×	×	1
6	1	1	0	0	0	1	×	1	×	1	1	×
7	1	1	1	0	1	0	×	1	×	0	×	1

Solution

The state table is given in Table 8.12. The state maps are

$$K_A = Q_B + Q_C$$

$$J_A = Q_B Q_C$$

$$K_B = \bar{Q}_A$$

$$J_B = 1$$

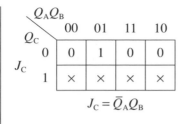

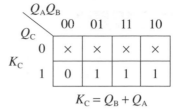

$$J_C = \bar{Q}_A Q_B$$

$$K_C = Q_B + Q_A$$

(Ans.)

PRACTICAL EXERCISE 8.3

Aim: to build and check the operation of a self-starting divide-by-5 counter that follows the counting sequence 2, 4, 6, 1, 3, 2, etc.

Components and equipment: two 74HC74 dual D flip-flop ICs, one 74HC32 quad 2-input OR gate IC, two 74HC11 triple 3-input AND gate ICs, three LEDs and three 270 Ω resistors. Pulse generator. Breadboard. Power supply.

Procedure:

(a) The design equations for the required counter are given in the answer to Example 8.2. The pinouts for the three ICs are given in Fig. 8.15. Draw the required circuit.

Fig. 8.15 *Pinouts for (a) 74HC74, (b) 74HC32 and (c) 74HC11 ICs*

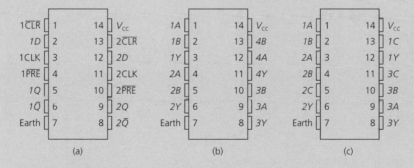

(b) Build the circuit with an LED in series with a 270 Ω resistor connected between each Q output and earth.

(c) Set the pulse generator to about 10 Hz and 5 V and apply it to the input of the circuit. Note the logical states of the LEDs and hence determine
 (i) the modulus
 (ii) the counting sequence of the counter.

(d) Comment on the complexity of the circuit and suggest how a simpler circuit might be obtained. Could the circuit be obtained using a 4-bit MSI synchronous counter?

Design a synchronous counter using D flip-flops to follow the counting sequence 0, 8, 4, 2, 9, 12, 14, 15, 7, 3, 1, 0, etc. using D flip-flops.

Table 8.13

Count	Present state				Next state				Required J-K inputs							
	Q_D	Q_C	Q_B	Q_A	Q_A^+	Q_B^+	Q_C^+	Q_D^+	J_D	K_D	J_C	K_C	J_B	K_B	J_A	K_A
0	0	0	0	0	1	0	0	0	1	×	0	×	0	×	0	×
8	1	0	0	0	0	1	0	0	×	1	1	×	0	×	0	×
4	0	1	0	0	0	0	1	0	0	×	×	1	1	×	0	×
2	0	0	1	0	1	0	0	1	1	×	0	×	×	1	1	×
9	1	0	0	1	1	1	0	0	×	0	1	×	0	×	×	1
12	1	1	0	0	1	1	1	0	×	0	×	0	1	×	0	×
14	1	1	1	0	1	1	1	1	×	0	×	0	×	0	1	×
15	1	1	1	1	0	1	1	1	×	1	×	0	×	0	×	0
7	0	1	1	1	0	0	1	1	0	×	×	1	×	0	×	0
3	0	0	1	1	0	0	0	1	0	×	0	×	×	1	×	0
1	0	0	0	1	0	0	0	0	0	×	0	×	0	×	×	1

Fig. 8.16

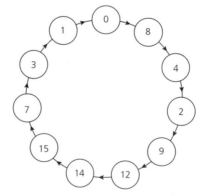

Solution

The state table is given in Table 8.13. The state diagram is given in Fig. 8.16. From the state table the state maps for the D inputs are

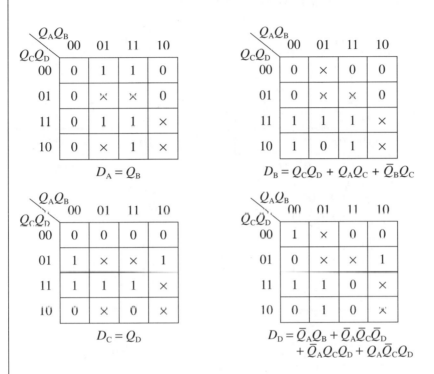

$Q_C Q_D$ \ $Q_A Q_B$	00	01	11	10
00	0	1	1	0
01	0	×	×	0
11	0	1	1	×
10	0	×	1	×

$$D_A = Q_B$$

$Q_C Q_D$ \ $Q_A Q_B$	00	01	11	10
00	0	×	0	0
01	0	×	×	0
11	1	1	1	×
10	1	0	1	×

$$D_B = Q_C Q_D + Q_A Q_C + \bar{Q}_B Q_C$$

$Q_C Q_D$ \ $Q_A Q_B$	00	01	11	10
00	0	0	0	0
01	1	×	×	1
11	1	1	1	×
10	0	×	0	×

$$D_C = Q_D$$

$Q_C Q_D$ \ $Q_A Q_B$	00	01	11	10
00	1	×	0	0
01	0	×	×	1
11	1	1	0	×
10	0	1	0	×

$$D_D = \bar{Q}_A Q_B + \bar{Q}_A \bar{Q}_C \bar{Q}_D + \bar{Q}_A Q_C Q_D + Q_A \bar{Q}_C Q_D$$

The required circuit is shown in Fig. 8.17 (*Ans.*)

Fig. 8.17 *Synchronous counter using D flip-flops has a count sequence of 0, 8, 4, 2, 9, 12, 14, 15, 7, 3, 1, 0, etc.*

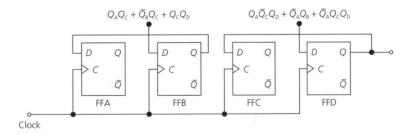

EXAMPLE 8.5

Repeat Example 8.4 using J-K flip-flops.

Solution

The state table is given in Table 8.12. From the state table the state maps are

J_A

Q_CQ_D \ Q_AQ_B	00	01	11	10
00	0	1	×	×
01	0	×	×	×
11	0	1	×	×
10	0	×	×	×

$$J_A = Q_B$$

K_A

Q_CQ_D \ Q_AQ_B	00	01	11	10
00	×	×	0	1
01	×	×	×	1
11	×	×	0	×
10	×	×	0	×

$$K_A = \bar{Q}_B$$

J_B

Q_CQ_D \ Q_AQ_B	00	01	11	10
00	0	×	×	0
01	0	×	×	0
11	×	×	×	1
10	1	×	×	×

$$J_B = Q_C$$

K_B

Q_CQ_D \ Q_AQ_B	00	01	11	10
00	×	1	1	×
01	×	×	×	×
11	×	0	0	×
10	×	×	0	×

$$K_B = \bar{Q}_C$$

J_C

Q_CQ_D \ Q_AQ_B	00	01	11	10
00	0	0	0	0
01	1	×	×	1
11	×	×	×	×
10	×	×	×	×

$$J_C = Q_D$$

K_C

Q_CQ_D \ Q_AQ_B	00	01	11	10
00	×	×	×	×
01	×	×	×	0
11	0	0	0	×
10	1	×	1	×

$$K_C = \bar{Q}_D$$

J_D

Q_CQ_D \ Q_AQ_B	00	01	11	10
00	1	1	0	0
01	×	×	×	×
11	×	×	×	×
10	0	×	0	×

$$J_D = \bar{Q}_A\bar{Q}_C$$

K_D

Q_CQ_D \ Q_AQ_B	00	01	11	10
00	×	×	×	×
01	1	×	×	0
11	0	0	1	×
10	×	×	×	×

$$K_D = Q_A Q_C + \bar{Q}_A\bar{Q}_C$$

Fig. 8.18 *Synchronous counter using J-K flip-flops has a count sequence of 0, 8, 4, 2, 9, 12, 14, 15, 7, 3, 1, 0, etc.*

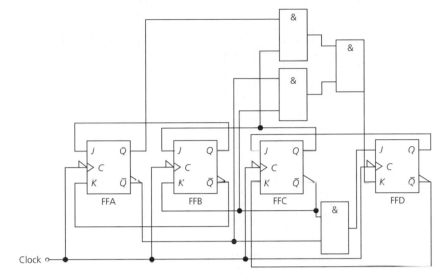

The designed circuit is shown in Fig. 8.18 (*Ans.*)

Table 8.14 *State table for a decade counter*

Count	Present state				Next state				Required inputs			
	Q_D	Q_C	Q_B	Q_A	Q_D^+	Q_C^+	Q_B^+	Q_A^+	T_D	T_C	T_B	T_A
0	0	0	0	0	0	0	0	1	0	0	0	1
1	0	0	0	1	0	0	1	0	0	0	1	1
2	0	0	1	0	0	0	1	1	0	0	0	1
3	0	0	1	1	0	1	0	0	0	1	1	1
4	0	1	0	0	0	1	0	1	0	0	0	1
5	0	1	0	1	0	1	1	0	0	0	1	1
6	0	1	1	0	0	1	1	1	0	0	0	1
7	0	1	1	1	1	0	0	0	1	1	1	1
8	1	0	0	0	1	0	0	1	0	0	0	1
9	1	0	0	1	0	0	0	0	1	0	0	1

Use of T flip-flops

If the *J* and *K* inputs of a J-K flip-flop are connected together the device acts as a T flip-flop. Each stage in the counter will then toggle at each clock transition.

Decade counter

The design of a decade counter using T flip-flops follows the same steps as for the D and J-K flip-flops. The state table is given in Table 8.14. From the state table, clearly $T_A = 1$. The state maps are

$$T_B = Q_A Q_B + Q_A \bar{Q}_D$$

$$T_C = Q_A Q_B$$

$$T_D = Q_A Q_D + Q_A Q_B Q_C$$

State checking table

The design of a counter can be checked with the aid of a *state checking table* (see Table 8.15). Consider the J-K flip-flop decade counter designed on p. 225.

Table 8.15 *State checking table*

	Present state												Next state				
	Q_D	Q_C	Q_B	Q_A	J_D	K_D	J_C	K_C	J_B	K_B	J_A	K_A	Q_D^+	Q_C^+	Q_B^+	Q_A^+	
0	0	0	0	0	0	0	0	0	0	0	1	1	0	0	0	1	1
1	0	0	0	1	0	1	0	0	1	1	1	1	0	0	1	0	2
2	0	0	1	0	0	0	0	0	0	0	1	1	0	0	1	1	3
3	0	0	1	1	0	1	1	1	1	1	1	1	0	1	0	0	4
4	0	1	0	0	0	0	0	0	0	0	1	1	0	1	0	1	5
5	0	1	0	1	0	1	0	0	1	1	1	1	0	1	1	0	6
6	0	1	1	0	0	0	0	0	0	0	1	1	0	1	1	1	7
7	0	1	1	1	1	1	1	1	1	1	1	1	1	0	0	0	8
8	1	0	0	0	0	0	0	0	0	0	1	1	1	0	0	1	9
9	1	0	0	1	0	1	0	0	0	1	1	1	0	0	0	0	0

Fig. 8.19 *State diagram for a 2-bit up-down counter*

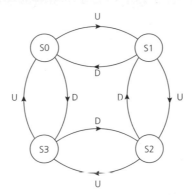

- The count starts at 0, so write $Q_A = Q_B = Q_C = Q_D = 0$.
- Determine from the circuit the J and K values of each flip-flop at this time. Since $Q_A = 0$, $J_B = K_B = J_C = K_C = J_D = K_D = 0$.
- Now determine the next state. Since K_A and J_A are at 1, only the first stage will toggle at the next clock transition so that the next state is 0001.
- Write down the next state of the previous line as the present state of the next line and find the values of each J and K input. Now, $K_B = K_D = 1$ since $Q_A = 1$. Also, since $\bar{Q}_D = 1$, both inputs to the left-hand AND gate are 1 and thus $J_B = 1$. All other J and K inputs are at 0.
- Determine the next state. Both flip-flops A and B toggle so that the next state is 0010.
- This procedure is repeated line-by-line, until a previous count is obtained. In this case this is 0010 which occurs on the tenth line.

Up-down counters

The design of an up-down counter follows the same procedure as the design of an up-counter except that the up/down control must be included. Figure 8.19 shows the state diagram for a 2-bit up-down counter. The lines connecting the state circles are labelled as 1 or as 0, the lines labelled 1 represent the logical state of the U/$\overline{D}$ control; when U/$\overline{D}$ is HIGH (1) the circuit counts up, and when U/$\overline{D}$ is LOW the circuit acts as a down-counter.

EXAMPLE 8.6

Design a divide-by-7 up-down counter using J-K flip-flops.

Solution

The state table of the required counter is given in Table 8.16. From the state table the state maps for each of the six J and K inputs can be written down (U denotes the state of the count-up line).

Table 8.16 *State table for a divide-by-7 up-down counter*

Present state			Count-up line	Next state			Required J-K inputs					
Q_C	Q_B	Q_A	U	Q_C^+	Q_B^+	Q_A^+	J_C	K_C	J_B	K_B	J_A	K_A
0	0	0	0	0	0	1	0	×	0	×	1	×
0	0	0	1	1	1	0	1	×	1	×	0	×
0	0	1	0	0	1	0	0	×	1	×	×	1
0	0	1	1	0	0	0	0	×	0	×	×	1
0	1	0	0	0	1	1	0	×	×	0	1	×
0	1	0	1	0	0	1	0	×	×	1	1	×
0	1	1	0	1	0	0	1	×	×	1	×	1
0	1	1	1	0	1	0	0	×	×	0	×	1
1	0	0	0	1	0	1	×	0	0	×	1	×
1	0	0	1	0	1	1	×	1	1	×	1	×
1	0	1	0	1	1	0	×	0	1	×	×	1
1	0	1	1	1	0	0	×	0	0	×	×	1
1	1	0	0	0	0	0	×	1	×	1	0	×
1	1	0	1	1	0	1	×	0	×	1	1	×

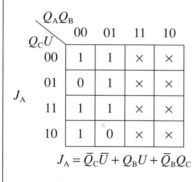

J_A

$Q_C U$ \ $Q_A Q_B$	00	01	11	10
00	1	1	×	×
01	0	1	×	×
11	1	1	×	×
10	1	0	×	×

$$J_A = \bar{Q}_C \bar{U} + Q_B U + \bar{Q}_B Q_C$$

K_A

$Q_C U$ \ $Q_A Q_B$	00	01	11	10
00	×	×	1	1
01	×	×	1	1
11	×	×	×	1
10	×	×	×	1

$$K_A = 1$$

J_B

$Q_C U$ \ $Q_A Q_B$	00	01	11	10
00	0	×	×	1
01	1	×	×	0
11	1	×	×	0
10	0	×	×	1

$$J_B = \bar{Q}_A U + Q_A \bar{U}$$

K_B

$Q_C U$ \ $Q_A Q_B$	00	01	11	10
00	×	0	1	×
01	×	1	0	×
11	×	1	×	×
10	×	1	×	×

$$K_B = Q_C + \bar{Q}_A U + Q_A \bar{U}$$

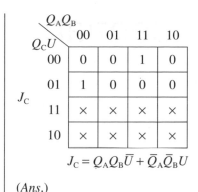

$$J_C = Q_A Q_B \bar{U} + \bar{Q}_A \bar{Q}_B U \qquad\qquad K_C = Q_B Q_C \bar{U} + \bar{Q}_A \bar{Q}_B U$$

(*Ans.*)

EXERCISES

8.1 Design a synchronous counter to follow the counting sequence 011, 100, 101, 111, 011, etc. Entry into an unwanted state is to change at the next clock transition to entry into state 111. Use D flip-flops.

8.2 Repeat exercise 8.1 using J-K flip-flops.

8.3 Design a self-starting counter to follow the counting sequence 0, 2, 4, 3, 6, 7, 0, etc. Use D flip-flops.

8.4 Design a synchronous counter that counts from 0 to 14 inclusive and then resets to 0. Use J-K flip-flops.

8.5 Design a synchronous counter to count down from 9 to 0 and then reset to 0. Use D flip-flops.

8.6 Design a self-starting divide-by-5 counter to follow the sequence 0, 1, 3, 6, 4, 0, etc. Use D flip-flops.

8.7 Draw the state diagram for a 3-bit Gray code counter. Hence, write down the state table if D flip-flops are to be employed. Draw the state maps and from them obtain expressions for D_A, D_B and D_C.

8.8 Design a counter to follow the count sequence 1, 2, 5, 7, 1, etc. Use J-K flip-flops.

8.9 Design a BCD counter that has each unwanted state just one clock pulse away from clearing all stages. Use T flip-flops.

8.10 Design a divide-by-5 counter to follow the sequence 1, 3, 2, 4, 6, 1, etc. Unwanted states 0, 5 and 7 are to lead to state 2.

8.11 Show how a counter that uses J-K flip-flops can be made to take up state 0 when it is first switched ON.

8.12 A divide-by-4 counter will operate only when its enable input is HIGH. Draw its state diagram.

9 Design of sequential systems

After reading this chapter you should be able to:

- Draw the block diagrams of synchronous and non-synchronous sequential systems.
- Write down the state diagram of a sequential system.
- Derive the state table for a sequential system.
- Use a flip-flop excitation map.
- Design a synchronous sequential system.
- Obtain the flow matrix for a non-synchronous sequential system.
- Simplify a flow matrix using row merging.
- Design a non-synchronous sequential system.
- Understand the causes of hazards and races in non-synchronous sequential systems.

A sequential system is one whose next output state depends not only on the input variables but also on the present state of the output(s). This means that the output(s) of the system is fed back to one, or more, of the inputs of the circuit. A sequential system is also known as *finite state* machine (FSM) since the sequential logic employed can have only a finite number of states.

In a synchronous sequential system all changes of state within the circuit take place at a clock transition. The present states of each flip-flop and/or gate in the circuit are the internal states at the instant new external data is applied to the input terminals of the circuit. When a clock transition occurs the next states are generated and are fed back to the input(s) to become the new present state. The internal states can be accurately defined and generated at each clock transition. The periodic time of the clock wave-form must be longer than the propagation delays in the system. Synchronous sequential systems always include at least one flip-flop and they are easier to design than are non-synchronous systems. The basic block diagram of a synchronous sequential system is shown in Fig. 9.1.

A non-synchronous sequential system operates at its own speed; the output of one stage provides the input to the next and the system may be implemented using combinational logic only. A non-synchronous circuit is unable to distinguish between consecutive

Fig. 9.1 *Synchronous sequential system*

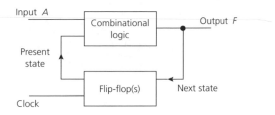

Table 9.1					
Present state	*Next state*				
Q	*J-K*	*00*	*01*	*11*	*10*
0		0	0	1	1
1		1	0	0	1

input signals that are at the same logic level. It is harder to design a non-synchronous system because of the presence of both stable and unstable states.

State tables

The operation of a sequential system can be described by a state diagram and this can be employed to derive the state table for the circuit. For the J-K flip-flop whose state diagram is shown in Fig. 8.5, the state table is given in Table 9.1. The state table has a number of rows equal to the number of possible states that the circuit is able to enter, and a number of columns equal to the number of possible combinations of the input variables. In each cell of the state table the next state of the circuit is entered when it is in the state represented by the row label, and the input variables are those specified by the column heading.

Consider a synchronous sequential circuit that has a single input terminal and a single output terminal and is required to detect the occurrence of two consecutive 1 bits in an input bit stream. Detection of two consecutive 1 bits is to be indicated by the output of the circuit going HIGH. The output is then to remain HIGH as long as further consecutive 1 bits are received. The circuit should be reset when a 0 bit arrives at its input terminals.

The state diagram for the circuit is shown in Fig. 9.2. State S0 is the initial state of the circuit; the circuit must remain in state S0 until a 1 bit is received. When a 1 bit occurs the circuit is required to move into state S1. If the next bit is another 1 the circuit

Fig. 9.2 *State diagram for a circuit that detects an 11-bit sequence*

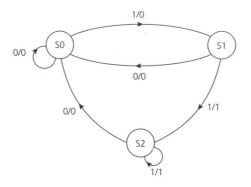

	Table 9.2				

		Next state		Output	
Present state	Input x = 0	= 1		= 0	= 1
S0	S0	S1		0	0
S1	S0	S2		0	1
S2	S0	S2		0	1

must move into state S2 and its output must go HIGH to indicate that two consecutive 1 bits have been received; but if the next bit is a 0 the circuit must return to state S0. Once the circuit is in state S2 it must remain there until a 0 bit is received, when it must return to its initial state S0. The state table for the circuit can be derived from the state diagram and is given in Table 9.2.

EXAMPLE 9.1

The circuit just considered is to be modified so that a third consecutive 1 bit arriving at the input terminals will take the output of the circuit LOW and a fourth consecutive 1 bit will take it HIGH again. Draw the modified state diagram and derive the new state table.

Solution

When the circuit is in state S2 another input 1 bit must take the circuit into a new state S3 in which the output is LOW. When the circuit is in state S3 another 1 bit must return the circuit to state S2. Hence the modified state diagram is as shown in Fig. 9.3. The state table of the modified circuit is given in Table 9.3 (*Ans.*)

Fig. 9.3 *Modified version of Fig. 9.2*

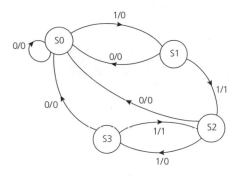

Table 9.3

Present state	Next state Input $x = 0$	$= 1$	Output $= 0$	$= 1$
S0	S0	S1	0	0
S1	S0	S2	0	1
S2	S0	S3	0	0
S3	S0	S2	0	1

Fig. 9.4

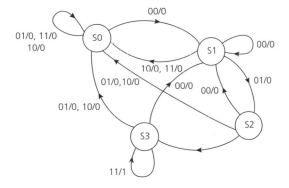

EXAMPLE 9.2

A circuit has two input terminals A and B and a single output F. The output of the circuit is to go HIGH only when inputs $AB = 00$, $= 01$, $= 11$ are received in that order. Once the output of the circuit has been taken HIGH it remains HIGH as long as consecutive $AB = 11$ bits are received. Obtain the state diagram and the state table for the circuit.

Solution

The state diagram for the circuit is shown in Fig. 9.4 and the state table is given in Table 9.4 (*Ans.*)

Table 9.4

Present state	Next state				Output F
	AB = 00	01	11	10	
S0	S1	S0	S0	S0	0
S1	S1	S2	S0	S0	0
S2	S1	S0	S3	S0	1
S3	S1	S0	S3	S0	0

Fig. 9.5

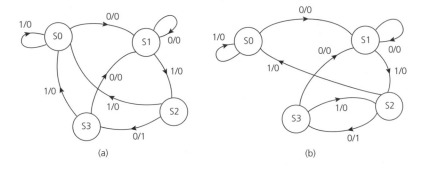

(a)   (b)

Table 9.5

Present state	Next state		Output
	Input x = 0	= 1	
S0	S1	S0	0
S1	S1	S2	0
S2	S3	S0	1
S3	S1	S0	0

EXAMPLE 9.3

Draw the state diagram and the state table for a single-input circuit that will detect the occurrence of the bit sequence 010 in a bit stream:

(a) With no overlaps (e.g. 10101010 contains two separate 010 groups of bits).
(b) With overlaps (e.g. 10101010 contains three occurrences of 010).

Solution

(a) The state diagram is shown in Fig. 9.5(a) and the state table is given in Table 9.5 (*Ans.*)

Table 9.6

Present state	Next state Input $x = 0$	$= 1$	Output
S0	S1	S0	0
S1	S1	S2	0
S2	S3	S0	1
S3	S1	S2	0

(b) The state diagram is shown in Fig. 9.5(b) and the state table is given in Table 9.6 (*Ans.*)

Excitation maps

Synchronous sequential circuits are operated under the control of a system clock and they always employ at least one flip-flop. The flip-flops may be of either the J-K, D or T types. An *excitation map* is used to determine the necessary inputs to the flip-flops for the required next state to occur at the next clock transition. The variables in the circuit are the present states and the external inputs and the axes of the excitation map must be labelled using the Gray code. The flip-flop equations are obtained from the excitation map in the same way as combinational logic equations can be obtained from a Karnaugh map. The equations for the output F of the circuit and for the required flip-flop inputs are derived separately. In writing down an excitation map it is necessary to have available the transition table for the type of flip-flop to be employed. Table 9.7 gives the transition tables for the J-K, D and T flip-flops.

Table 9.5 has been re-written in Table 9.8 to show the states of the Q_A and Q_B outputs of the two flip-flops that are necessary. Using Tables 9.7 and 9.8, the excitation maps for the circuit may be drawn. When the circuit is in its initial state S0 and the input x to the circuit is LOW (0) the circuit should move into its state S1. This means that it should change from $Q_A = Q_B - 0$ to $Q_A - 1$ and $Q_B - 0$. From Table 9.7 it can be seen that this requires the following flip-flop inputs: $J_A = 1$, $K_A = \times$ (don't care), $J_B = 0$, $K_B = \times$, $D_A = 1$, $D_B = 0$, $T_A = 1$ and $T_B = 0$. These values are entered in the $Q_A Q_B = 00$, $x = 0$ cell

Table 9.7

Present state Q	Next state Q^+	J-K		D	T	S-R	
0	0	0	×	0	0	0	×
0	1	1	×	1	1	1	0
1	0	×	1	0	1	0	1
1	1	×	0	1	0	×	0

Table 9.8

	Present state		Next state Input $x = 0$		1		Output
State	Q_B	Q_A	Q_B	Q_A	Q_B	Q_A	
S0	0	0	0	1	0	0	0
S1	0	1	0	1	1	1	0
S2	1	1	1	0	0	0	1
S3	1	0	0	1	0	0	0

of each excitation map. Similarly, when the input $x = 1$ the circuit should remain in the state S0 for both the present and next states. Again, from Table 9.7, this requires that $J_A = J_B = 0$, $K_A = K_B = 0$, $D_A = D_B = 0$, and $T_A = T_B = 0$. These values are entered in the $Q_A Q_B = 00$, $x = 1$ cell of each excitation map. The rest of the entries have been obtained in the same way.

(a) J-K flip-flops.

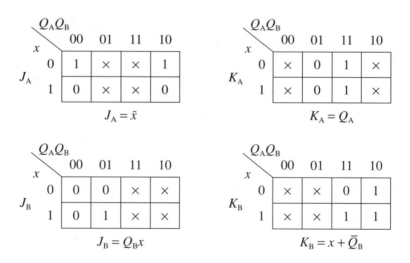

(b) D flip-flops.

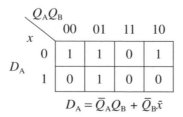

(c) T flip-flops.

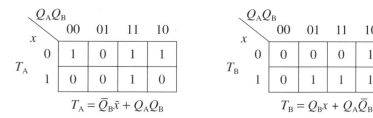

$$T_A = \bar{Q}_B \bar{x} + Q_A Q_B$$

$$T_B = Q_B x + Q_A \bar{Q}_B$$

Design of a synchronous sequential circuit

The steps to be taken in the design of a synchronous sequential circuit are as follows:

- Draw the state diagram of the required circuit.
- Determine the number of flip-flops needed and decide which kind to use.
- Use the state diagram to obtain the state table for the circuit.
- Draw the excitation maps.
- Use the excitation maps to obtain equations for the flip-flop inputs.
- Determine an expression for the output F of the circuit.
- Using these expressions draw the required circuit and then decide whether SSI/MSI devices or a PLD are to be used. If the former, decide which ICs to use.

The procedure is illustrated by the following design. A circuit is required that will detect the bit sequence 1001 in an input bit stream, including any overlaps. Detection of the required bit sequence is to be indicated by the output of the circuit going HIGH.

The state diagram of the required circuit is shown in Fig. 9.6. Initially, the circuit is in state S0 and here it will remain until a 1 bit is received, then the circuit will move to its state S1. It will then remain in state S1 until a 0 bit arrives at the input to the circuit when it will move into state S2. Once in state S2 a 0 bit followed by a 1 bit will move the circuit through state S3 into state S4 and when it is in state S4 the output of the circuit will go HIGH to indicate that the bit sequence 1001 has been detected.

Fig. 9.6 *State diagram for a circuit that detects a 1001 bit sequence*

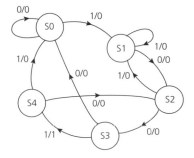

Table 9.9

State	Q_C	Q_B	Q_A	Input $x = 1$	Q_C	Q_B	Q_A	$= 0$	Q_C	Q_B	Q_A	F
		Present state									Next state	Output
S0	0	0	0	S1	0	0	1	S0	0	0	0	0
S1	0	0	1	S1	0	0	1	S2	0	1	1	0
S2	0	1	1	S1	0	0	1	S3	0	1	0	0
S3	0	1	0	S4	1	1	0	S0	0	0	0	0
S4	1	1	0	S0	0	0	0	S2	0	1	1	1

When the circuit is in state S2 an input 1 bit means that the required bit sequence is not present; however, this 1 bit could be the start of a new required sequence and hence the circuit must move back to state S1.

When the circuit is in state S3 an input 0 bit means that the wanted bit sequence cannot now occur and hence the circuit must go back to its initial state S0.

When the circuit is in final state S4 an input 1 bit means that the next sequence is not the required one and so the circuit must revert back to state S0. If, however, the next input bit is a 0 it could be the second bit of a following required bit sequence, i.e. 1001 001, and hence the circuit moves to state S2.

There are five different states and hence three flip-flops are necessary. The five states are best numbered so that only 1 bit changes at a time in the state table, and one possible numbering scheme is shown in Table 9.9.

D flip-flops

When the circuit is in its initial state S0 it should stay there if the input $x = 0$ or should go to state S1 if $x = 1$. The first state requires that $D_A = 0$ and the second state requires that $D_A = 1$. Also, $D_B = D_C = 0$ for both $x = 0$ and $x = 1$. Hence, the excitation maps for the three flip-flops have these values entered in the $\bar{Q}_A\bar{Q}_B\bar{Q}_C\bar{x}$ and $\bar{Q}_A\bar{Q}_B\bar{Q}_C x$ cells.

Next, when the circuit is in state S1 it should move to state S2 if $x = 0$ but remain in state S1 if $x = 1$. This requires that when $x = 0$, Q_A stays at 1 and so $D_A = 1$, Q_B changes from 0 to 1 which needs $D_B = 1$. Q_C does not change so $D_C = 0$. Also, when $x = 1$, Q_A remains at 1 so that $D_A = 1$, while Q_B and Q_C both stay at 0. Hence, $D_B = D_C = 0$. These values are entered into cells $Q_A\bar{Q}_B\bar{Q}_C\bar{x}$ and $Q_A\bar{Q}_B\bar{Q}_C x$ in all three maps.

D_A

$Q_C x$ \ $Q_A Q_B$	00	01	11	10
00	0	0	0	1
01	1	0	1	1
11	×	0	×	×
10	×	1	×	×

D_B

$Q_C x$ \ $Q_A Q_B$	00	01	11	10
00	0	0	1	1
01	0	1	0	0
11	×	0	×	×
10	×	1	×	×

| | Q_AQ_B | | | |
Q_Cx	00	01	11	10
00	0	0	0	0
01	0	1	0	0
11	×	0	×	×
10	×	0	×	×

D_C

| | Q_AQ_B | | | |
Q_Cx	00	01	11	10
00	0	0	0	0
01	0	0	0	0
11	×	1	×	×
10	×	1	×	×

F

Fig. 9.7 *D flip-flop implementation of the state diagram given in Fig. 9.6*

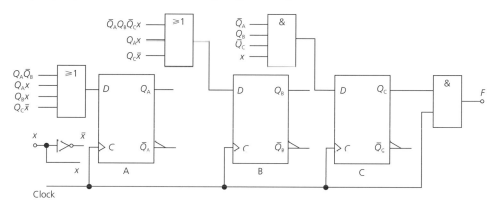

Similarly, the circuit should go from state S2 to state S3 if $x = 0$, requiring Q_A to change from 1 to 0 and so $D_A = 0$; Q_B stays at 1 and Q_C stays at 0 so that $D_B = 1$ and $D_C = 0$. These values are entered into cells $Q_AQ_BQ_C\bar{x}$ and $Q_AQ_B\bar{Q}_Cx$ in all three maps.

In a similar manner, the entries for S3 and S4 have been made and are shown in the three maps. The remainder of the cells are 'don't cares'. From the excitation maps,

$$D_A = Q_A\bar{Q}_B + Q_Ax + Q_Bx + Q_C\bar{x},$$

$$D_B = Q_A\bar{x} + \bar{Q}_AQ_B\bar{Q}_Cx + Q_C\bar{x} \text{ and } D_C = \bar{Q}_AQ_B\bar{Q}_Cx$$

$$F = Q_C$$

The circuit is shown in Fig. 9.7.

J-K flip-flops

The same procedure as for D flip-flops is followed but now it is necessary to determine the required inputs for both the J and K terminals of each flip-flop for a required change in the state of the circuit to take place. When the circuit is in state S0, $x = 0$ requires the circuit to remain in S0 and so Q_A, Q_B and Q_C must all remain at 0. Hence, the required inputs are $J = 0$ and $K = \times$ (don't care). These values are entered into the $\bar{Q}_A\bar{Q}_B\bar{Q}_C\bar{x}$ cell of each map. If $x = 1$, the circuit must move from S0 to S1 and this requires that Q_A only changes state to 1. Hence, the necessary inputs to flip-flop A are $J_A = 1$ and $K_A = \times$. These values are entered into cells $\bar{Q}_A\bar{Q}_B\bar{Q}_Cx$ in the J_A and K_A maps. Continuing in this way gives the state maps shown.

Q_AQ_B

Q_Cx	00	01	11	10
00	0	0	×	×
01	1	0	×	×
11	×	0	×	×
10	×	1	×	×

J_A

Q_AQ_B

Q_Cx	00	01	11	10
00	×	×	1	0
01	×	×	0	0
11	×	×	×	×
10	×	×	×	×

K_A

Q_AQ_B

Q_Cx	00	01	11	10
00	0	×	×	1
01	0	×	×	×
11	×	×	×	×
10	×	×	×	×

J_B

Q_AQ_B

Q_Cx	00	01	11	10
00	×	1	0	×
01	×	0	1	0
11	×	1	×	×
10	×	0	×	×

K_B

Q_AQ_B

Q_Cx	00	01	11	10
00	0	0	0	0
01	0	1	0	0
11	×	×	×	×
10	×	×	×	×

J_C

Q_AQ_B

Q_Cx	00	01	11	10
00	×	×	×	0
01	×	×	×	×
11	×	1	×	×
10	×	0	×	×

K_C

From the maps: $J_A = \bar{Q}_Bx + Q_C\bar{x}$, $K_A = Q_B\bar{x}$, $J_B = Q_A$, $K_B = \bar{Q}_C\bar{Q}_B\bar{x} + Q_Cx + Q_AQ_Bx$, $J_C = \bar{Q}_AQ_Bx$ and $K_C = x$.

The circuit is shown in Fig. 9.8. It can be seen that the extra flexibility given by the J-K flip-flop having two input terminals leads to simpler combinational logic.

Fig. 9.8 *J-K flip-flop implementation of the state diagram given in Fig. 9.6*

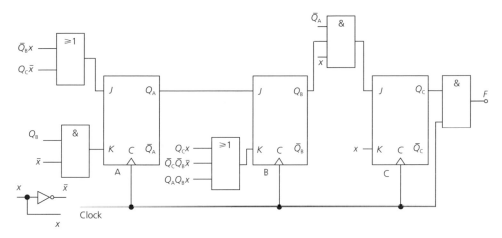

T flip-flops

The state maps obtained using the transition table for a T flip-flop are

T_A

Q_Cx \ Q_AQ_B	00	01	11	10
00	0	0	1	0
01	1	0	0	0
11	×	1	×	×
10	×	0	×	×

T_B

Q_Cx \ Q_AQ_B	00	01	11	10
00	0	1	0	1
01	0	0	1	0
11	×	0	×	×
10	×	1	×	×

T_C

Q_Cx \ Q_AQ_B	00	01	11	10
00	0	0	0	0
01	0	1	0	0
11	×	1	×	×
10	×	1	×	×

From the maps: $T_A = \bar{Q}_A\bar{Q}_Bx + Q_Cx + Q_AQ_B\bar{x}$, $T_B = \bar{Q}_AQ_B\bar{x} + Q_AQ_Bx + Q_A\bar{Q}_B\bar{x}$ and $T_C = Q_C + \bar{Q}_AQ_Bx$. The required circuit is shown in Fig. 9.9.

Fig. 9.9 *T flip-flop implementation of the state diagram given in Fig. 9.6*

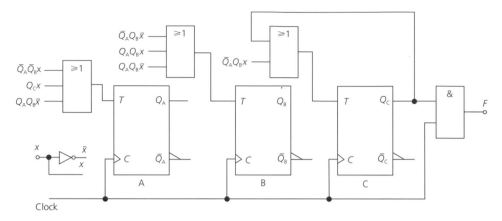

EXAMPLE 9.4

Design a circuit to detect the presence of three consecutive 1 bits in a bit stream. The output of the circuit is to go HIGH to indicate when this condition occurs and then stay HIGH until the next 0 bit arrives. Use D flip-flops.

Fig. 9.10

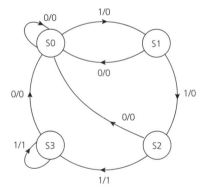

Solution

The state diagram for the circuit is shown in Fig. 9.10. In this figure S0 is the initial state of the circuit, S1 is the state for one received 1 bit, S2 is the state for two consecutive 1 bits, and S3 is the state for three consecutive 1 bits. There are four different states and hence two D flip-flops are required. The state table for the required circuit is obtained from the state diagram and is given in Table 9.10. The state maps are

---

Table 9.10

State	Q_B	Q_A	Next state Input $x = 0$ Q_B^+	Q_A^+	Required inputs D_B	D_A	Next state Input $x = 1$ Q_B^+	Q_A^+	Required inputs D_B	D_A	Output F
S0	0	0	0	0	0	0	0	1	0	1	0
S1	0	1	0	0	0	0	1	0	1	0	0
S2	1	0	0	0	0	0	1	1	1	1	0
S3	1	1	0	0	0	0	1	1	1	1	1

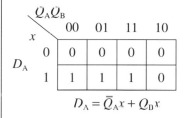

$$D_A = \bar{Q}_A x + Q_D x$$

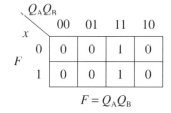

$$D_B = Q_B x + Q_A x \qquad F = Q_A Q_B$$

The required circuit is shown in Fig. 9.11 (*Ans.*)

Fig. 9.11

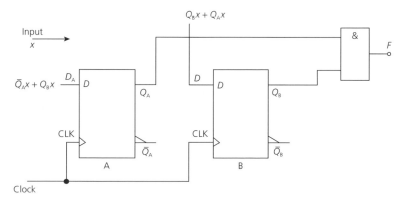

Fig. 9.12

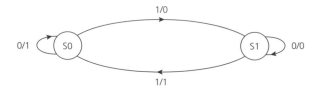

Table 9.11

Present state	Input	Next state	Output
S0	0	S0	1
S0	1	S1	0
S1	0	S1	0
S1	1	S0	1

Fig. 9.13

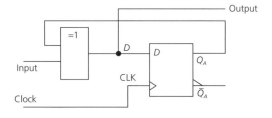

EXAMPLE 9.5

Design a circuit whose output goes HIGH whenever the number of 1 bits arriving at its input terminal is even.

Solution

The state diagram for the required circuit is shown in Fig. 9.12. S0 is the state when an even number of bits have been inputted and S1 is the state for an odd number of bits received. The state table is given in Table 9.11.

If S0 is denoted by 0 and S1 by 1 it is clear that:

next state output = (present state) $\oplus$ (input)

The required circuit is shown in Fig. 9.13 (*Ans.*)

Aim: to build and confirm the correct operation of a circuit which detects the presence of three consecutive 1 bits in a bit stream.

Components and equipment: one 74HC74 dual D flip-flop IC, one 74HC08 quad 2-input AND gate IC, one 74HC32 quad 2-input OR gate IC, one LED and one 270 Ω resistor. Pulse generator. Breadboard. Power supply.

Procedure:

(a)   Build the circuit shown in Fig. 9.11.

(b)   Set the clock frequency to 1 kHz and input the bit sequence 1010110111 and see whether the three 1-bit sequence is indicated by the LED glowing visibly.

(c)   Next input the bit sequence 1011111. Does the LED remain ON?

(d)   Input the bit sequence 1011110. Does the LED dim when the 0 following three or more 1s is entered?

(e)   Have any errors been detected? Try operating the circuit at several higher clock frequencies. Explain how any errors arise.

EXAMPLE 9.6

Design a circuit using D flip-flops that will compare two 4-bit numbers A and B and signal which of them is the larger.

Fig. 9.14

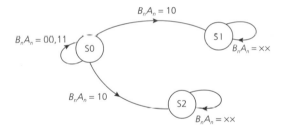

Table 9.12

Present state	Next state			
	$B_n A_n = 00$	01	11	10
S0	S0	S2	S0	S1
S1	S1	S1	S1	S1
S2	S2	S2	S2	S2

Solution

If $A > B$, $F_1 = 1$, $F_2 = 0$, if $A < B$, $F_2 = 1$ and $F_1 = 0$. If $A = B$, $F_1 = F_2 = 0$. The state diagram of the circuit is shown in Fig. 9.14. From the state diagram the state table can be produced and is shown in Table 9.12.

Let S0 = 00, S1 = 10 and S2 = 01, then the state table can be re-written as shown in Table 9.13. The state maps are

Table 9.13

Present state	Next state								Required inputs								Outputs	
	$A_nB_n = 00$		$= 01$		$= 11$		$= 10$											
Q_B Q_A	Q_B^+	Q_A^+	Q_B^+	Q_A^+	Q_B^+	Q_A^+	Q_B^+	Q_A^+	D_B	D_A	D_B	D_A	D_B	D_A	D_B	D_A	F_1	F_2
0 0	0	0	1	0	0	0	0	1	0	0	1	0	0	0	0	1	0	0
1 0	1	0	1	0	1	0	1	0	1	0	1	0	1	0	1	0	0	1
0 1	0	1	0	1	0	1	0	1	0	1	0	1	0	1	0	1	1	0

D_A

A_nB_n \ Q_AQ_B	00	01	11	10
00	0	1	×	0
01	0	1	×	0
11	0	1	×	0
10	1	1	×	0

$$D_A = Q_B + \bar{Q}_A\bar{B}_nA_n$$
$$F_1 = Q_A$$

A_nB_n \ Q_AQ_B	00	01	11	10
00	0	0	×	1
01	1	0	×	1
11	0	0	×	1
10	0	0	×	1

$$D_B = Q_A + \bar{Q}_BB_n\bar{A}_n$$
$$F_2 = Q_B$$

(*Ans.*)

EXAMPLE 9.7

Design a synchronous sequence detector to detect the presence of 101 in a bit stream. Use J-K flip-flops.

Fig. 9.15

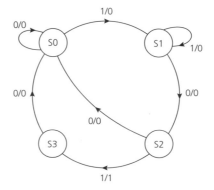

Table 9.14

	Present state		Next state x = 0		Required inputs				Next state x = 1		Required inputs			
	Q_B	Q_A	Q_B^+	Q_A^+	J_B	K_B	J_A	K_A	Q_B^+	Q_A^+	J_B	K_B	J_A	K_A
S0	0	0	0	0	0	×	0	×	0	1	0	×	1	×
S1	0	1	1	1	1	×	×	0	0	1	0	×	×	0
S2	1	1	0	0	×	1	×	1	1	0	×	0	×	1
S3	1	0	0	0	×	1	0	×	0	1	×	1	1	×

Solution

The state diagram is shown in Fig. 9.15. There are four states so two flip-flops are required. The state table with S0 = 00, S1 = 01, S2 = 11 and S3 = 10 is given in Table 9.14. The state maps are

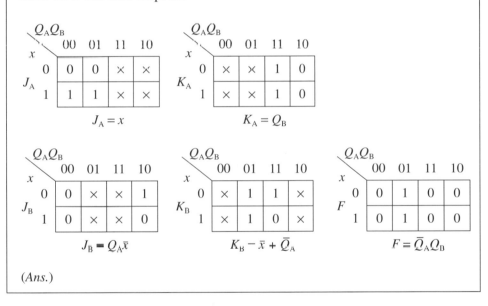

$J_A = x$

$K_A = Q_B$

$J_B = Q_A\bar{x}$

$K_B = \bar{x} + \bar{Q}_A$

$F = \bar{Q}_A Q_B$

(*Ans.*)

Aim: to build the circuit shown in Fig. 9.8. To check that the output of the circuit goes HIGH only when the bit sequence 1001 is applied to the input of the circuit. Components and equipment: two 74HC112 dual J-K flip-flop ICs, one 74HC08 quad 2-input AND gate IC, one 74HC11 triple 3-input AND gate IC, one 74HC32 quad 2-input OR gate IC, one 74HC04 hex inverter IC, one LED and one 270 Ω resistor. Pulse generator. Breadboard. 5 V d.c. voltage supply. Power supply.

Procedure:

(a) Build the circuit given in Fig. 9.8.
(b) Set the clock frequency to 1 kHz and 5 V at 50 per cent duty cycle. Apply the bit stream 1001 to the input terminal x of the circuit and note whether the LED glows visibly when the fourth bit is entered.
(c) Enter, in turn, by operating the switch, the bit streams 10011, 10010 and 1001001 and each time note what happens.
(d) Apply a random bit stream to the circuit and observe if the LED glows at any point.
(e) Repeat (b), (c) and (d) at clock frequencies of
   (i) 1 MHz
   (ii) 20 MHz
   (iii) 30 MHz.
   If a logic analyser is available use it to observe the input and output waveforms of the circuit.
(f) Comment on the operation of the circuit. Did it correctly indicate the presence of the bit stream 1001
   (i) on its own
   (ii) as a part of a longer bit stream?

Non-synchronous sequential systems

A non-synchronous sequential system does not have a clock to regulate the events that take place in the circuit. The control inputs to a non-synchronous sequential circuit are pulses that are applied to one, or more, input terminals. The circuit will respond immediately to any input change instead of waiting for the next clock tran-sition like a synchronous circuit. If an unchanging sequence of bits, say 11111, is applied to the input of a synchronous circuit this will be taken as a series of 1 inputs at each clock transition, but if the same bit sequence is applied to a non-synchronous circuit the input signal would be taken as a single 1 bit. Timing may therefore be difficult to design unless it can be arranged that the input variables never change state simultaneously.

A simple example of a non-synchronous sequential circuit is shown in Fig. 9.16. A 2-input OR gate has a variable A applied to one of its inputs and its output F is applied to the other input. The logical state of the gate is time dependent. The fed-back signal is labelled as f, instead of F, to indicate that, because the operation of the gate is not instantaneous, they might not be the same at all instants in time. When an input change occurs the new state of the output F will take some time to propagate through the circuit. During this time the new value of F and the present value f may differ from one another. The state table of the circuit is given in Table 9.15. The state map is

Fig. 9.16 *Non-synchronous sequential circuit*

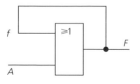

Table 9.15

Present state		Next state
A	f	F
0	0	0
0	1	1
1	0	1
1	1	1

	A 0	1
f 0	0	1
1	1	1

and hence $F = A + f$.

Another example of a non-synchronous circuit is shown in Fig. 9.17. The circuit has two inputs, A and B, and a single output F. When the input variables change the next output state of the circuit is a function not only of the input variables but also of the present state of the output. The output fed back to the input is labelled as f and not as F; this is because there is always a propagation delay through the circuit. Hence there is always a short time delay between f being applied to the input terminal and F appearing at the output terminal. This means that for a short time f may not be equal to F. Therefore, when a non-synchronous circuit is designed the next state and the present state must be considered separately.

The state table for the circuit is shown in Table 9.16. The state map is

AB	00	01	11	10
f 0	0	1	1	0
1	0	1	1	1

Hence, $F = B + Af$.

Fig. 9.17 *Non-synchronous sequential circuit*

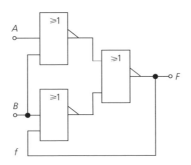

Table 9.16

Present state					Next state
B	A	f	B + f	A + B	F = B + f + A + B
0	0	0	1	1	0
0	0	1	0	1	0
0	1	0	1	0	0
0	1	1	0	0	1
1	0	0	0	0	1
1	0	1	0	0	1
1	1	0	0	0	1
1	1	1	0	0	1

PRACTICAL EXERCISE 9.3

Aim: to determine the operation of a simple non-synchronous sequential circuit.
Components and equipment: one 74HC00 quad 2-input NAND gate IC, one LED
and one 270 Ω resistor. Two pulse generators. CRO. Breadboard. Logic analyser (if
available). Power supply.

Procedure:

(a) Build the circuit shown in Fig. 9.18.

(b) Apply all the possible combinations of 1 and 0 to the inputs A and B of the
circuit. Each time note the logical states of the LEDs and hence determine the
Boolean expression that describes the operation of the circuit.

(c) Remove the switched connections to 1 and 0 and the inputs A and B and,
instead, connect a pulse generator to each input. Set the frequency of one
generator to 10 Hz and the frequency of the other generator to 15 Hz. Use
the CRO to observe the input and output waveforms of the circuit. Increase

Fig. 9.18

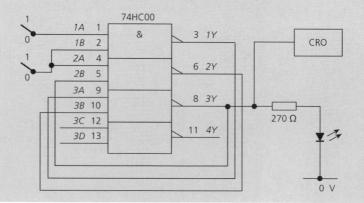

the frequencies of the generators to several different pairs of frequencies up to the region of the highest operating frequency of the IC. Each time observe the output waveform.

(d) Connect a logic analyser to the input and output terminals of the circuit and use it to observe the circuit waveforms at different input frequencies.

(e) Comment on the results. Did the propagation delays of the gates have a noticeable effect on the output waveform? If so, at what frequency did it first become evident?

Stable and unstable states

A stable state is one for which the entry in the state map for the output F of the circuit and for the fed-back value f are the same. The stable states are usually ringed. An unstable state exists when the circuit is in the process of changing from one stable state to another; for an unstable state the F entry differs from the condition for f for that row. Unstable states are not ringed.

The stable and unstable states for the circuit in Fig. 9.17 are shown in the next map. If, now, the stable states are numbered using the Gray code the *flow matrix* of the circuit will be obtained. This is shown by the map below left, which is the map for Table 9.16 again.

AB
f	00	01	11	10
0	(0)	1	1	(0)
1	0	(1)	(1)	(1)

AB
f	00	01	11	10
0	(1)	2	3	(4)
1	1	(2)	(3)	(5)

If there are two stable states in the one column then the state in the $f = 0$ row should be numbered first. The unstable states are then numbered with the same number as the stable state in the same column to which each leads. The flow matrix for Fig. 9.17 is shown by the map above right. The information given by the flow matrix shows that, for example, when $A = B = 0$ the circuit will take up its stable state 1, i.e. $F = 0$, and when $A = B = 1$ the circuit will be in its stable state 3, i.e. $F = 1$.

As a further example of stable and unstable states consider the S-R latch. This circuit is in a stable state when $S = 1$, $R = 0$ and $Q = 1$. When both the input variables change so that $S = 0$, $R = 1$, there will be a short time delay before the latch changes state to have $Q = 0$. For a short time, therefore, an unstable state exists in which $S = 0$, $R = 1$ and $Q = 1$. A circuit with two S-R latches could experience a change in the input variables that result in the circuit changing state from $Q_A = Q_B = 1$ to $Q_A = Q_B = 0$. This change in state could, possibly, occur simultaneously but it is much more likely that there will be an unstable state, with either $Q_A = 1$ and $Q_B = 0$, or $Q_A = 0$ and $Q_B = 1$, because of the different switching speeds of the two latches. If the outputs Q_A and Q_B are applied as inputs to other circuitry, incorrect system operation is very likely to occur. To avoid such problems non-synchronous sequential circuits are usually designed so that only one input variable is allowed to change at any instant in time.

Design of a non-synchronous sequential circuit

The design procedure for a non-synchronous sequential system is as follows:

(a) Draw the state diagram.
(b) Obtain the *primitive state (flow) table*.
(c) Simplify the primitive state table if possible by merging rows and obtain the merged state table.
(d) Allocate secondary variables to identify all the rows in the merged state table.
(e) Draw the flow matrix.
(f) Draw the excitation map(s).
(g) Derive the logic equations.
(h) Draw the designed circuit.

Consider a circuit that has two inputs A and B and a single output F. The output of the circuit is to go HIGH if the sequence $A = B = 0$, $A = 0$ and $B = 1$, $A = B = 1$ occurs at the inputs. The output should then remain HIGH as long as $A = 1$. If input A should change from 1 to 0 and then back to 1, while input B remains at 1, the output should go HIGH.

State diagram

The initial state S0 of the circuit is when $A = B = 0$. If B changes to 1 the circuit moves to its state S1. When the circuit is in state S1 a change in input B from 1 to 0 moves the circuit back to state S0, but if input A changes from 0 to 1 (when $A = B = 1$) the circuit goes into its state S2 and output goes HIGH.

Now a change in A from 1 to 0 ($AB = 01$) will move the circuit back into state S1 and the output of the circuit will go LOW; but a change in B from 1 to 0 ($AB = 10$) moves the circuit to state S3 and the output remains HIGH. When the circuit is in state S1 *after* being in state S2 a move to state S2 will also take the output HIGH. If, in either state S1 or S3, the next input change gives $A = B = 0$ the circuit will revert to its initial state S0. The circuit will go from state S3 to state S2 if B changes from 0 to 1 and the output will then remain HIGH.

If, when in state S0, A changes from 0 to 1 before B has changed from 0 to 1 a sequence of states is entered none of which will lead to the output of the circuit going HIGH.

The state diagram is shown in Fig. 9.19.

Primitive state (flow) table

The primitive state, or flow, table can be obtained from the state diagram. This table must have one row for each stable state, and so for this circuit there will be seven rows. An entry must be made in the table for each unstable state; these entries indicate the next stable state that will be taken up by the circuit for the particular input variables AB for that column in the table. If a particular condition cannot occur – in this circuit

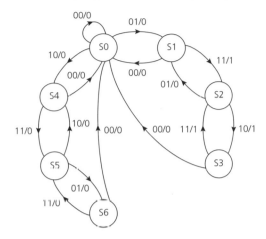

because simultaneous changes in *A* and *B* are not permitted – it is represented in the table by a dash.

The initial stable state of the circuit is S0 and this is represented in the first row of the table under the column heading *AB* = 00. Since this is a stable state it is ringed. If, then, *A* = 0 and *B* = 1 the circuit will go into an unstable state S1 until it moves into stable state S1. Thus, unstable state S1 is entered in the first row under the column heading *AB* = 01. If, instead, *AB* = 10 the circuit enters unstable state S4 before it goes to its stable state S4. Hence, S4 is entered in the first row under the column heading 10. Since *A* and *B* cannot change simultaneously the condition *A* = *B* = 1 does not happen and a dash is entered at the intersection of the first row and the 11 column. The first row entries in the primitive state table, therefore, are as shown below.

Row	AB				Output
	00	01	11	10	F
a	(S0)	S1	–	S4	0

The output of the circuit is 0 since the conditions for the output to go HIGH have not yet occurred. The rest of the entries into the primitive state table are determined in a similar manner leading to Table 9.17.

Row merging

Once the primitive state table has been obtained it will often contain more stable states than are necessary and so the next step should be to attempt to reduce the number of rows in the table. Two or more rows may be merged together if they have the same numbered states, stable or unstable, in each column of the table. Dashed entries are treated as 'don't cares' and may be assumed to have any state number. Each row in the

Table 9.17 _Primitive state table_

| Row | Input variables AB | | | | Output |
	00	01	11	10	F
a	Ⓢ0	S1	–	S4	0
b	S0	Ⓢ1	S2	–	0
c	–	S1	Ⓢ2	S3	1
d	S0	–	S2	Ⓢ3	1
e	S0	–	S5	Ⓢ4	0
f	–	S6	Ⓢ5	S4	0
g	S0	Ⓢ6	S5	–	0

Table 9.18

Rows	AB	00	01	11	10
a/b		Ⓢ0	Ⓢ1	S2	S4
c/d		S0	S1	Ⓢ2	Ⓢ3
e/f/g		S0	Ⓢ6	Ⓢ5	Ⓢ4

primitive state table may only appear in _one_ merged row. The output F of the circuit is ignored; it can later be obtained from the primitive state table.

Thus, rows a and b in Table 9.17 can be merged to give the merged row

　Ⓢ0　Ⓢ1　S2　S4

and rows c and d can be merged to give the merged row

　S0　S1　Ⓢ2　Ⓢ3

The merged state table is given in Table 9.18.

Alternatively, rows b, c and d could be merged together leaving row a on its own. The merged state table would then be as shown in Table 9.19.

Other possibilities also exist, e.g. rows a and e could be merged together.

Secondary variables

Since there are only three rows – in either of the merged state tables – only two secondary variables x and y will be required to identify them. These are variables that

Table 9.19

Rows	AB	00	01	11	10
a		(S0)	S1	–	S4
b/c/d		S0	(S1)	(S2)	(S3)
e/f/g		S0	(S6)	(S5)	(S4)

Fig. 9.20 *Secondary variables x and y*

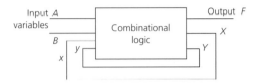

are provided by the circuit itself (see Fig. 9.20). The rows should be allotted x and y values that place adjacently rows that contain similar numbered stable and unstable states. In Table 9.18 there is movement from row a/b to row e/f/g [S4 to S4 and S0 to S0] and also between rows a/b and c/d, but none between rows c/d and e/f/g. Hence, a suitable numbering would be

row c/d $xy = 00$
row a/b $xy = 01$
row c/f/g $xy = 11$

The xy combination 10 is then not used.

Flow matrix

The flow matrix maps the stable and unstable states with 'don't cares' indicated by a dash. The flow matrix for Table 9.18, using the suggested xy row allocations, is shown below.

AB / xy	00	01	11	10
00	S0	S1	(S2)	(S3)
01	(S0)	(S1)	S2	S4
11	S0	(S6)	(S5)	(S4)
10	–	–	–	–

The flow matrix is minimized using the Karnaugh map looping technique. Hence, the next step is to allocate suitable state numbers to all of the 'don't cares'. A possible choice is illustrated by the flow matrix shown below.

xy \ *AB*	00	01	11	10
00	S0	S1	(S2)	(S3)
01	(S0)	(S1)	S2	S3
11	S0	(S6)	(S5)	(S4)
10	S0	S6	S5	S4

Excitation map

The excitation map for the circuit is drawn by writing down the values of *X* and *Y* for each stable state that agrees with the *xy* entry in that row. Thus, in the position occupied by stable states S2 and S3 in the top row of the flow matrix 00 is entered in the excitation map. Similarly, S0 and S1 in the second row are represented by 01. Next, the entries for the stable states are copied for the unstable states of the same number. This is shown by the next maps

XY

xy \ *AB*	00	01	11	10
00			00	00
01	01	01		
11		11	11	11
10				

XY

xy \ *AB*	00	01	11	10
00	01	01	00	00
01	01	01	00	11
11	01	11	11	11
10	01	11	11	00

Lastly, separate excitation maps can be drawn for both *X* and *Y* as shown below.

X

xy \ *AB*	00	01	11	10
00	0	0	0	0
01	0	0	0	1
11	0	1	1	1
10	0	1	1	0

Y

xy \ *AB*	00	01	11	10
00	1	1	0	0
01	1	1	0	1
11	1	1	1	1
10	1	1	1	0

Logic equations

From the excitation maps,

$$X = Bx + A\bar{B}y$$

$$Y = \bar{A} + \bar{B}y + Bx$$

From the primitive state table (Table 9.17) it can be seen that the output F is HIGH when the circuit is in either of its stable states S2 or S3. From the flow matrix a mapping for F can be derived, by writing 1 for each S2 and S3 entry and a 0 for all other stable and unstable states. Thus, the output map is shown below. From this map $F = AB\bar{x} + A\bar{B}\bar{y}$

	AB 00	01	11	10
xy 00	0	0	1	1
01	0	0	1	0
11	0	0	0	0
10	0	0	0	1

F

The circuit

The circuit is shown, using AND and OR gates, in Fig. 9.21. It may be implemented using random logic, probably using either NAND gates or NOR gates only, or, more likely, one of the other methods that are available such as a multiplexer or a PLD.

Use of flip-flops

The design of a non-synchronous sequential circuit may also be carried out using flip-flops. The merged state table given in Table 9.18 has three rows so that two flip-flops are needed. Suppose that J-K flip-flops are to be employed. Four state maps are then required. Referring to the excitation map for the circuit and to Table 9.9, for the present states $Q_A Q_B$ of the flip-flop outputs to become the required next states $Q_A^+ Q_B^+$ it is necessary to compare the row assignments and the table entries to determine the conditions necessary to set and reset each J-K flip-flop. For example, in the top left-hand cell of the J_A and K_A maps the present and next states of Q_A are both 0 and hence, from Table 9.9, $J_A = 0$ and $K_A = X$. Similarly, for the top left-hand cell of the J_B and K_B maps, Q_B is to change from 0 to 1 and hence $J_B = 1$ and $K_B = X$. The remainder of the cell entries have been obtained in the same way.

Fig. 9.21

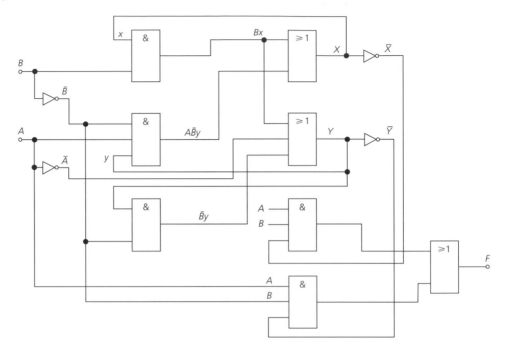

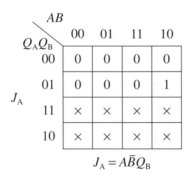

$$J_A = A\bar{B}Q_B$$

$$K_A = \bar{A}\bar{B} + \bar{B}\bar{Q}_B$$

$$J_B = \bar{A} + BQ_A$$

$$K_B = AB\bar{Q}_A$$

EXAMPLE 9.8

Design a circuit that has two inputs *A* and *B* and a single output. The output is to go HIGH whenever the sequence *A* = 1, *B* = 11, in that order, occurs. The output should then stay HIGH until another 1 pulse is received at the *B* input. *A* and *B* never go HIGH simultaneously.

Solution

The state diagram is given in Fig. 9.22. The initial state of the circuit is S0 and the circuit will remain in this state until a 1 pulse appears at input *A*, no matter how many pulses appear at the *B* input. When the first pulse appears at input *A* the circuit moves to state S1 and here it will remain until a pulse appears at *B*. Then the circuit will move into state S2. While the circuit is in state S2 another pulse at input *B* will move the circuit to state S3 and then the output will go HIGH. If, while the circuit is in state S2, another pulse arrives at input *A* the circuit will revert to state S1.

The circuit will remain at state S3 with a HIGH output until another pulse arrives at input *B* when the circuit will go back to state S0.

The primitive state table, obtained from the state diagram, is given in Table 9.20. No row merging is possible. The flow matrices are

Fig. 9.22

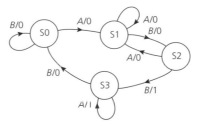

Table 9.20

Row	AB	00	01	11	10	Output F
a		(S0)	S0	–	S1	0
b		–	S1	–	(S1)	0
c		–	(S2)	–	S1	0
d		–	(S3)	–	S3	1

AB \ xy	00	01	11	10
00	(S0)	S0	–	S1
01	–	S1	–	(S1)
11	–	(S2)	–	S1
10	–	(S3)	–	S3

X

AB \ xy	00	01	11	10
00	(S0)	S0	S1	S1
01	S1	S1	S1	(S1)
11	S2	(S2)	S2	S1
10	S3	(S3)	S3	S3

Y

The excitation maps are

AB \ xy	00	01	11	10
00	0	0	0	0
01	0	0	0	0
11	1	1	1	0
10	1	1	1	1

X

AB \ xy	00	01	11	10
00	0	0	1	1
01	1	1	1	1
11	1	1	1	1
10	0	0	0	0

Y

From the maps, $X = \bar{A}x + Bx + x\bar{y}$ and $Y = A\bar{x} + y$

The output of the circuit is HIGH only when the circuit is in stable state S3 and, from the flow matrix, $F = xy$. The circuit is shown in Fig. 9.23 (*Ans.*)

Fig. 9.23

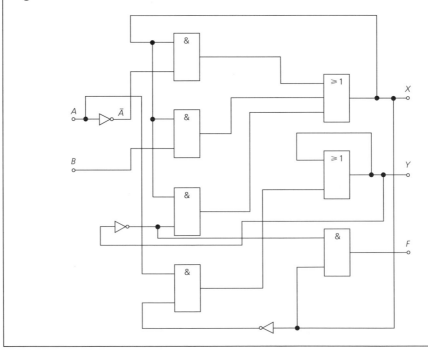

Aim: to build a non-synchronous sequential circuit with two inputs *A* and *B* that is able to detect the bit sequence *A* = 1, *B* = 11 in that order. The output is to go HIGH to signal the receipt of that bit sequence. *A* and *B* never go HIGH simultaneously.

Components and equipment: two 74HC08 quad 2-input AND gates ICs, one 74HC32 quad 2-input OR gate IC, one 74HC04 hex inverter IC, three LEDs and three 270 Ω resistors. Breadboard. 5 V d.c. voltage source. Power supply.

Procedure:

(a)   Build the circuit given in Fig. 9.23. Connect the LEDs between the inputs and outputs of the circuit and earth with the 270 Ω resistors in series to limit current flow.

(b)   Set
  (i) *A* = *B* = 0
  (ii) *A* = 0, *B* = 1
  (iii) *A* = 0, *B* = 11
  and each time note the logical state of the LEDs.

(c)   Set
  (i) *A* = 1, *B* = 0
  (ii) *A* = *B* = 1
  (iii) *A* = 1, *B* = 11
  and each time note the logical states of the LEDs.

(d)   Comment on the operation of the circuit.

Hazards and races

Because there is no clock to regulate the operation of a non-synchronous sequential circuit, different ICs, gates or flip-flops may introduce differing time delays when a circuit changes from one stable state to another. This may lead to the occurrence of incorrect outputs. The causes of the problem are known as *hazards* and as *races*.

Races

When a non-synchronous sequential circuit moves from one stable state to another it may go through one or more unstable states. If there is only one possible path between the two states the sequence that is followed is known as a *cycle*. The flow table below is for a circuit that has four stable states and uses flip-flops. If the circuit is in its stable state 3 when the input variables change from 11 to 00 the circuit will immediately move to its unstable state 1 in row 11 and column 00. It then moves to unstable state 1 in row 01 and, finally, to stable state 1 in row 00. The cycle is indicated by the arrows.

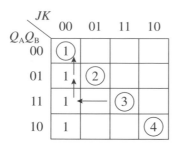

When there are two or more possible paths for the sequence of unstable states, but the circuit always gets into the correct state, the sequence followed is called a *race*. The move from stable state 3 to stable state 1 requires both Q_A and Q_B to change from 1 to 0 but, almost certainly, one of the flip-flops will be faster to change state than the other. If flip-flop A is the faster to switch then Q_A will change before Q_B does, and so the circuit moves to row 01 first and then to row 00 when Q_B changes. If, on the other hand, flip-flop B is the faster then the circuit will move to row 10 first and then to row 00. The race between the two flip-flops is said to be *non-critical* since the circuit always reaches its correct next state regardless of the path taken.

A race is said to be *critical* if, when different paths are taken, because of inherent time delays, the circuit may end up in any one of two or more different stable states. This situation could occur if the stable state 2 were in the 00 column instead of the 01 column as in the flow table shown below. When the circuit is in stable state 3 and the input variables change to 00 then, depending upon which flip-flop is the faster to change state, the circuit could move (i) to row 10 and then to its correct stable state 1, or (ii) move to row 01 and thence to stable state 2. If the circuit does get into stable state 2 it will remain there because the present and next states are the same. Critical races can be avoided if the assignment of rows is made so that transitions are made between adjacent rows only (see p. 264).

(see p. 264)

```
  JK
       00   01   11   10
Q_AQ_B
  00  (1)
  01  (2)
  11           (3)
  10                (4)
```

Hazards

A *static hazard* is said to exist if a change in the input variables that ought not to alter the output does in fact cause the output to change momentarily. A static hazard may arise if any two 1 entries in an excitation map are not looped together. There is, for example, a static hazard in the excitation map which is shown below.

```
     AB
          00   01   11   10
xy
   00     0    0    0    0

   01     0    0    0   (1)

   11     0   (1   (1)  1)

   10     0   (1    1)   0
```

The minimal solution is $X = Bx + A\bar{B}y$. To remove the static hazard the adjacent squares $AB = xy = 11$, and $AB = 10\ xy = 11$, must also be looped together. The equation for the circuit is then

$$X = Bx + A\bar{B}y + Axy$$

EXERCISES

9.1 Design a 3-bit circuit that will measure the length of an input pulse.

9.2 Design a circuit to follow the sequence given in Table 9.21 to generate a 3-phase voltage.

9.3 Design a synchronous circuit that will detect the bit sequence 1101 in an input bit stream. Detection of the sequence is to be indicated by the output of the circuit going HIGH.

9.4 Draw the state diagram for a circuit that has two inputs A and B and one output F. The output F is to go HIGH whenever the input A is HIGH and input B follows the sequence 101. The output must then stay HIGH until A goes LOW.

9.5 Design a synchronous code generator that produces the bit sequence 000, 110, 011, 101, 111, 001, 000, etc. Use J-K flip-flops.

9.6 Repeat exercise 9.5 using D flip-flops.

9.7 Design a synchronous sequence detector having two inputs A and B and a single output F. The output is to go HIGH after the input sequence $A = 0\ B = 1$, $A = B = 1$, $A = 1\ B = 0$. The output must stay HIGH until $A = B = 0$. Use D flip-flops.

Table 9.21

	0	1	2	3	4	5
A	0	1	1	1	0	0
B	0	0	1	1	1	0
C	0	0	0	1	1	1

Fig. 9.24

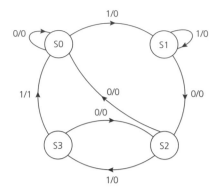

9.8 Design a non-synchronous sequential circuit that indicates by its output going HIGH, each time a second leading-edge transition occurs on an input pulse train. The output should stay HIGH for the duration of the second pulse.

9.9 Design a sequential circuit whose output will go HIGH when the bit sequence 010 occurs in a bit stream unless the sequence 100 has previously occurred. Use D flip-flops.

9.10 The state diagram of a circuit is given in Fig. 9.24.
(a) State, in words, what the function of the circuit is.
(b) Obtain the state maps and from them determine the equations for the inputs to the J and K terminals of the J-K flip-flops that are to be employed.

9.11 A vending machine is to deliver a bar of chocolate after 40p has been placed into its money slot. The 40p may be made up of any combination of 20p and 10p coins. No change is given. Design the logic for the machine. Assume that coin sensors can recognize deposited coins and reject any 2p or 1p coins that may be inserted.

10 Programmable logic devices

After reading this chapter you should be able to:

- Outline the differences between the main types of PLD.
- Describe the structure of a PAL and explain how it may be use to implement a logic function.
- Describe the structure of a PLA and explain how it may be used to implement a logic function.
- Describe the structure of a GAL and explain how it may be used to implement a logic function.
- Describe the architecture of a GAL and explain how its OLMC operates.
- Understand the use of software in the design of circuits to be implemented using a GAL.
- Implement simple sequential circuits using a GAL.

Some complex digital systems may require the use of tens of SSI/MSI ICs and these would occupy a large physical board area and take a long time to insert, interconnect and test. The number of ICs required can be considerably reduced if *programmable logic devices* (PLDs) are employed. Reducing the number of ICs employed means less board space is occupied, the power requirements are lowered, assembly is quicker, cheaper and less error-prone, the reliability is greater and, lastly, fault finding is easier. The use of LSI/VLSI circuits, such as read-only memories (ROMs) and microprocessors, does, of course, achieve the same ends but for many circuit functions there are no LSI/VLSI solutions available.

The need to design more compact digital systems than is possible using SSI/MSI devices has led to the use of much more complex ICs. An *application specific integrated circuit* (ASIC) is an IC that has been designed for a specific application. A single ASIC can replace a large number of standard SSI/MSI devices and in some cases a complete board of ICs may be replaced by a single IC. A *full-custom* ASIC is very costly and time consuming to get to its final design stage, although once this point has been reached production costs are low. The use of a full-custom ASIC can be economically justified only when high-volume production is anticipated, generally in excess of about 200 000 items. A cheaper and quicker alternative is the use of a *semi-custom* ASIC which employs a cell library; these devices are either *gate array* ASICs or they are *standard cell* ASICs.

A gate array consists of a large number of blocks that are formed within the silicon chip but are left unconnected. Each block is a functional logic unit. A required system is produced by depositing a metal interconnection network to link the blocks to the customer's specification. Further metal connects appropriate points in the system to the IC package pins. The production of a gate array ASIC is subject to considerable wastage because of difficulties in designing the gate interconnections. As a result many gates remain unused. The interconnections introduce propagation delays so that the final circuit is usually not as good as could be obtained from a full-custom design.

An alternative approach to semi-custom ASIC design is to use standard cells. A standard cell ASIC is made from a number of building blocks that are listed by the manufacturer in a *cell library*. A designer selects the combinations of standard cells required for the design and specifies, using a *computer-aided design* (CAD) package, how the selected cells are to be interconnected to produce the required system. The information is then sent to the manufacturer who completes the design by using the same masking techniques that are employed for a full-custom design. Only required modules are used in the design so that there are no unused blocks. This method of developing an ASIC has the advantage that, since each standard cell is known to work correctly, it is very likely that the system will work correctly.

Gate array ASICs are cheaper than standard cell ASICs and they are economical to use for production runs of between about 5000 and 50 000 items. Standard cell ASICs do not become economic until the required production run is greater than about 50 000.

Programmable logic devices

Programmable logic provides designers with an alternative to both standard IC devices and full-custom and semi-custom ASICs. A programmable logic device (PLD) offers a general architecture which the user can customize by programming the IC to perform a particular logic function. Some PLDs can be programmed only once but others may be erased and re-programmed as often as required. PLDs are economical to use in relatively small numbers, i.e. from 1 to about 5000, and they are used extensively in modern digital systems. PLDs are smaller, cheaper, and easier to use than gate arrays.

A PLD contains an AND array whose output is connected to an OR array as shown by the basic block diagram given in Fig. 10.1. The input variables A, B and C are

Fig. 10.1 *Programmable logic device*

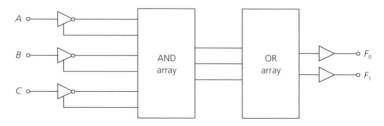

Fig. 10.2 *(a) Bipolar PLD uses transistors to make connections in an AND array and (b) CMOS PLD uses E²CMOS cells to make connections in an AND array*

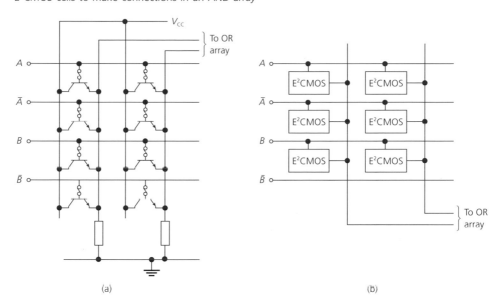

(a)

(b)

applied to input buffer amplifiers whose outputs give each variable in both true and complemented form. The input variables and their complements are applied to the AND array where the products of the input variables are produced. The products are then passed on to the OR array where the sum-of-product equations are formed. Lastly, these SOP equations are amplified before they appear at the output terminals of the device; in some devices the outputs can be programmed to be either active-LOW or active-HIGH. Either the AND array only, or both the AND array and the OR array, are user programmable to implement a required logic function. The logic function may be just combinational or it may include one, or more, D flip-flops.

Bipolar PLDs use bipolar transistors in series with fuses to make the connections in the AND (and perhaps the OR) array or matrix as shown in Fig. 10.2(a). A bipolar device is supplied with all its links intact and the user must break, by blowing the fuse, all links that are not required for the circuit to be implemented. A PLD has programming inputs that can be used to break certain links to implement a required logic function. This process is known as *programming* the PLD. The programming inputs are used only during programming and are left unused thereafter.

In a CMOS PLD the connections in the AND matrix are established by EPROM cells. The PLD is supplied with all connections in the matrix broken (see Fig. 10.2(b)) and programming the device results in the required connections being established. Erasable cells allow a programmed device to be tested fully during manufacture to guarantee reliability, whereas bipolar devices cannot be tested. CMOS PLDs also consume less power than bipolar devices and so they generate less heat. This further increases reliability and allows a simpler power supply to be used. Generic Array Logic (GAL) devices are made using high-speed electrically erasable CMOS technology which offers a high degree of testability and quality as well as allowing erasure of a programmed pattern. Some GAL devices can directly replace PALs in nearly all applications. *Output logic macrocells* are used that allow the user to configure the output circuitry in different ways so that specified PALs can be emulated.

Several different kinds of PLD are available and these include the following:

- Programmable logic array (PLA).
- Programmable array logic (PAL).
- Generic Array Logic (GAL).
- *Field programmable gate arrays* (FPGAs).

An FPGA contains a programmable AND array with each product term produced available at an output terminal; it does not have an OR array. An FPGA may contain thousands of gates and have programmable inputs and outputs. Lattice Semiconductor Corporation offers *programmable large-scale integration* (pLSI) and *in-system plSI* (ispLSI) GAL devices that combine the performance and ease of use of PLDs with the high density and flexibility of FPGAs. The isp technology allows a device to be programmed or re-programmed while it is still connected in a circuit. *Programmable electrically erasable logic* (PEEL) and *erasable PLD* (ELPD) devices allow a user to move up to more complex circuits without any need to change either technology or technique. The PEEL family includes devices with both programmable AND and programmable AND/OR arrays. A PEEL array complex PLD combines a PLA with FPGA-like logic cells to provide an alternative to GALs. PEEL devices can be direct replacements for both PAL and PLA devices. An ELPD device combines the CMOS and EEPROM technologies to provide a much higher density than a PLA but they are both more expensive and slower. The internal architecture of an ELPD device is the same as that of an FPLA device. A *complex programmable device* (CPLD) is an IC that combines a number of PAL devices on the same chip.

The dimensions of a PLD are $m \times n \times p$, where m is the number of inputs, n is the number of product terms that are generated, and p is the number of outputs. Thus, a $7 \times 4 \times 2$ PLD has seven inputs, four product terms and two outputs. The three kinds of programmable logic device differ from one another in that:

- A programmable read-only memory (PROM) has a fixed array and a programmable OR array.
- A programmable array logic (PAL) device has a programmable AND array and a fixed OR array.
- A field programmable logic array (FPLA) device has both its AND and OR arrays programmable.

PROMs and FPLAs are not often employed since PALs are easier to program and to use. Any circuit design that uses a PAL can be emulated by a GAL.

The AND and OR arrays consist of a matrix of rows and columns. A connection between a row and a column is permanent in a fixed array and is made by a transistor in a programmable array. Permanent connections are indicated by a dot and programmed

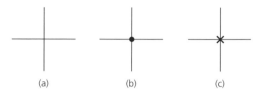

(a)   (b)   (c)

connections are indicated by a ×. This convention is illustrated in Fig. 10.3 which shows (a) no connection, (b) a fixed connection, and (c) a programmed connection.

Programmable array logic

A *programmable array logic* (PAL) device consists of a programmable AND array or matrix followed by a fixed, non-programmable, OR array which allows Boolean equations in sum-of-product form to be implemented. The block diagram of a PAL is shown in Fig. 10.4. Most PALs have several inputs and outputs. The PAL was the most common type of 'once-only' programmable PLD but since it uses bipolar technology it has largely been replaced in new designs by the GAL. The GAL (Generic Array Logic) is a re-programmable device that uses CMOS technology. The basic block diagram of a GAL is shown in Fig. 10.5.

Figure 10.6 shows a $2 \times 4 \times 2$ PAL, i.e. one with two inputs and two outputs. With all the links in the AND matrix made, the output of each column is $A\bar{A}B\bar{B} = 0$. The product terms generated by the AND array depend upon its programming. Before a logic function is implemented by a PAL the equation must be expressed in its sum-of-products form. Suppose the Boolean equations $F_0 = A\bar{B} + \bar{A}B$ and $F_1 = AB + \bar{A}\bar{B}$ are to be implemented. The programming of the AND array should then be as shown in Fig. 10.7.

Fig. 10.4 *Block diagram of a PAL*

Fig. 10.5 *Block diagram of a GAL*

Fig. 10.6 *A 2 × 4 × 2 PAL*

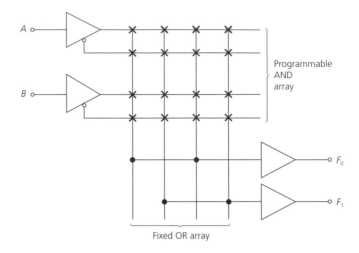

Programmable AND array

Fixed OR array

Fig. 10.7 *PAL implementation of $F_0 = A\bar{B} + \bar{A}B$ and $F_1 = AB + \bar{A}\bar{B}$*

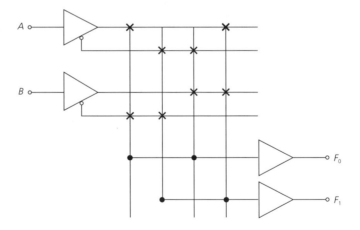

Representation of a PAL

The drawing of a PAL (or a PLA) rapidly becomes complicated as its dimensions increase and hence it is customary to show only one input to a gate. This method of drawing is shown in Fig. 10.8 which is for a $4 \times 8 \times 3$ PAL. The programmed and fixed connections are shown by × and dots, respectively, and each column acts like a 4-input

Fig. 10.8 *A 4 × 8 × 3 PAL*

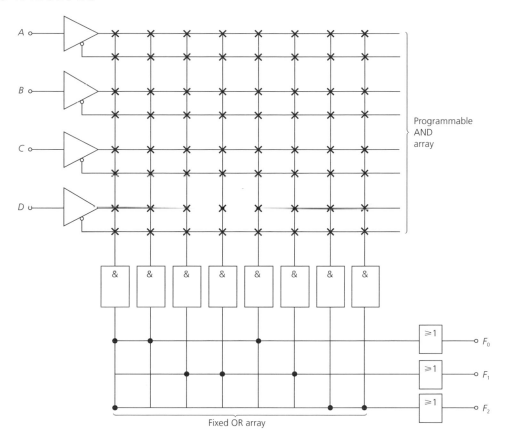

AND gate. The output of each AND gate is connected to an input of an OR gate. Each AND gate has an input for each input variable *and* its complement and this means that the number of inputs per AND gate is equal to twice the number of input variables. Here, each line into an AND gate symbol represents eight inputs.

EXAMPLE 10.1

Implement the logic functions $F_0 = C\bar{D} + \bar{C}D$ and $F_1 = \bar{C}\bar{D} + AB + \bar{A}D$ using a PAL.

Solution

The required connections are shown in Fig. 10.9 (*Ans.*)

Fig. 10.9 *PAL implementation of $F_0 = C\bar{D} + \bar{C}D$ and $F_1 = AB + \bar{A}D + \bar{C}\bar{D}$*

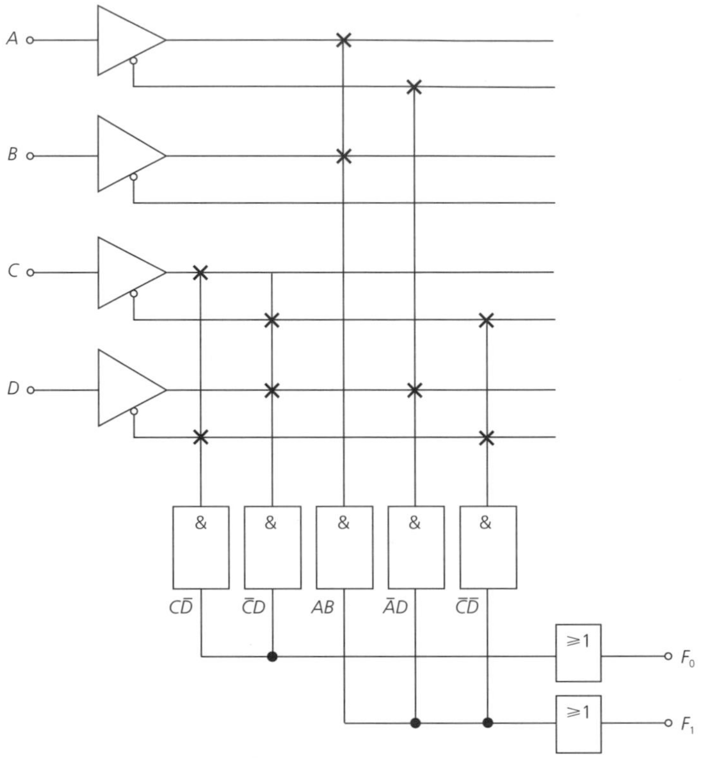

PAL devices are available in a variety of dimensions and specifying the OR connections is a task of device selection and not one of programming. Some examples of available PALs are given in Fig. 10.10. Each PAL is labelled as mHn, mLn or mRn, where m = number of inputs, n = number of outputs and H = active-HIGH output, L = active-LOW output and R = registered output (from a D flip-flop). The 16L8, for example, has 16 inputs and eight active-LOW outputs. It can be seen from the figure that six of these outputs are the outputs of a 7-input OR gate buffered and inverted by a three-state amplifier. The buffer amplifier is enabled or disabled by the product term on the eighth line from the AND matrix. Each of these outputs can be fed back to the AND matrix via an amplifier which also has true and complemented outputs. The other two active-LOW outputs are also derived from a 7-input OR gate but for these no feedback path is provided. Some PALs have input/output pins that can be programmed to act as either an input or as an output.

Outputs

The PALs listed in Fig. 10.10 have three types of output logic which allow the outputs to be configured for specific applications.

Fig. 10.10 *Some available PALs*

PAL	No. of data inputs	No. of outputs and configurations

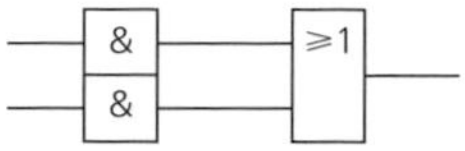

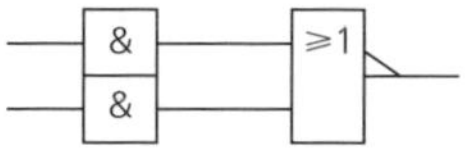

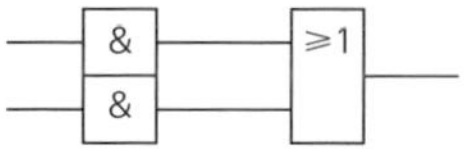

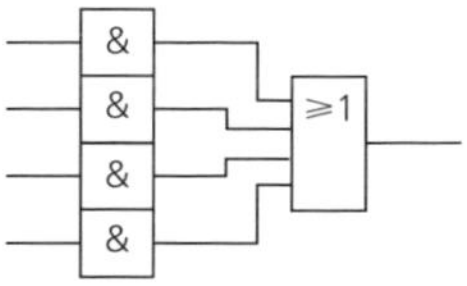

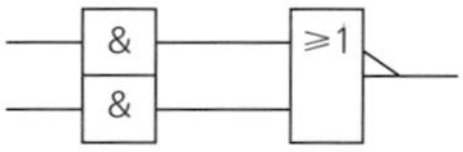

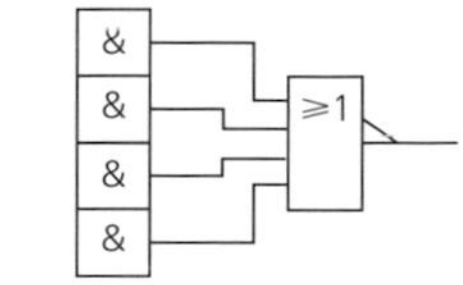

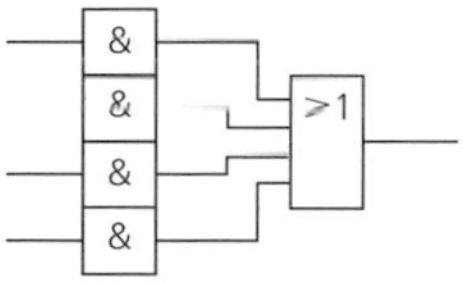

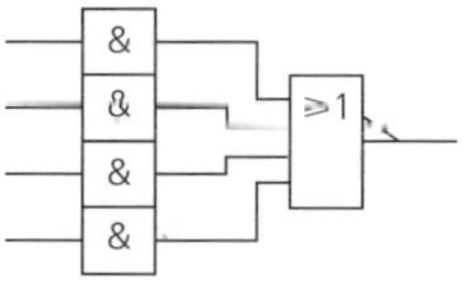

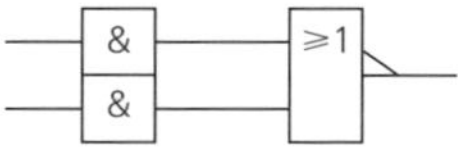

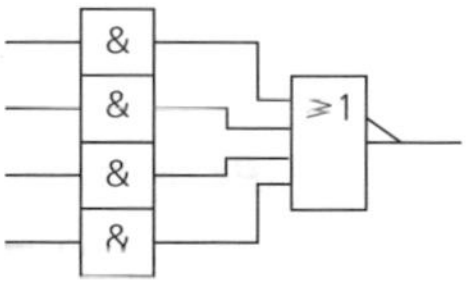

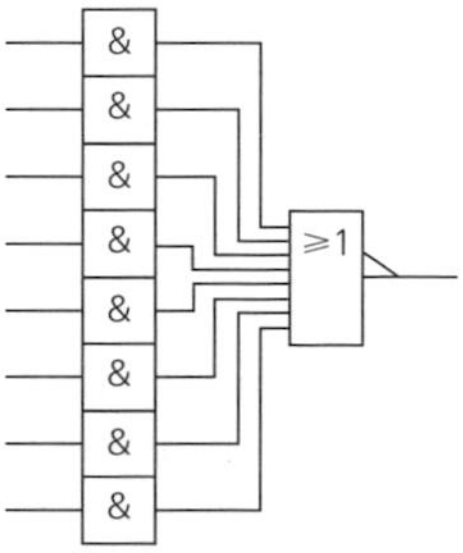

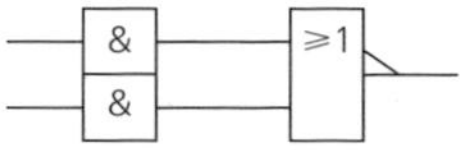

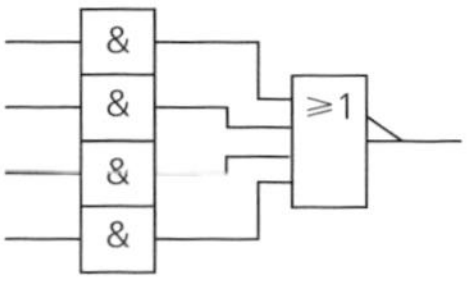

Fig. 10.11 *PAL: combinational output*

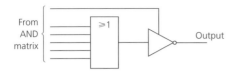

Fig. 10.12 *PAL: combinational input/output*

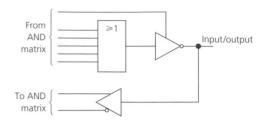

Fig. 10.13 *PAL: programmable polarity output*

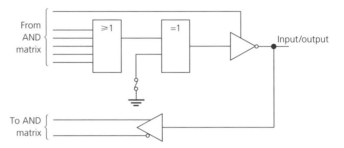

Combinational output

A combinational output is used for the implementation of Boolean SOP equations. Figure 10.11 shows the circuit used.

Combinational input/output

This configuration is employed when the output function is to be fed back to the AND matrix, or the input/output pin is to be used as an input only. The circuit involved is shown in Fig. 10.12.

Programmable polarity output

This feature allows an output function to be presented in either its true or its complemented form. An exclusive-OR gate is programmed to act as either an inverter or as a non-inverter. For the gate to invert its input signal the fusible link at its other input must be blown; conversely, for a true output the fusible link is left intact. The basic arrangement is shown in Fig. 10.13.

The logic diagram of the 16L8 PAL is given in Fig. 10.14. The device has 16 input pins each of which is connected to a buffer amplifier. Each amplifier has two outputs, one inverting and the other non-inverting, so both the true and the complemented forms of the input variable are made available to the AND matrix. Ten of these inputs (pins 1 to 9 and pin 11) are dedicated inputs, i.e. they cannot be used as outputs, and the other six (pins 13 through to 18) can act as either inputs or outputs. There are eight active-LOW outputs, two of which (pins 12 and 19) are dedicated outputs while the others (pins 13 through to 18) can be configured as either inputs or outputs. All outputs can be switched into a high output impedance condition by the product term on the eighth OR gate input.

Fig. 10.14 *Logic diagram of a 16L8 PAL*

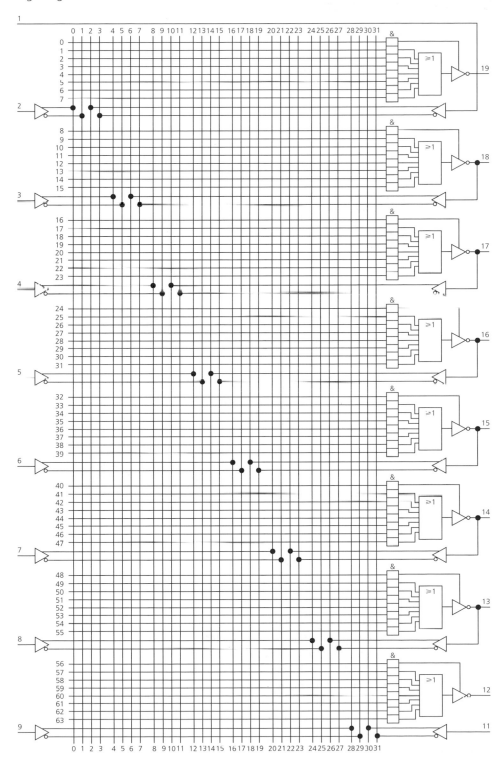

EXAMPLE 10.2

The Boolean equations $F = ABCD + \bar{A}BCD + A\bar{B}CD + AB\bar{C}D + \bar{A}\bar{B}CD + ABC\bar{D}$ and $G = \bar{A}\bar{B}\bar{C}\bar{D} + \bar{A}\bar{B}C\bar{D} + \bar{A}B\bar{C}D + \bar{A}B\bar{C}\bar{D} + ABCD + ABC\bar{D} + A\bar{B}CD + A\bar{B}C\bar{D}$ are to be implemented on a 16L8 PAL. Determine the required connections.

Solution

The mappings for F and G are

F

CD \ AB	00	01	11	10
00	0	0	1	0
01	0	0	0	0
11	1	1	1	1
10	0	0	1	0

G

CD \ AB	00	01	11	10
00	1	1	0	0
01	1	1	0	0
11	0	0	1	1
10	0	0	1	1

From the map, $F = CD + AB\bar{D}$ and $\bar{F} = \overline{CD + AB\bar{D}}$, $G = \bar{A}\bar{C} + AC$, $\bar{G} = \overline{\bar{A}\bar{C} + AC}$. The implementation of $\bar{F}$ and of $\bar{G}$ requires the use of two of the OR gates. The connections for $\bar{F}$ are shown in Fig. 10.15 (*Ans.*)

Fig. 10.15

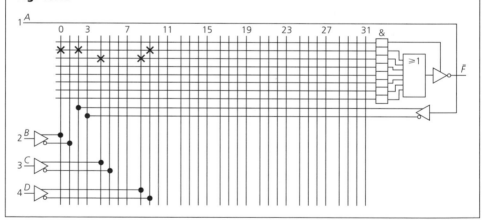

If the equation to be implemented will fit into the dimensions of the PAL there is no need to attempt to minimize the Boolean expression. If, however, the equation is too large for the PAL, minimization will be necessary until the total number of product terms required is no longer larger than the number the PAL can handle. Often the number of product terms needed can be reduced by the use of an external inverter. As an example, the function $F_0 = B\bar{C}\bar{D} + AB\bar{C}D + A\bar{B}\bar{C}D + \bar{B}CD + \bar{A}C\bar{D} + \bar{A}BC$, with don't cares $\bar{A}\bar{B}C\bar{D}$, $A\bar{B}\bar{C}D$ and $A\bar{B}C\bar{D}$, has been mapped

AB CD	00	01	11	10
00	×	1	1	×
01	0	0	1	1
11	1	1	0	1
10	1	1	0	×

Looping the 1 cells gives $F_0 = A\bar{C} + A\bar{B} + \bar{A}C + \bar{C}\bar{D}$. This equation has four product terms. If, instead, the 0 cells are looped to give $\bar{F}_0 = \bar{A}CD + ABC$ only two product terms are required.

Field programmable logic array

A *field programmable logic array* (FPLA) device is more flexible than a PAL because both its AND array and its OR array can be programmed. All possible product terms can be generated by the AND array and any, or all, of these terms can be summed by the OR array to give sum of product terms. Because of this inherent flexibility an FPLA is able to implement functions that a PAL could not. When an FPLA includes flip flops it is often known as a *field programmable logic sequencer* (FPLS). FPLAs have never been widely used and now very few are listed in distributor's catalogues. The basic circuit of an FPLA is shown in Fig. 10.16.

Fig. 10.16 *Basic circuit of an FLPA*

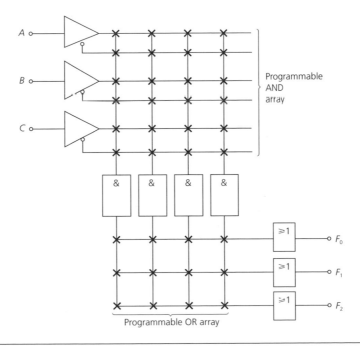

Use an FPLA to implement the logic functions $F_0 = A\bar{B}C\bar{D} + \bar{A}B\bar{C}D + \bar{A}\bar{B}CD$, $F_1 = A + \bar{B}C$ and $F_2 = B\bar{C} + \bar{B}C$.

Solution

The programmed PLA is shown in Fig. 10.17 (*Ans.*)

Fig. 10.17

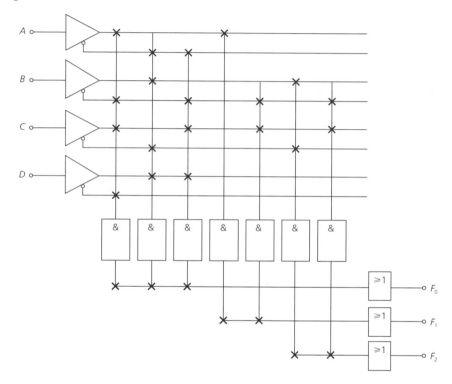

Implement the circuit in Example 9.4 using a PLA.

Solution

The state table of the required circuit is shown in Table 9.8 which is reproduced here with the headings slightly modified. The table is re-written in modified form in Table 10.1. The AND terms are read from the present state values of A, B and x. Thus, $P_1 = \bar{A}\bar{B}\bar{x}$, $P_2 = \bar{A}B\bar{x}$, $P_3 = \bar{A}Bx$, $P_4 = AB\bar{x}$ and $P_5 = A\bar{B}\bar{x}$.

The OR terms are read from the next state values of A and B. Thus, $A^+ = P_1 + P_2 + P_3$ and $B^+ = P_2 + P_3 + P_5$.

The output F of the circuit is HIGH in only one row in the table and hence $F = P_4$. The PLA implementation of the circuit is shown in Fig. 10.18 (*Ans.*)

Table 9.8

| | Present state | | Next state | | | | |
| | A | B | x = 0 | | x = 1 | | Output |
			A^+	B^+	A^+	B^+	
S0	0	0	1	0	0	0	0
S1	0	1	0	1	1	1	0
S2	1	1	1	0	0	0	1
S3	1	0	0	1	0	0	0

Table 10.1

| | Present state | | | Product term | Next state | | Output |
	A	B	x		A^+	B^+	
S0	0	0	0	P_1	1	0	0
	0	0	1		0	0	0
S1	0	1	0	P_2	0	1	0
	0	1	1	P_3	1	1	0
S2	1	1	0	P_4	1	0	1
	1	1	1	–	0	0	0
S3	1	0	0	P_5	0	1	0
	1	0	1	–	0	0	0

Fig. 10.18

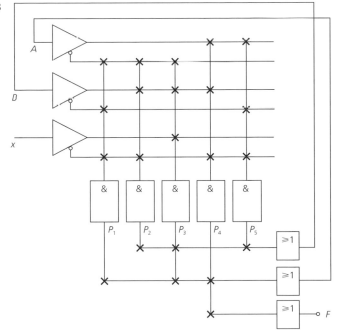

Table 10.2 *PALs emulated by the 16V8A GAL*
10L8, 10H8, 12L6, 12L8, 14L4, 14H4, 16L2, 16H2, 16R8, 16L8 and 16H8

Table 10.3 *PALs emulated by the 20V8 GAL*
14L8, 14H8, 16L6, 16H6, 18L2, 18H2, 20L2, 20H2, 20L8 and 20H8

Generic Array Logic

Generic Array Logic (GAL) devices are electrically erasable CMOS ICs that combine reconfigurable logic, CMOS low-power dissipation and TTL high-speed performance. A GAL may be re-programmed as often as is required. The internal architecture of a GAL is very similar to that of a PAL and it also has a programmable AND matrix to which a number of fixed OR gates are connected. The GAL family of devices was originally developed by Lattice Semiconductor Corporation but several other manufacturers, e.g. National Semiconductor, have been licensed to produce the devices which have become the industry standard for erasable programmable logic devices.

The three main devices in the GAL family are the 22V10B, the 16V8A and the 20V8; the last two have been designed so that they can be programmed to emulate most PAL devices. The first two digits indicate the number of inputs, *including* those outputs that may be configured to act as inputs, V means variable, and the output digits state the number of outputs. Any 20-pin package PAL can be emulated by the 16V8A GAL and any 24-pin package PAL can be emulated by the 20V8 GAL. Tables 10.2 and 10.3 show the PAL devices that can be emulated by each GAL device.

There are also low-voltage versions of many GAL devices which are able to operate from a 3.3 V power supply and interface with 5 V circuitry. Such devices are indicated by the letters LV, e.g. 16LV8.

GAL 22V10B

The block diagram of the GAL 22V10B is given in Fig. 10.19; other GALs have similar block diagrams, differences exist in the circuitry of the OLMCs and in the numbers of inputs and outputs. The 22V10B has 12 dedicated inputs that are applied to an input buffer amplifier that has true and complemented outputs which are inputted to the AND matrix. There are ten *output logic macrocells* (OLMCs) all of which can be configured to act either as an input or as an output. Each OLMC can be programmed so that its logic circuitry is configured as either a combinational input or output, or as a registered output. [Registered means that the output is taken from the Q terminal of a D flip-flop.] The logic diagram of the OLMC of a 22V10B GAL is shown in Fig. 10.20. The signals

Fig. 10.19 *Block diagram of a GAL 22V10B. (Courtesy of Lattice Semiconductor Corporation)*

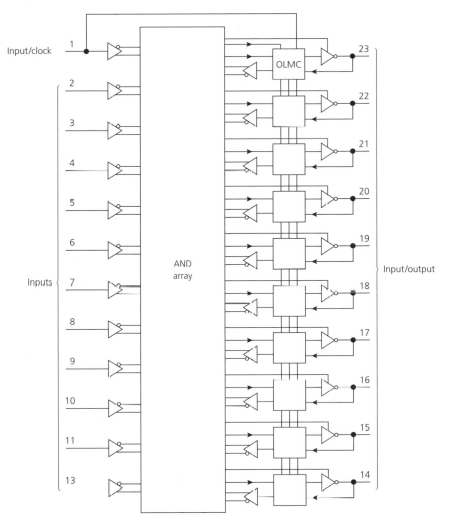

S1 and S0 determine which of the four inputs, D, $\bar{D}$, Q or $\bar{Q}$, to the upper multiplexer is connected to the three-state buffer amplifier. The use of Q and $\bar{Q}$, or D and $\bar{D}$, allows the output of the OLMC to be either active-HIGH or active-LOW in any mode. The buffer amplifier is enabled or disabled by the product term applied from the AND matrix. The signal S1 determines which of the two inputs, Q or the output of the buffer, is fed back to the AND matrix. S1 and S2 are held in some dedicated locations in the AND matrix.

The OLMC can be configured to act in any one of four ways:

(a) Combinational mode with active-LOW output.
(b) Combinational mode with active-HIGH output.
(c) Registered mode with active-LOW output.
(d) Registered mode with active-HIGH output.

Fig. 10.20 *Logic diagram of a GAL 22V10B OLMC*

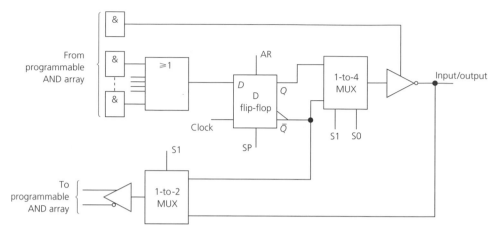

Fig. 10.21 *GAL 22V10B OLMC: (a) combinational mode with active-LOW output and (b) with active-HIGH output*

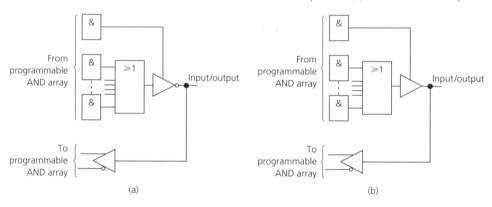

(a)   (b)

Combinational mode with active-LOW output

The output of the OR gate is connected to the buffer amplifier to give the circuit shown in Fig. 10.21(a). The three-state buffer amplifier is enabled by the line from the AND matrix being held HIGH. The signal from the multiplexer is selected by programming S0 = 0, S1 = 1 so that the input to the buffer is the OR gate output inverted.

Combinational mode with active-HIGH output

The combinational mode with active-HIGH output is selected by setting S0 = S1 = 1. The circuit is similar to that for active-LOW output but now the buffer amplifier output is not inverted (see Fig. 10.21(b)).

If the buffer amplifier is disabled and put into its high output impedance state the input/output pin becomes an input pin. An input signal applied to the pin is directed to the AND matrix via the input amplifier. The control line is taken LOW by the product term on the line from the AND matrix.

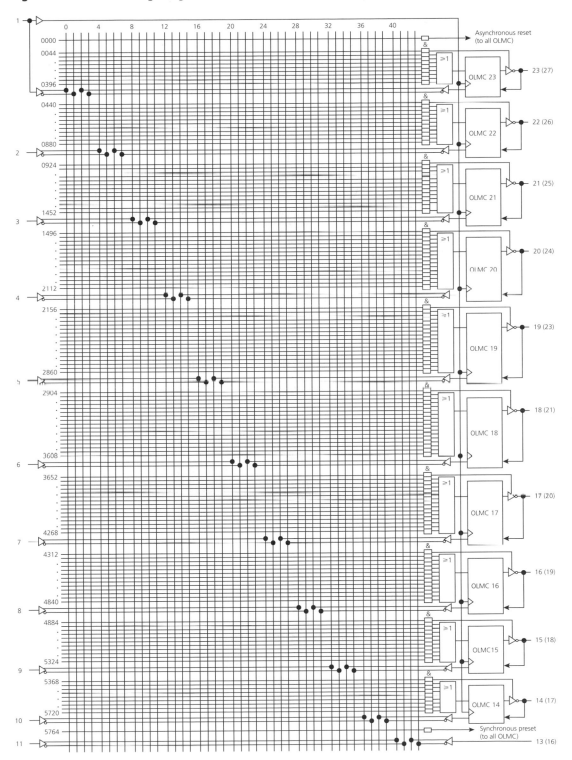

Fig. 10.22 GAL 22V10B logic diagram

Registered mode with active-LOW output

To obtain the registered mode of operation with an active-LOW output set signals $S0 = S1 = 0$ so that the Q output of the D flip-flop is selected. The other multiplexer feeds back either the $\bar{Q}$ output or the output of the buffer depending on the logical state of S1. Input AR non-synchronously resets the flip-flop to $Q = 0$ and input SP sets the flip-flop to $Q = 1$ at the leading edge of a clock pulse. Both AR and SP are derived from product terms in the AND matrix.

Registered mode with active-HIGH output

Now the $\bar{Q}$ output of the D flip-flop is selected by the multiplexer, which means that S0 $= 1$ and S1 $= 0$. The GAL 22V10B has the capacity for up to ten separate SOP Boolean equations to be implemented, the largest of which may contain up to 16 product terms.

Logic diagram

The logic diagram of the GAL 22V10B is shown in Fig. 10.22.

EXAMPLE 10.5

From the logic diagram of the 22V10B determine the locations in the AND matrix at which the bits S0 and S1 for:

(a)   OLMC 20
(b)   OLMC 14

are held.

Solution

(a)   S0: 5814, S1: 5815 (*Ans.*)
(b)   S0: 5826, S1: 5827 (*Ans.*)

EXAMPLE 10.6

The Boolean equation implemented by the GAL 22V10B shown in Fig. 10.23 is $F = ABCD + \bar{A}\bar{B}\bar{C}\bar{D} + \bar{A}B\bar{C}D + A\bar{B}C\bar{D} + AB\bar{C}\bar{D}$. State:

(a)   The product terms on each of the eight inputs.
(b)   The values of control signals S0 and S1.
(c)   The logical state of the line from AND matrix to buffer amplifier.
(d)   What would the output be if the control signals were $S0 = 0$, $S1 = 1$?

Fig. 10.23

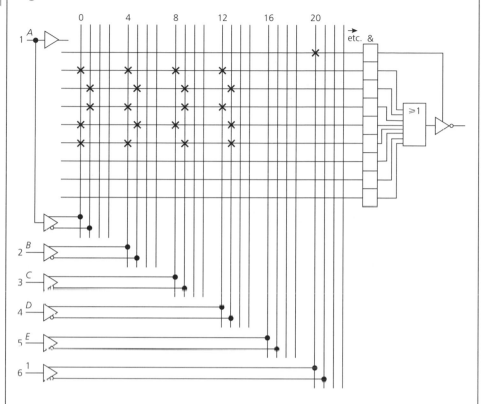

Solution

(a) The product terms on the eight inputs are the same as given in the example (*Ans.*)

(b) S0 = S1 = 1 (*Ans.*)

(c) HIGH (*Ans.*)

(d) The OLMC is now active-LOW so the output is
$$F = \overline{\overline{A}BCD + \overline{A}\overline{B}\overline{C}\overline{D} + \overline{A}B\overline{C}D + A\overline{B}C\overline{D} + AB\overline{C}\overline{D}} \ (Ans.)$$

EXAMPLE 10.7

The function $F = ABC\overline{D}\overline{E} + \overline{A}\overline{B}CDE + \overline{A}B\overline{C}D\overline{E} + A\overline{B}\overline{C}DE + A\overline{B}C\overline{D}E + \overline{A}B\overline{C}DE + ABCDE$ is to be implemented using a GAL 22V10B. Draw the necessary connections in the AND matrix.

Solution

The terms in the equation can be implemented using one OLMC. Figure 10.24 shows the necessary connections in the AND matrix (*Ans.*)

Fig. 10.24

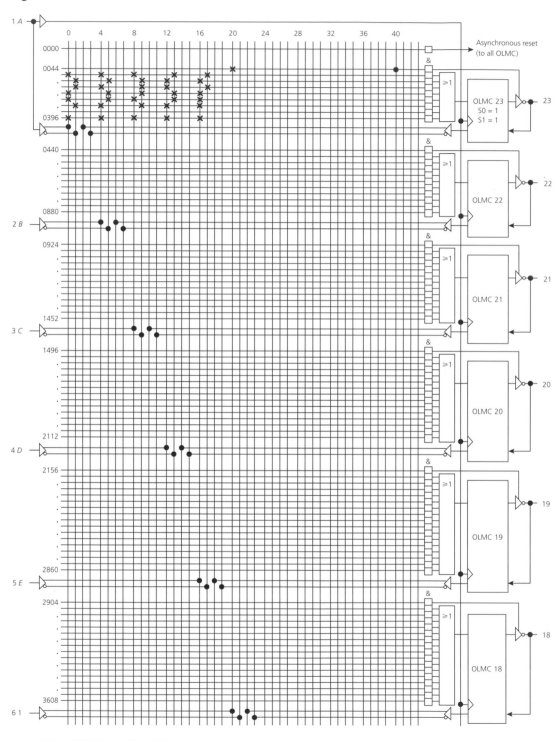

GAL 16V8A

The logic diagram of a GAL 16V8A is shown in Fig. 10.25. Each AND gate has 32 inputs. The device has eight dedicated input pins (2 through to 9) and eight programmable OLMCs which allow the user to configure each input/output pin (12 through to 19) in a number of different ways. There are two special function pins (1 and 11) and, finally, pin 10 is connected to earth and pin 20 is connected to V_{CC}. All the inputs are available at the AND matrix in both true and inverted form. The AND matrix contains 64 input rows and 32 output columns. The eight dedicated inputs and their complements are each connected to a different column in the AND matrix. Also, a logic level and its complement is fed back from each OLMC to a separate AND matrix column. Whether or not these feedback lines have a signal on them depends upon the programmed configuration of the associated OLMC. In total, therefore, there are $4 \times 8 = 32$ input columns in the AND matrix.

Programming the GAL consists of making the chosen connection between the 64 rows and the 32 columns of the matrix. Each made connection applies that input to the AND gate at the end of the row. Some locations in the AND matrix store bits that are used for programming the device. There is no access to these locations when the device is in use. Because its OLMC can be programmed to have different functions the GAL 16V8 is able to emulate many of the common PALs including the 16R8 and 16L8 (see Table 10.2).

Output logic macrocells

Figure 10.26 shows the circuit of an OLMC in the GAL 16V8A; clearly it differs from the OLMC used in the GAL 22V10B. Eight different product terms (AND gate outputs) are applied to each of the eight OLMCs. Inside an OLMC seven of these products are ORed together to give the sum of the input product terms. The eighth product term is connected to the inputs of both a 2-input multiplexer and a 4-input multiplexer. The SOP equation at the output of the OR gate is either routed to the input/output pin of the OLMC to implement a Boolean equation, or it is clocked into a D flip-flop to implement a circuit with a registered output.

The OLMC contains four multiplexers all of which are under the control of programmable bits that are labelled as SYN, AC0 and AC1. SYN and AC0 are both applied to all OLMCs simultaneously but the AC1 bit is individual to each OLMC.

TSMUX
The output of this multiplexer controls the input to the buffer amplifier.

(a) If the V_{CC} input to the multiplexer is selected by the programmable bits the output inverter is enabled and the OLMC provides a combinational logic circuit.
(b) When the earth input is selected the output inverter is put into its high output impedance state and this allows the input/output pin to be used as an input.
(c) Selection of the OE input allows the output pin to be either enabled or disabled by an external signal applied to the OE input at pin 11.
(d) When a product term from the AND array is selected that term is used to enable or disable the output pin.

Fig. 10.25 *GAL 16V8 logic diagram. (Courtesy of Lattice Semiconductor Corporation)*

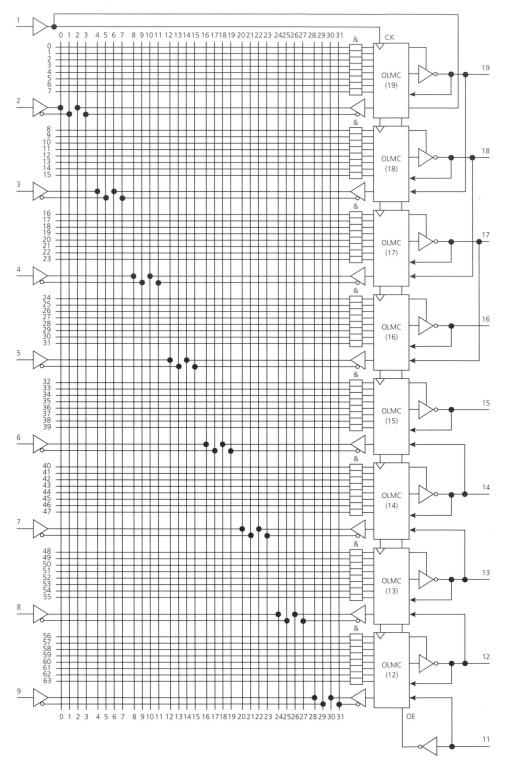

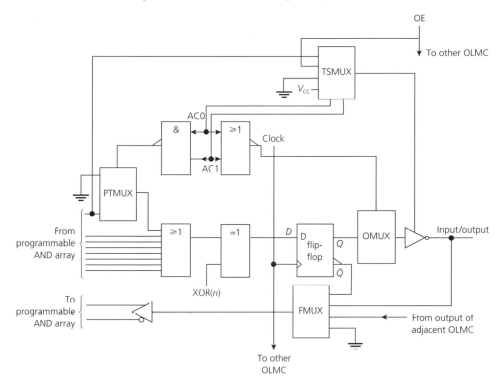

FMUX

The feedback multiplexer selects the signal that is fed back to the AND matrix under the control of the AC0 and AC1 bits.

(a) If either the adjacent OLMC output or the input/output pin is selected an existing output state can be fed back to the AND matrix in some of the operating modes. This feature allows sequential circuits like an S-R latch to be implemented or the input/output pin to be used as a dedicated input terminal.

(b) When the $\bar{Q}$ output of the D flip-flop is selected the present state of the flip-flop is fed back to the AND matrix to allow a synchronous sequential circuit to be implemented.

OMUX

This multiplexer selects either the output of the OR gate (combinational) or the Q output of the D flip-flop (registered), and applies it to the input of the three-state buffer amplifier. The selection is determined by the logical state of $\overline{AC0} + AC1$.

PTMUX

The eighth product term from the AND matrix is one of the inputs to this multiplexer and the other is earth. One of the inputs is selected by the signal appearing at the output of the NAND gate. When $\overline{AC0 \cdot AC1} = 0$ the signal applied to the eighth OR gate input is 0, but when $AC0 \cdot AC1 = 1$ the eighth row of the AND matrix becomes the last OR gate input.

Fig. 10.27 *GAL 16V8: OLMC operated in simple mode: (a) dedicated input, (b) dedicated output and (c) output with feedback to AND array*

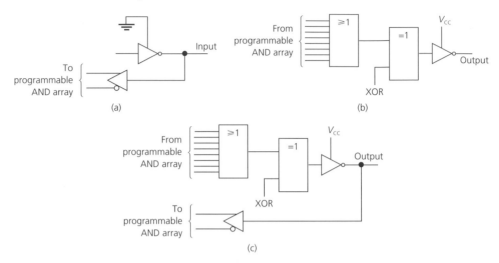

Exclusive-OR gate

The exclusive-OR gate can also be programmed. The XOR input voltage level determines whether the output is active-HIGH or LOW. When the XOR input is HIGH the input from the OR gate is inverted to give an active-HIGH output. When the XOR input is LOW the input from the OR gate is not inverted and the output is then active-LOW.

Output logic macrocell modes

The configuration of each OLMC is determined by the programming of the GAL. The OLMCs can be configured in any one of three different modes. These are known as:

(a) *Simple mode*, which is used for combinational logic circuits whose outputs are always active.
(b) *Complex mode*, which also implements combinational logic circuits but with outputs that can be disabled by a product term from the AND matrix.
(c) *Registered* mode, in which an OLMC can either provide the same function as the complex mode or implement a circuit that includes a clocked D flip-flop.

Two bits, SYN and AC0, select the mode of operation for all the OLMCs.

Row 60 of the AND array stores the *architecture control word*, the bits of which determine the mode and configuration of the OLMC. The device is programmed by setting SYN, AC0 and AC1 to 0 and/or 1.

Simple mode

To program the OLMCs to operate in the simple mode set SYN = 1, AC0 = 0, and AC1 = 0 for a dedicated combinational output and AC1 = 1 for a dedicated input. Each of the OLMCs is independently programmed to be in one of these configurations. In the dedicated input configuration, shown in Fig. 10.27(a), the buffer amplifier is disabled by being in its high output impedance state. The input/output pin is connected to the AND matrix by routing it through the adjacent OLMC. Pins 11 and 1 can be used as inputs that are routed via OLMCs 12 and 19, respectively. Pins 15 and 16 cannot be used as dedicated inputs because they do not have a connection to an adjacent OLMC. If AC1 = 0 the buffer amplifier is enabled and the OLMC can act as an output circuit (see Fig. 10.27(b)). Alternatively (see Fig. 10.27(c)), the output of the buffer amplifier can be fed back to the AND matrix. OLMCs 15 and 16 are always in the dedicated output configuration with zero feedback when in simple mode.

EXAMPLE 10.8

Implement the function $F = AB + \bar{B}C + A\bar{B}\bar{C}$ on a 16V8 GAL.

Solution

The simple mode is required: hence, SYN = 1 and AC0 = 0. If OLMCs 18 and 19 are used (the top two) then 19 must be configured to act as a dedicated output. Hence, AC1(19) = 0. AC1(18) = 0. The required connections to be programmed into the device are shown in Fig. 10.28. Note that rows with no ✗s have all their columns connected so that the output of the AND gate is 0 (*Ans.*)

Fig. 10.28

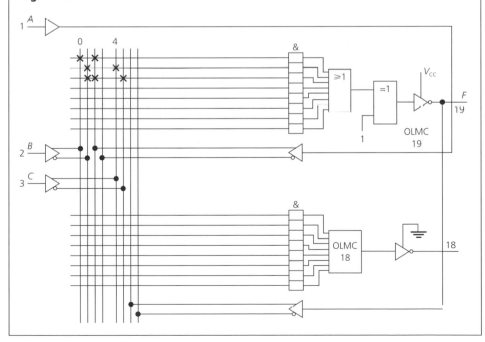

Fig. 10.29 *GAL 16V8: OLMC operated in complex mode: (a) input and (b) Input/output*

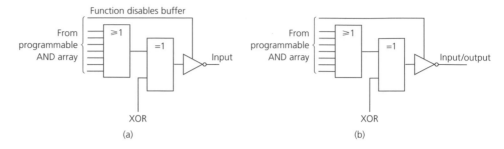

(a)   (b)

Fig. 10.30 *GAL 16V8: operated in registered mode: (a) output with feedback to AND arrray and (b) registered output with feedback*

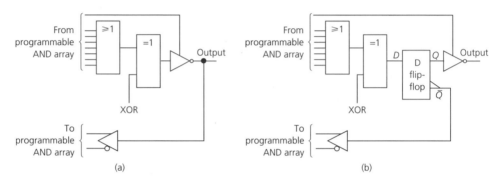

(a)   (b)

Complex mode

The complex mode is selected by setting SYN = 1 and AC0 = 1. The AC1 bit in each OLMC is also set to 1. The three-state inverter at the input/output pin is enabled, or disabled, by a product term in the top row of the AND matrix entering the OLMC. If that expression disables the inverter then that OLMC acts as a dedicated input circuit (see Fig. 10.29(a)). If the inverter is not disabled the OLMC acts as an input/output circuit (see Fig. 10.29(b)). OLMCs 12 and 19 can be used only as outputs because the feedback path is used to allow pins 1 and 11 to be used as inputs. The logical states of outputs 13 through to 18 are fed back through the same OLMC to the AND matrix.

Registered mode

The registered mode is entered by programming the OLMCs with SYN = 0 and AC0 = 1. In this mode there are two possible configurations:

(a) Registered when AC1 = 0.
(b) Combinational when AC1 = 1.

The combinational configuration is shown in Fig. 10.30(a) and is similar in its operation to the complex mode. The output of an OLMC is controlled by a product term from

---

the AND matrix. The output signal is fed back to the AND matrix via the same OLMC. In the registered configuration, shown in Fig. 10.30(b), OLMCs 13 through to 17 and OLMC 19, the D flip-flop is used to synchronize all the registered outputs to a common clock. The clock input is applied to pin 1. All the three-state inverters in OLMCs that are configured as registered outputs are controlled by the OE signal applied to pin 11. The registered outputs have up to eight product terms for the SOP logic that is applied to the D inputs of the flip-flops.

EXAMPLE 10.9

Implement a 2-bit synchronous down-counter using the GAL 16V8.

Solution

The 2-bit synchronous counter must use D flip-flops when implemented using a PAL or a GAL and requires $D_A = \bar{Q}_A$ and $D_B = Q_A Q_B + \bar{Q}_A \bar{Q}_B$. Implementation of these equations using the GAL 16V8 is shown in Fig. 10.31 (*Ans.*)

Fig. 10.31 *Two bit down counter*

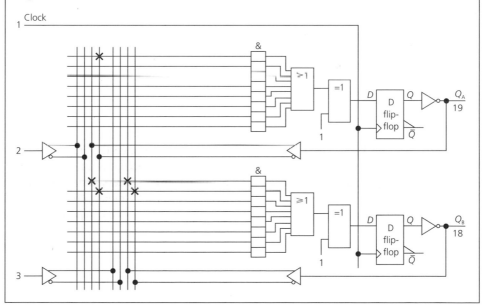

Programming a PAL/GAL

Once a PAL/GAL design has been finalized it must be programmed into the target device. GAL modes of operation cannot be mixed; all OLMCs must be programmed to

Fig. 10.32 *GAL programming steps*

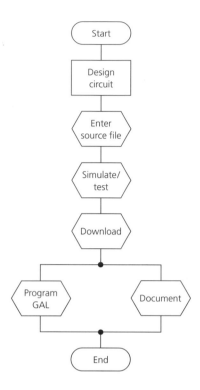

the same mode. Hardware logic programmers are available which are able to program either a PAL or a GAL. The device to be programmed is plugged into a ZIF socket on the hardware programmer and that, in turn, is connected to the computer by a cable. All PLD programming software and hardware programmers conform to a standard established by the Joint Electronic Device Engineering Council (JEDEC). Before programming starts the hardware programmer must be told what type of device it is about to program. A bipolar PAL that has fusible links cannot be re-programmed if an error is made, or if the design is to be altered. A GAL does not suffer from this disadvantage and may be re-programmed many times.

GAL devices can be used to emulate a wide variety of PLD devices. The GAL 16V8 and the GAL 20V10B between them can replace most 20-pin and 24-pin PAL devices. The GAL devices can be programmed directly from an existing JEDEC file or from a master PAL device. Programming a GAL device automatically configures its architecture to emulate the source PAL device. The resulting GAL device is 100 per cent compatible with the original PAL device.

The steps involved in the programming of a PAL or GAL are shown in Fig. 10.32. The text editor is employed to create a *source file* that will contain the details of the design that is to be programmed into the target device. Boolean equations, truth tables and state tables can all be used to enter the circuit design into the computer. Some newer software packages also allow entry using schematic diagrams, macro library, timing waveforms or *hardware description language* (HDL). The source file is inputted to a logic compiler which generates a JEDEC file version of the design. All programming languages produce a standard JEDEC file.

Test vectors can also be entered so that the design can be tested by the software before any programming begins. This means that the required logic can be designed, simulated, and (if necessary) de-bugged using software before any hardware is involved. The compiler configures the bits in the JEDEC file that make/break links in the AND matrix and also configures the programmable features of the output circuitry (OLMC in a GAL). The information in the JEDEC file is downloaded to the programmer in the form of a *fuse map*. The fuse map or plot indicates which connections in the AND matrix are to be broken and which are to be left intact. The fuse plot is changed into a form suitable for the hardware programmer. The programming software then communicates with the hardware programmer and informs the programmer what kind of PAL/GAL is to be programmed before the JEDEC file is downloaded. Once the JEDEC file has been produced the operation of the required circuit can be simulated using the test vectors. If the circuit is found to function correctly a start command can then be transmitted by the computer to the hardware programmer to commence the programming of the PLD according to the fuse map. The fuse map indicates to the programmer which connections in the AND matrix are to be left broken and which are to be made.

The programming software must allow the user to specify the type of target device to be programmed, i.e. PAL or GAL, and the model, and allow declarations and pin assignments to be made before information about the required circuit is entered. If an existing PAL JEDEC file is available the programming process is even simpler. Just select the GAL code and download the file to the hardware programmer and then insert the GAL device (one that is able to emulate the PAL). The hardware programmer will then automatically configure the GAL to emulate the PAL.

Programming languages

Several programming languages exist, but perhaps the two most popular are *ABEL* (Advanced Boolean Expression Language) and *CUPL* (Universal Compiler for Programmable Logic). Both languages are comprehensive and only a brief introduction to ABEL will be presented here. Both ABEL and CUPL employ the same logic operators to represent a Boolean equation. These are: AND &, OR #, NOT !, and exclusive-OR $. The arithmetic operators employed are + (addition), − (subtraction), * (multiplication), and / (division). The relational operators are == (equal), != (not equal), < (less than), > (greater than), <= (less than or equal to), and >= (greater than or equal to). The assignment operators are = (combinational output) and := (registered output).

All equations must end with a semi-colon. If an entered Boolean equation is too large for the target device ABEL will automatically minimize the equation.

EXAMPLE 10.10

Write the equations:

(a) $F = AB + \bar{B}\bar{C}$.
(b) $F = A\bar{B}\bar{C} + \bar{A}BC$.
(c) $F = (A + B + \bar{C})(\bar{A} + C)$

in ABEL/CUPL.

Solution

(a) F = A&B#!B&!C (*Ans.*)
(b) F = A&!B&!C#!A&B&C (*Ans.*)
(c) F = (A#B#!C)&(!A#C) (*Ans.*)

Fields are provided into which optional information such as the designer's name, the date, an outline description of the circuit, etc. can be entered. Declarations must then be made. These define the type of device, assign pin numbers, label pins and specify any internal variables. Pins may be given any desired labels. All assignments must end with a semi-colon as must the Boolean equations that specify the required circuit. There is no need for input Boolean equations to be simplified before entry. If an entered equation contains too many terms for the target GAL the software will automatically carry out a simplification of the equation. If there are still too many terms an error message will be generated.

Comments can be included to make a program clearer and can be written anywhere so long as they are preceded by /* and followed by */ in CUPL and preceded by " in ABEL. Comments are ignored by the software and are not implemented. Comments can be used to provide headings within a program to make it clearer; two examples are:

(a) ABEL; "inputs.
(b) CUPL; /* inputs */ .

An ABEL program has several segments:

- MODULE: all ABEL programs must start with a MODULE statement and an optional title statement. The MODULE and title statements give the program a name and an indication of its function. If required, the programmer's name, the date of writing and any other information can be entered after a comment sign.
- Description: the next segment of the program gives information about the type of PLD to be programmed, e.g. device 'P22V10'. This is known as the device declaration; device is a keyword and must be used. Either lower case or capitals can be employed. The letter P is used for both PAL and GAL devices. Next in the description segment are written the pin assignments. The pin declaration should be in the form:

A, B, C, D, PIN 1, 2, 3, 4;

- PIN is a keyword that must be employed. The pin declaration assigns the input and output of the circuit to specific pins on the PAL/GAL package, input *A* to pin 1, input *B* to pin 2 and so on. The programmer will need to look at the logic diagram of the device to be programmed to decide which package pins to assign to which input/output. Headings may be used to make it clear what the assignments are, e.g. "input pins" and "output pins". Unused pins do not have to be labelled. Lastly, the programmer must specify if the outputs of the circuit are to be active-LOW or active-HIGH if the device to be programmed gives a choice (when a GAL is to emulate a PAL that PAL will have either active-HIGH or active-LOW outputs).
- Logic description: the Boolean equations that describe the operation of the required logic circuit are entered into this segment of the program. Equations are entered using the syntax:

    variable = expression, e.g. F = A&B&!C.

If the programmer prefers, the information can be entered in the form of a truth table, or, for a sequential circuit, a state table. The header format for a truth table entry is:

    TRUTH TABLE ([A, B, C] → [F_0, F_1])

- Extensions: an expression may be required to control the operation of an embedded device such as the buffer amplifier in an OLMC. As an example of this consider the buffer amplifier. For it to be enabled permanently the OE line should be HIGH. This is indicated by: Output.OE = 'b'1. If the amplifier is to be disabled permanently the expression should be: Output.OE = 'b'0. Output is the label given to the input/output pin and 'b' indicates a binary number.
- Test vectors: in the last part of the program a list can be entered that gives details of possible input states and the expected outputs. This information can be used to simulate and test the programmed circuit before it is entered into the IC.

    An example of declarations, pin assignments, etc. is:

ABEL   encoder device 'P16V8A'; " encoder is a description of the circuit; this word can be anything the user wants. Device is a keyword and must be used. The letter P is required for both PALs and GALs.
INPUTS
A, B, C, D, PIN 1, 2, 3, 4;   " PIN is another keyword, the numbers following
                                     are the chosen pin numbers.
OUTPUTS
F1, F2, PIN 18, 19;
    Specification of the circuit   " Enter by Boolean equation or by truth table
    Test vectors   " Same format as truth table

If a ! is placed in front of one, or more, of the input variables that variable is active-LOW. A typical example of the appearance of an ABEL program is:

MODULE   Type of circuit
Title   'Brief description'
          "Device declaration
Circuit device 'P22V16';
     "Input pins
     A, B, PIN 1, 2;
     "Output pins
     F, PIN 3;
     G, PIN 4;
equations (or truth table or state table)
     END

EXAMPLE 10.11

Write the source file for the Boolean equation $F = ABCD + \bar{A}\bar{B}CD + A\bar{B}CD$ $+ \bar{A}\bar{B}\bar{C}\bar{D}$ using:

(a) ABEL.
(b) CUPL.

Solution

(a) MODULE
       Title 'Combinational_logic_circuit
       Device 'P16V8';
"input pins
       A, B, C, D PIN 2, 3, 4, 5;
       "output pin
       F PIN 18;
       equations F = A&B&C&D#!A&!B&C&D#A&!B&C&D#!A&!B&!C&!D;
END
(*Ans.*)
(b) Logic_circuit device P16V8A;
Applied Digital Electronics
          /* combinational logic circuit */
          /* inputs*/
          PIN 2 = A
          PIN 3 = B
          PIN 4 = C
          PIN 5 = D
          /* output*/
          PIN 18 = F
          /* Boolean equations */

F = A&B&C&D#!A&!B&C&D#A&!B&C&D#!A&!B&!C&!D
END
(*Ans.*)

EXAMPLE 10.12

A full-adder is to be implemented using a 22V16 GAL. Write the source file for the sum *S* using:

(a) Equation entry.
(b) Truth table entry.

Solution

(a) The Boolean equations for a full-adder are given in equations (5.5) and (5.6) and are:

Sum $S = (\overline{A \oplus B})C_{in} + (A \oplus B)\bar{C}_{in}$ and carry-out $= C_{out} = AB + (A \oplus B)C_{in}$. Hence, the ABEL program is:

```
MODULE full_adder
Title 'full-adder
Circuit device 'P22V10';
A, B, Cin, S, Cout, PIN 1, 2, 3, 4, 5;
'Equations
    S = (A$B)&!Cin#!(A$B)&Cin
END full-adder
```
(*Ans.*)

(b)
```
MODULE full_adder
Title 'full-adder
device 'P22V10';
A, B, Cin, S, Cout PIN 1, 2, 3, 4, 5;
TRUTH_TABLE [A, B, Cin] → [S, Cout];
            [0, 0, 0] → [0, 0];
            [1, 0, 0] → [1, 0];
            [0, 1, 0] → [1, 0];
            [1, 1, 0] → [0, 1];
            [0, 0, 1] → [1, 0];
            [1, 0, 1] → [0, 1];
            [0, 1, 1] → [0, 1];
            [1, 1, 1] → [1, 1];
END
```
(*Ans.*)

Registered mode

A GAL 16V8 can be used to implement counters with up to eight stages but the number of available input pins may pose a problem.

When a GAL is used in registered mode some more operators are employed:

PIN ISTYPE 'reg specifies an output as registered.
PIN ISTYPE 'com specifies an output as combinational.
P ISTYPE 'reg, buffer means output not inverted.
Q ISTYPE 'reg, buffer means output inverted.

In the registered mode the .D extension is used to tell the programmer what is to be connected to the D input terminal of the D flip-flop:

QA.D = !QA means that the D_A input is to be connected to the $\bar{Q}_A$ output, and
QB.D = QA means that the D_B input is to be connected to the Q_A output.
Q0 := D0 means D_0 will become the same as D_0 at the next clock transition. It must be followed by
Q0.CLK = clock.
.CLK means clocked.
.x. means don't care.

As an example of the use of the registered symbols consider a 4-bit shift register:

MODULE shift register
device declaration 4-bit shift register
device 'P16V8';
clock, clear, data_in PIN 1, 2, 3;
QA, QB, QC, QD PIN 15, 16, 17, 18 ISTYPE 'reg, buffer;
QA := data_in
[QB, QC, QD] := [QA, QB, QC];
[QA, QB, QC, QD].CLK = clock;
[QA, QB, QC, QD] .R = clear;

EXAMPLE 10.13

Write down the Boolean equations for:

(a) A 2-bit counter.
(b) A 3-bit counter.

Solution

(a) The least significant flip-flop must toggle at each clock edge. Hence, QA = !QA&C. The other flip-flop must toggle only when Q_A is HIGH. Hence, QB = !QB&QA&C. Q_B should go LOW whenever this statement is not true but also it must remain HIGH when $Q_A = 0$ and $Q_B = 1$, i.e. as the count moves from 2 to 3. Hence, QB = !QB&QA&C#QB&QA&C (*Ans.*)
(b) Q_C must go from LOW to HIGH when the count moves from 3 to 4 and it must then remain HIGH for the next three counts. Hence, QC = !QC&QB&QA&C#QC&!QB&C#QC&!QA&C (*Ans.*)

EXAMPLE 10.14

Implement a 4-bit SISO shift register using the GAL 22V10.

Solution

D flip-flops must be used. The *D* input of each flip-flop is connected to the *Q* output of the preceding flip-flop. The serial data input is applied to the *D* input of the left-hand flip-flop. Hence, the ABEL source file is:

        MODULE Shift_register
        Title '4-bit SISO shift register'
        "device declaration
        Shift register device 'P22V10';
        "Pin declaration
        clock, clear PIN 1, 2;
        data_in PIN 3;
QA := data_in;
[QB, QC, QD] := [QA, QB, QC];
[QA, QB, QC, QD].CLK = Clock;
[QA, QB, QC, QD].AR = Clear;
END

Test vectors can be added to check the correct operation of the circuit. These are:

    TEST_VECTORS

([Clock, Clear, Data_in] → [QA, QB, QC, QD])
[0, 0, 1] → [0, 0, 0, 0];
[.c., 1, 1] → [1, 0, 0, 0];
[.c., 1, 0] → [0, 1, 0, 0];
[.c., 1, 1] → [1, 0, 1, 0];
[.c., 1, 0] → [0, 1, 0, 1];
[.c., 1, 1] → [1, 0, 1, 0];
[.c., 0, 1] → [0, 0, 0, 0];
END
(*Ans.*)

EXAMPLE 10.15

Implement a 4-bit binary counter with count 0 through to 15 using the GAL 22V10.

Solution

The state diagram is given in Fig. 10.33. From the state diagram the state table in Table 10.4 has been obtained. The state maps are

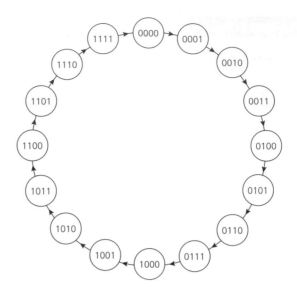

Fig. 10.33

Table 10.4

Present state				Next state				Required inputs			
Q_D	Q_C	Q_B	Q_A	Q_D^+	Q_C^+	Q_B^+	Q_A^+	D_D	D_C	D_B	D_A
0	0	0	0	0	0	0	1	0	0	0	1
0	0	0	1	0	0	1	0	0	0	1	0
0	0	1	0	0	0	1	1	0	0	1	1
0	0	1	1	0	1	0	0	0	1	0	0
0	1	0	0	0	1	0	1	0	1	0	1
0	1	0	1	0	1	1	0	0	1	1	0
0	1	1	0	0	1	1	1	0	1	1	1
0	1	1	1	1	0	0	0	1	0	0	0
1	0	0	0	1	0	0	1	1	0	0	1
1	0	0	1	1	0	1	0	1	0	1	0
1	0	1	0	1	0	1	1	1	0	1	1
1	0	1	1	1	1	0	0	1	1	0	0
1	1	0	0	1	1	0	1	1	1	0	1
1	1	0	1	1	1	1	0	1	1	1	0
1	1	1	0	1	1	1	1	1	1	1	1
1	1	1	1	0	0	0	0	0	0	0	0

D_A

$Q_C Q_D$ \ $Q_A Q_B$	00	01	11	10
00	1	1	0	0
01	1	1	0	0
11	1	1	0	0
10	1	1	0	0

D_B

$Q_C Q_D$ \ $Q_A Q_B$	00	01	11	10
00	0	1	0	1
01	0	1	0	1
11	0	1	0	1
10	0	1	0	1

Q_CQ_D \ Q_AQ_B	00	01	11	10
00	0	0	1	0
01	0	0	1	0
11	1	1	0	1
10	1	1	0	1

D_C

Q_CQ_D \ Q_AQ_B	00	01	11	10
00	0	0	0	0
01	1	1	1	1
11	1	1	0	1
10	0	0	1	0

D_D

From the maps: $D_A = \bar{Q}_A$, $D_B = \bar{Q}_A Q_B + Q_A \bar{Q}_B$, $D_C = \bar{Q}_A Q_C + \bar{Q}_B Q_C + Q_A Q_B \bar{Q}_C$,
$$D_D = \bar{Q}_C Q_D + \bar{Q}_A \bar{Q}_C + \bar{Q}_B Q_D + Q_A Q_B Q_C \bar{Q}_D.$$

"Because the Q output takes up the value of the D input at the next clock pulse, equations are expressed using Q outputs instead of D inputs.

```
    MODULE FOUR_BIT COUNTER
    Title '4-bit binary counter'
    "device declaration
    device 'P22V10';
    Clock, Clear PIN 1, 2,
    QA, QB, QC, QD PIN 20, 21, 22, 23 ISTYPE 'reg, buffer';
Equations
    QA := !QA;
    QB := !QA&QB#QA!QB;
    QC := !QA&QC#!QB&QC#QA&QB&!QC;
    QD := !QC&QD#!QA&!QC#QB&QD#QA&QB&QC&!QD;
    [QA, QB, QC, QD].CLK = Clock;
    [QA, QB, QC, QD].AR = !Clear;
END
```

(*Ans.*)

EXAMPLE 10.16

Repeat Example 10.15 using state table entry.

Solution

```
    MODULE FOUR_BIT COUNTER
    Title '4-bit binary counter'
    Counter device 'P22V10';
    Clock, Clear PIN 1,2;
    QA, QB, QC, QD PIN 20, 21, 22, 23 ISTYPE 'reg, buffer';
"State definitions
```

◀ QSTATE = [QD, QC, QB, QA];
        A = [0, 0, 0, 0];
        B = [0, 0, 0, 1];
        C = [0, 0, 1, 0];
        D = [0, 0, 1, 1];
        E = [0, 1, 0, 0];
        F = [0, 1, 0, 1];
        G = [0, 1, 1, 0];
        H = [0, 1, 1, 1];
        I = [1, 0, 0, 0];
        J = [1, 0, 0, 1];
        K = [1, 0, 1, 0];
        L = [1, 0, 1, 1];
        M = [1, 1, 0, 0];
        N = [1, 1, 0, 1];
        O = [1, 1, 1, 0];
        P = [1, 1, 1, 1];
Equations
QSTATE.CLK = clock;
QSTATE.AR = !Clear;
State_diagram QSTATE
    State A: GOTO B;
    State B: GOTO C;
    State C: GOTO D;
    State D: GOTO E;
    State E: GOTO F;
    State F: GOTO G;
    State G: GOTO H;
    State H: GOTO I;
    State I: GOTO J;
    State J: GOTO K;
    State K: GOTO L;
    State L: GOTO M;
    State M: GOTO N;
    State N: GOTO O;
    State O: GOTO P;
    State P: GOTO A;
END
(*Ans.*)

EXERCISES

10.1 Use a PLD to implement a circuit that has three inputs *A*, *B*, and *C* and whose output goes HIGH only when two, or more, of the inputs are HIGH.

Fig. 10.34

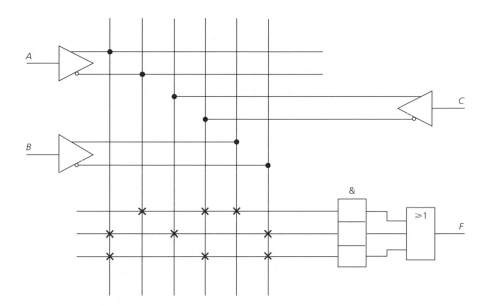

10.2 (a) How many inputs and outputs has
(i) a PAL 14H4
(ii) a PAL 16L2
(iii) a PAL 16R4?
(b) How many inputs and outputs has
(i) a GAL 22V10
(ii) a GAL 20V8
(iii) a GAL 16V8?

10.3 Figure 10.34 shows a part of a GAL. Determine the logic expression that is implemented.

10.4 Show how a PAL or GAL is programmed to implement each of the following Boolean equations:
(a) $F = \bar{A}BC + AB\bar{C} + A\bar{B} + AC$.
(b) $F = A\bar{B}C + ABC + \bar{A}\bar{B} + \bar{A}\bar{C}$.
(c) $F = ABC + \bar{A}\bar{B}\bar{C}$.

10.5 Show how each of the expressions in exercise 10.4 would be implemented using a 12H6 PAL.

10.6 Write a source file in ABEL to program a GAL 16V8 to implement the equations $F = A + B$, $G = CD$, $H = \bar{A}\bar{B}$, $I = \bar{C} + \bar{D}$ and $J = \bar{A} \oplus \bar{C}$.

10.7 Write an ABEL source file to program a GAL 16V8 with the Boolean expression $F = AB\bar{C}\bar{D} + A\bar{B}C\bar{D} + ABCD + \bar{A}BCD + A\bar{B}CD + \bar{A}\bar{B}CD + ABC\bar{D} + \bar{A}\bar{B}C\bar{D} + A\bar{B}\bar{C}D + \bar{A}BC\bar{D}$.

10.8 Produce the ABEL source file for $Y = \bar{A} + \bar{B} + \bar{C} + D + E + \bar{F} + \bar{G} + II + I$ using a GAL 16V8.

10.9 Write the ABEL representation of:
   (a) An exclusive-OR gate.
   (b) A half-adder.
   Use truth table entry.

10.10 Answer the following questions:
   (a) What is ABEL?
   (b) How is the exclusive-OR function indicated in ABEL?
   (c) What are test vectors and when are they used? Must they always be included in a program?
   (d) What is meant by a JEDEC file?
   (e) Must separate programs be used to program PAL and GAL devices?

10.11 Write programs for a GAL 16V8 to implement:
   (a) The AND logic function.
   (b) The exclusive-NOR logic function.

10.12 Write an ABEL source file for a 1-to-4 multiplexer.

11 Interfacing between digital and analogue systems

After reading this chapter you should be able to:

- Understand the use of digital-to-analogue and analogue-to-digital conversion.
- Describe some applications for analogue-to-digital and digital-to-analogue converters.
- Describe the operation of both weighted binary resistor and R/2R digital-to-analogue converters.
- Explain the operation of each of the different types of analogue-to-digital converter.
- Understand the meanings of the terms employed in the data sheets for both analogue-to-digital and digital-to-analogue converters.
- Explain the errors that may occur in a conversion.
- Perform calculations on the performance of both kinds of converter.
- State the relative merits of the different types of analogue-to-digital converter.
- Understand the need to employ a sample-and-hold amplifier in conjunction with an analogue-to-digital converter.
- Outline the performance of a data acquisition system.

There are many areas in electrical engineering where analogue signals must be processed in some way. The phenomena that occur in nature, such as temperature, pressure, liquid flow, are all analogue quantities and if these phenomena are to be processed by a digital system, or computer, they must first be converted from their original analogue forms into equivalent digital signals. This is the function of an *analogue-to-digital converter* (ADC). After processing, it may be necessary to convert the digital signal into its equivalent analogue form and this conversion is the function of a *digital-to-analogue converter* (DAC). *Data acquisition* is the process of collecting analogue *natural* phenomena to observe, to analyse and/or to process its parameters. A typical example of a data acquisition system is an industrial system that monitors the temperature of a process, compares it with the expected temperature range, and, if necessary, outputs control signals to either raise or lower the temperature of the process.

The function of an ADC is to convert an input analogue voltage into the equivalent digital word. The block diagram of an ADC is shown in Fig. 11.1. The conversion of an analogue voltage into a digital word is carried out by the use of a technique known as *quantization*, and involves sampling the analogue voltage at regular intervals of time

Fig. 11.1 *Principle of an analogue-to-digital converter*

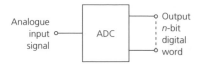

Fig. 11.2 *Principle of a digital-to-analogue converter*

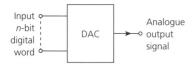

and producing digital signals that are the binary equivalent of each sampled voltage. The *sampling frequency* must be at least twice the highest frequency contained in the analogue signal otherwise an effect, known as *aliasing*, will occur. Aliasing causes severe distortion of the signal because frequency components produced during the sampling process interfere with the frequencies in the analogue signal. An analogue signal with a peak value of V_m can be divided, or *quantized*, into $V_m/2^n$ different values, where n is the number of bits employed by the ADC. Each quantization step or level represents a range of analogue voltages and hence it is only an approximation to the true value. The greater the number of quantization steps, and hence the larger the number of bits employed, the smaller will be the step size. The difference between the analogue signal and the quantized approximation to it produces an error in the signal when it is reconstituted and this is known as *quantization error* or *noise*.

The function of a DAC is to convert an input digital word into the equivalent analogue voltage. The block diagram of a DAC is given in Fig. 11.2. The DAC output is a stepped approximation to the required analogue voltage. By increasing the number of bits in the input digital word the number of different possible output values is increased and the approximation becomes closer to the true analogue voltage. Digital-to-analogue conversion is a simpler process than analogue-to-digital conversion and hence DACs will be considered before ADCs.

Digital-to-analogue converters

A *digital-to-analogue converter* (DAC) is a circuit that converts an input digital word into an equivalent analogue voltage or current. Essentially, a DAC consists of a voltage or current reference, a number of binary weighted resistors, some electronic switches and an op-amp summer circuit. Most DACs have a unipolar output voltage and employ normal binary coding; for such circuits the ideal transfer function is given by

$$V_{out} = V_{FS}(B_0/2 + B_1/4 + B_2/8 + B_3/16 + \text{etc.}) \tag{11.1}$$

Fig. 11.3 *Transfer function of a 3-bit DAC*

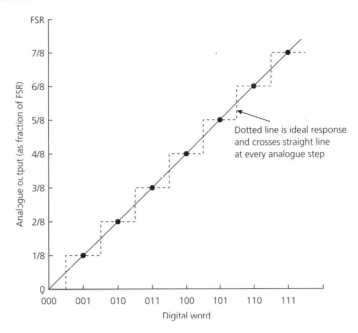

where V_{FS} is the maximum (or full-scale) output voltage, B_0 is the most significant bit (MSB), B_1 is the next most significant bit and so on. The most significant bit is equal to $V_{FS}/2$ and the least significant bit is equal to $V_{FS}/2^n$. Figure 11.3 shows the transfer function of a 3-bit DAC. The analogue output voltage is represented by a number of discrete voltages. The smallest increment in the analogue voltage is contributed by the least significant bit (LSB) of the input digital word and it is equal to $V_{FS}/2^n$. The *ideal* transfer function is obtained by drawing a straight line between the origin and the full-scale voltage and this is shown by the dotted line in the figure. The difference between the ideal and practical transfer functions shows that some error exists; this error can be reduced by increasing the number of bits used by the input digital words and hence increasing the number of possible analogue voltages. When 3 bits are employed only 2^3 or eight different analogue voltages can be outputted; if the number of bits is increased to 4, 2^4 or 16 analogue voltages are possible and so on for greater numbers of bits. This is illustrated by the transfer function of a 4-bit DAC given in Fig. 11.4. Clearly the error has been reduced; the *resolution* has been improved.

The analogue output voltage of a DAC can never reach V_{FS} since its maximum value is always smaller than V_{FS} by one LSB.

The maximum output voltage $V_{max} = [(2^n - 1)V_{FS}]/2^n$ (11.2)

The output voltage produced by any digital word is

$V_{out} = $ (decimal equivalent of digital word $\times V_{max})/(2^n - 1)$ (11.3)

Fig. 11.4 *Transfer function of a 4-bit DAC*

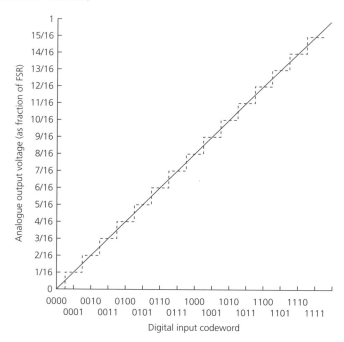

Analogue output voltage (as fraction of FSR)

Digital input codeword

EXAMPLE 11.1

The full-scale voltage of a DAC is 10 V. If the DAC is:

(a)   A 3-bit circuit
(b)   An 8-bit circuit
      calculate the output voltage represented by
        (i) the LSB
        (ii) the MSB
        (iii) V_m.

Solution

(a)   (i) LSB voltage = $10/2^3$ = 1.25 V (*Ans.*)
      (ii) MSB voltage = $10/2$ = 5 V (*Ans.*)
      (iii) V_m = 10 − 1.25 = 8.75 V (*Ans.*)
(b)   (i) LSB voltage = $10/2^8$ = 39.1 mV (*Ans.*)
      (ii) MSB voltage = $10/2$ = 5 V (*Ans.*)
      (iii) V_m = 10 − 0.0391 = 9.9609 V (*Ans.*)

◀
EXAMPLE 11.2

Calculate the output voltage of a 12-bit DAC with $V_{FS} = 10$ V when the input digital word is:

(a) 99H.
(b) 0AH.

Solution

$V_m = (2^{12} - 1)10/2^{12} = 9.976$ V.

(a) 99H = 1001 1001 = 153_{10}. Hence, $V_{out} = (153 \times 9.976)/4095 = 0.373$ V (Ans.)
(b) 0AH = 10_{10}. Hence, $V_{out} = (10 \times 9.976)/4095 = 0.0244$ V (Ans.)

Three other codes are sometimes employed for the input digital word of a DAC (or the output digital word of an ADC).

- The *offset binary code* is obtained by offsetting the natural binary code so that the half full-scale analogue voltage is 0 V. The transfer function of a DAC that uses this code is given in Fig. 11.5(a).
- The 2s complement code is obtained by inverting the MSB of the offset binary code. The transfer function of a DAC that uses this code is given in Fig. 11.5(b).
- The *sign magnitude code* uses the MSB to indicate the polarity of the analogue signal. MSB = 1 indicates a positive voltage and MSB = 0 indicates a negative voltage. Either natural binary or BCD may be used to indicate the magnitude of the analogue voltage. Offset binary, 2s complement and sign magnitude codes are shown in Table 11.1.

Fig. 11.5 *DAC transfer functions: (a) offset binary code and (b) 2s complement code*

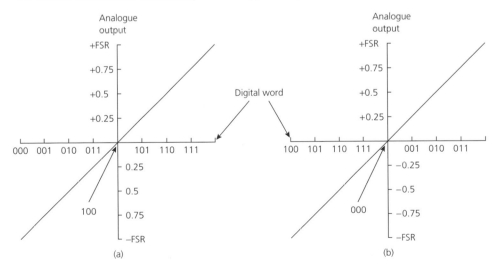

(a)   (b)

Table 11.1

Scale	Offset binary	Twos complement	Sign magnitude
+FS − LSB	11111111	01111111	11111111
+3/4 FS	11100000	01100000	11100000
+1/2 FS	11000000	01000000	11000000
+1/4 FS	10100000	00100000	10100000
+0	00000000	00000000	10000000
−0	00000000	00000000	00000000
−1/4 FS	01100000	11100000	00100000
−1/2 FS	01000000	11000000	01000000
−3/4 FS	00100000	10100000	01100000
−FS + LSB	00000001	10000001	01111111
−FS	00000000	10000000	–

EXAMPLE 11.3

Determine the maximum and minimum output voltages of a 12-bit DAC that uses:

(a) Unipolar binary.
(b) Offset binary.
(c) 2s complement binary code.

for the input digital word.
The nominal full-scale voltage is 10 V.

Solution

(a) $V_{max} = 10 - 10/2^{12} = 10 - 10/4096 = +9.9976$ V, $V_{min} = 0$ V (*Ans.*)
(b) $V_{max} = 5(1 - 1/2^{11}) = 5(1 - 4.883 \times 10^{-4}) = +4.976$ V, $V_{min} = -5$ V (*Ans.*)
(c) $V_{max} = 5(1 - 1/2^{11}) = +4.976$ V, $V_{min} = -5$ V (*Ans.*)

DAC data sheets

The data sheets for a DAC have a similar format to that employed with digital circuits. The data sheet commences with a description of the main features of the device as well as its pinout. Next come the absolute maximum ratings and the electrical characteristics. The data sheet usually ends with graphs, typical applications and operating instructions. A number of terms appear in DAC data sheets with which the user must be familiar if the information given is to be understood:

- Full-scale output. The full-scale output is the maximum voltage, or current, that can be outputted by the device. A DAC produces its full-scale output voltage when the digital input word is all 1s.
- Resolution. The resolution of a DAC is the smallest change in the input voltage that will cause a change to occur in the output voltage and it is determined by the

Table 11.2

Number of bits (n)	States 2^n	Resolution (%)	LSB weight in ppm	Bit weight for 10 V FSR	Dynamic range (dB)
0	1	100	10^6	10 V	0
1	2	50	500×10^3	5 V	6.02
2	4	25	250×10^3	2.5 V	12.04
4	16	6.25	62 500	0.625 V	24.08
8	256	0.39	3906	39 mV	48.16
10	1024	0.1	977	9.8 mV	60.21
12	4096	0.02	244	2.4 mV	72.25
14	16 384	0.006	61	610 µV	84.29
16	65 536	0.002	15	152 µV	96.33

Fig. 11.6 *Non-linearity and differential non-linearity*

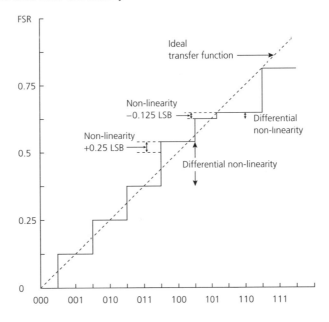

number n of bits in the digital input word. The resolution is equal to $1/2^n$. If, for example, $n = 8$ there are $2^8 = 256$ possible output voltages so that the smallest possible change in the output voltage is 1/256 times the full-scale voltage. Table 11.2 gives the resolutions for other bit numbers.

- Conversion time. The *conversion time* of a DAC is the time that elapses from the 'start conversion' command being given to the circuit and the analogue voltage appearing at the output terminals.
- Linearity. The *linearity* of a DAC is an indication of the maximum deviation of the slope of the transfer function from the average slope. It is illustrated in Fig. 11.6.

Fig. 11.7 *Settling time of a DAC*

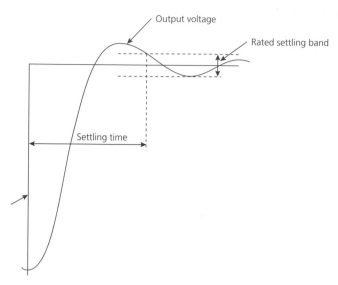

- Absolute accuracy. The absolute accuracy of a DAC is the largest difference between the actual output voltage and the output voltage predicted by the ideal transfer characteristic when a given input digital word is applied to the circuit. The absolute accuracy can be quoted as a percentage of the full-scale range, or as a percentage of the least significant bit. (For a DAC with a bipolar output range the percentage accuracy may be quoted in terms of the end-to-end range.) The absolute accuracy may be larger or smaller than the resolution.
- Relative accuracy. The relative accuracy of a DAC is the largest difference between the transfer function of the device and a straight line drawn between the output voltages 0 V and V_m.
- Monotonicity. A DAC is said to be monotonic if an increase in the input digital word always results in an increase in the output analogue voltage. If the size of a step should at any point be larger than one LSB the DAC will be non-monotonic at that point.
- Zero offset. The zero offset of a DAC is the difference between 0 V and the actual analogue voltage when the input digital word is all 0s.
- Settling time. When the input digital word changes suddenly over a wide range the analogue output voltage will not be able to change immediately to its new value. Figure 11.7 shows how the analogue output voltage will exhibit some overshoot and take some time to settle to its new value. The output voltage is said to have settled when it remains within a specified band of values, usually ±LSB/2, about its final value.
- Dynamic range. The dynamic range of a DAC is the ratio of the maximum output voltage V_m to the smallest change in the output voltage.

$$\text{Dynamic range} = 20 \log_{10} 2^n \text{ dB} \qquad (11.4)$$

- Maximum throughput rate. The maximum throughput rate is the maximum number of conversions that the DAC is able to carry out in 1s.

Table 11.3 *DACs*

DAC	Resolution (bits)	Settling time ± LSB (µs)	Linearity error ± LSB	Power dissipation (mW)	Microprocessor compatible
DAC 0808	8	0.15	0.5	305	No
ZN 425	8	1	0.5	190	No
AD 7524	8	0.4	0.5	10	Yes
AD 561	10	0.25	0.5	300	No
DAC 312	12	0.25	0.5	375	No
AD 767	12	4	1	400	Yes
DAC 16	16	0.5	2	1000	No

Fig. 11.8 *Basic DAC*

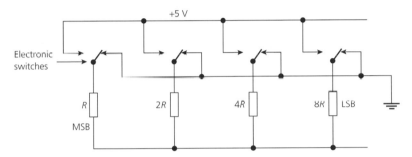

Short-form data

Short-form data for DACs is employed to make it easier for a user to choose possible devices for a particular application. Table 11.3 gives details of a small sample of DACs.

DAC circuits

Binary-weighted resistor DAC

Figure 11.8 shows the basic arrangement of a DAC. Four resistors are connected by electronic switches either to +5 V or to 0 V and have their other terminals connected together. The resistor values are binary weighted, i.e. their values are R, $2R$, $4R$ and $8R$, with the highest values of resistance being connected to the least significant switch.

Suppose that the input digital word is 1000. Then the circuit can be re-drawn as shown in Fig. 11.9(a); this circuit can, in turn, be re-drawn to give Fig. 11.9(b). From Fig. 11.9(b) the output voltage V_{out} of the circuit is

$$V_{out} = (5 \times 8R/7)/(R + 8R/7) = (5 \times 8)/15 = 2.667 \text{ V}$$

Similarly, when the input digital word is 0100, the circuit of Fig. 11.9(a) can be re-drawn to give first Fig. 11.10(a) and then Fig. 11.10(b). From Fig. 11.10(b),

$$V_{out} = (5 \times 8R/11)/(2R + 8R/11) = (5 \times 8)/30 = 1.333 \text{ V}$$

Fig. 11.9 *Figure 11.8 re-drawn for an input digital codeword of 1000*

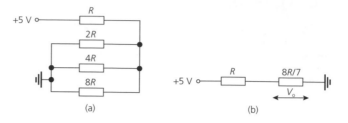

(a)   (b)

Fig. 11.10 *Figure 11.8 re-drawn for an input digital codeword of 0100*

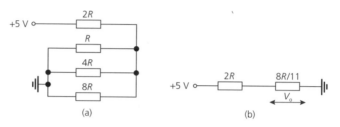

(a)   (b)

Table 11.4

Digital word	0000	0001	0010	0011	0100	0101	0110	0111
Output voltage (V)	0	0.333	0.667	1	1.333	1.667	2	2.333
Digital word	1000	1001	1010	1011	1100	1101	1110	1111
Output voltage (V)	2.667	3	3.333	3.667	4	4.333	4.667	5

Fig. 11.11 *Weighted resistor DAC*

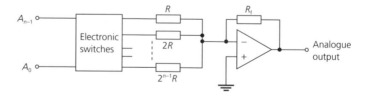

When the input signal is 1000, or decimal 8, the output voltage is $5/15 \times 8$, and when the digital word is 0100 (4) the output voltage is $5/15 \times 4$. Using this result the output voltages for other digital words have been calculated and are given in Table 11.4.

The output voltage of the resistor network is applied to the input of an op-amp connected as an inverting amplifier. This is shown in Fig. 11.11.

EXAMPLE 11.4

Calculate the output voltage of the circuit given in Fig. 11.11 if $R = R_f$ and the input digital word is

(a) 1010.
(b) 0110.

The voltage applied to each resistor is either +5 V or 0 V.

Solution

(a) $V_{out} = -5(1/R + 0/2R + 1/4R + 0/8R) = -6.25$ V *(Ans.)*
(b) $V_{out} = -5(0/R + 1/2R + 1/4R + 0/8R) = -3.75$ V *(Ans.)*

Switched current source DAC

Currents can be switched into, and out of, a circuit faster than is possible with voltages and hence integrated circuit DACs generally use current switching. A typical circuit is shown in Fig. 11.12. It consists of an array of switched current sources that are binary weighted by the emitter resistors whose values are R, $2R$, $4R$ and $8R$. The collector currents of the transistors are summed by the op-amp to produce the converted analogue voltage. When a digital input is HIGH its associated diode is turned OFF and the associated

Fig. 11.12 *Weighted current-switching DAC*

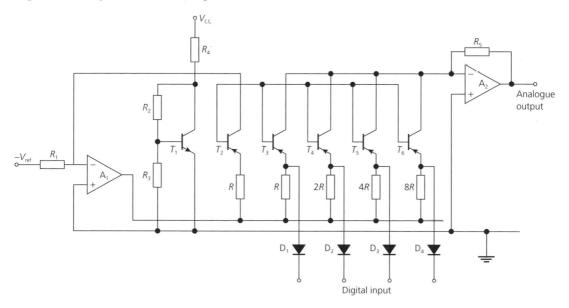

Fig. 11.13 *R/2R resistance network*

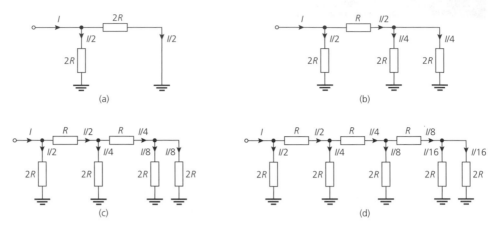

(a) (b) (c) (d)

current source is turned ON. Conversely, if a digital input is LOW its associated current source will be OFF.

R/2R DAC

The weighted resistor type of DAC requires the use of several precise-valued resistors which it is difficult, and hence expensive, to provide and so an alternative technique is more often employed. The *R/2R* type of DAC employs resistors of only two different values.

Figure 11.13(a) shows two resistors of value $2R$ connected in parallel. Since the resistors are of equal value the input current I will divide into two equal parts $I/2$. If the right-hand resistor is provided by a resistor R connected in series with two $2R$ resistors in parallel the circuit given in Fig. 11.13(b) is obtained. Now the $I/2$ current that flows into the right-hand resistor in Fig. 11.13(a) now splits into two equal $I/4$ parts.

If, now (see Fig. 11.13(c)), the right-hand resistor in Fig. 11.13(b) is provided by two $2R$ resistors connected in parallel, currents of $I/8$ will be obtained. The ideas can be further extended, and Fig. 11.13(d) shows how currents of $I/16$ can be produced. Clearly, the currents that flow in the resistors have a binary relationship to one another.

The circuit of a 4-bit DAC that uses an *R/2R* network is shown in Fig. 11.14. The lower end of each shunt resistor $2R$ is connected to an electronic switch that connects each resistor either to earth or to the current-summing line. The input digital word is applied to the PIPO shift register whose parallel outputs are used to operate the electronic switches. The most significant bit switches a current of $I/2$, the next most significant bit a current of $I/4$ and so on. The currents that are switched to the inverting terminal of the op-amp are summed and converted into the analogue output voltage.

Fig. 11.14 *R/2R DAC*

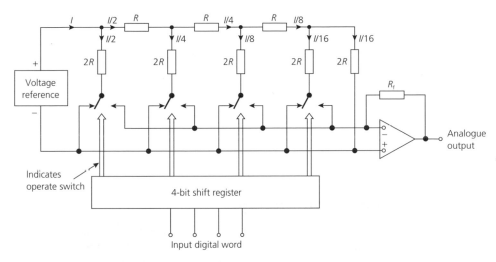

EXAMPLE 11.5

Calculate the output voltage of the circuit given in Fig. 11.14 when the input digital word is 1111, if $V_{REF} = 5$ V, $R_f = 10$ kΩ and $R = 5$ kΩ.

Solution

The current into the $R/2R$ network is $V_{REF}/R_{in} = V_{REF}/R = 5/5000 = 1$ mA. When the input digital word is 1111 all currents are switched to the output and hence the summed current is $I/2 + I/4 + I/8 + I/16 = 15/16$ mA. Therefore, $V_{out} = -15/16 \times 10^{-3} \times 10^4 = -9.375$ V *(Ans.)*

EXAMPLE 11.6

An 8-bit $R/2R$ DAC has a reference voltage of 3 V. The circuit is to have an output voltage of 0–6 V. Determine the necessary value for R. The feedback resistor is 39 kΩ.

Solution

When the input digital word is 1111 1111 the current flowing out of the network to the op-amp is equal to $255/256 \times I$, where I is the current that flows into the network. $I = 3/R$. Hence, $(255/256)I \times 39 \times 10^3 = 6$, and $I = 154.5$ μA. Therefore, $154.5 \times 10^{-6} = 3/R$, and $R = 19.417$ kΩ *(Ans.)*

Fig. 11.15 *Making the output of a DAC bipolar*

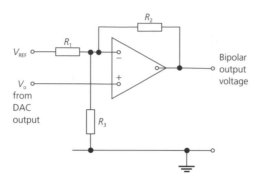

Bipolar output voltage

The analogue output voltage of an IC DAC is unipolar but some applications may require a bipolar output voltage. A bipolar voltage may be obtained by adding a negative offset voltage equal to $V_m/2$ so that when the DAC output is at mid-range the output of the circuit is 0 V. One commonly employed arrangement is shown in Fig. 11.15.

EXAMPLE 11.7

The 8-bit DAC in Example 11.6 is to be modified so that the output voltage is 0 V for an input digital word of 1000 0000. If the reference voltage applied to the op-amp is –5 V determine the necessary value of R_3 in Fig. 11.15.

Solution

In the basic circuit, when the input digital word is 1000 0000 the current flowing into the op-amp is $154.5 \times 10^{-6} \times 255/256 \approx 154$ μA. To obtain zero output voltage for this input word an offset current of the same value must flow into the other op-amp input terminal. Therefore, $154 \times 10^{-6} = 5/R3$, or $R_3 = 32.5$ kΩ (*Ans.*)

Integrated circuit DACs

Several manufacturers supply integrated circuit DACs and Table 11.2 (p. 323) lists a few commonly employed examples. The block diagram and pinout of the DAC 0808 are given in Fig. 11.16. A_1 is the most significant input and A_8 is the least significant input.

To use the DAC 0808 a reference voltage must be applied to pin 14, and pin 15 connected to earth. The output current of the DAC at pin 4 must be connected to an op-amp to be converted into a voltage; Fig. 11.17 shows a possible circuit. The compensating capacitor should be about 220 pF. The reference current is equal to V_{REF}/R_1 and it determines the full-scale output current which has 255/256 times the reference value. The value of R_2 is chosen to have the same value as R_1. The current flowing out of the output terminal is equal to $(V_{REF}/R_1)(A_1/2 + A_2/4 + A_3/8 + \ldots + A_8/256)$. The output voltage of the circuit is then I_{out} times R_f.

Fig. 11.16 *DAC 08080: (a) block diagram and (b) pinout*

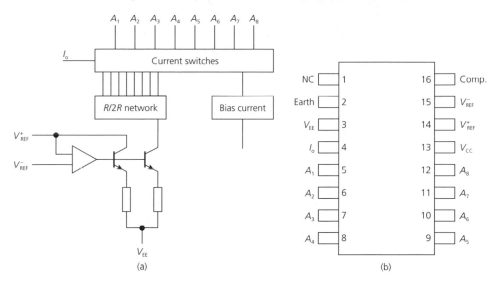

(a)

(b)

Fig. 11.17 *Circuit using the DAC 0808*

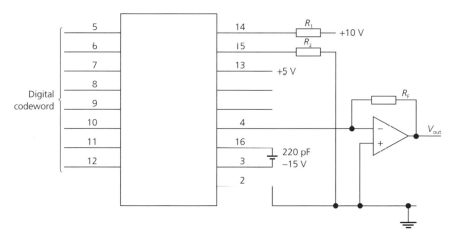

EXAMPLE 11.8

Calculate the output voltage of the circuit shown in Fig. 11.17 if the input digital word is 1011 0001 and $R_1 = R_2 = 4.7$ kΩ and $R_f = 6.2$ kΩ.

Solution

$I_{out} = [12/(4.7 \times 10^3)](1/2 + 1/8 + 1/16 + 1/256) = 1.764$ mA

$V_{out} = 1.764 \times 10^{-3} \times 6.2 \times 10^3 = 10.937$ V (*Ans.*)

Aim: to check the operation of a DAC.

Components and equipment: one 0808 DAC, one 741 op-amp, varied resistors and one 10 kΩ variable resistor. Pulse generator. CRO. Breadboard. Digital voltmeter. Power supply.

Procedure:

(a) Build the circuit shown in Fig. 11.18.

(b) Connect the DVM to the output of the DAC and set all the digital inputs to the DAC to logic 0. Note the reading of the DVM. It ought to be zero; is it? If it is not, explain why.

(c) If the output voltage is not zero then there is a zero offset in the circuit. Devise an op-amp circuit to remove the offset and then build this circuit. Connect the output of the DAC to the input of the op-amp circuit and connect the op-amp output to the DVM. Vary the value of the variable resistor until the DVM indicates 0 V. The zero offset has then been reduced to zero.

(d) Set all the digital inputs to the DAC to the logical 1 voltage level and adjust the variable resistor R_3 until the DVM reads 5 V. Then the FS is 5 V.

(e) Connect the DAC inputs to either 1 or 0 to obtain outputs of
   (i) 50 per cent FSR
   (ii) 25 per cent FSR
   (iii) 10 per cent FSR.
   Each time note the DVM reading. Use the results to draw the transfer function of the DAC, obtaining any more points that may be required.

(f) Connect all the digital inputs together and connect their junction to the pulse generator. Connect the op-amp output to the CRO. Vary the frequency of the pulse generator to a number of different values and each time note the waveforms displayed by the CRO. Measure the settling time of the DAC at each frequency.

Fig. 11.18

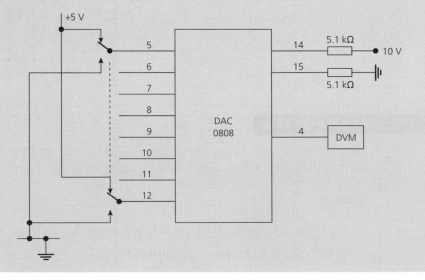

PRACTICAL EXERCISE 11.2

Aim: to confirm the operation of a DAC.
Components and equipment: one 74HC161 4-bit synchronous counter, and one DAC 0808. Pulse generator. CRO. Breadboard. Power supply.
Procedure:

(a) Build the circuit shown in Fig. 11.19.
(b) Start the counter and then ensure that it repeatedly cycles through a count of 0 to 7. Then connect the output of the counter to the input of the DAC and the output of the pulse generator to the clock terminal of the counter.
(c) Set the clock frequency to 1 kHz and note the waveform displayed on the CRO.
(d) Repeat (c) for clock frequencies of
   (i) 1 MHz
   (ii) 2 MHz
   (iii) 25 MHz.
(e) For each waveform state
   (i) whether any differential non-linearity was observed
   (ii) whether there was any offset error
   (iii) if any other errors were observed.

Fig. 11.19

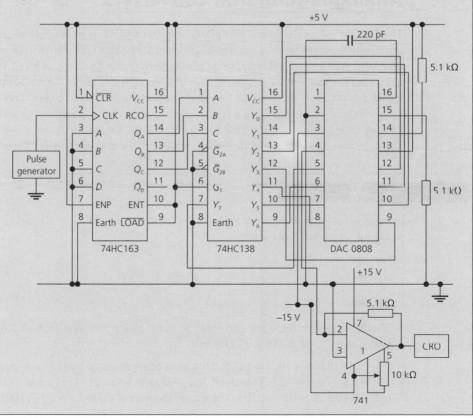

Fig. 11.20 *Transfer function of a 3-bit ADC*

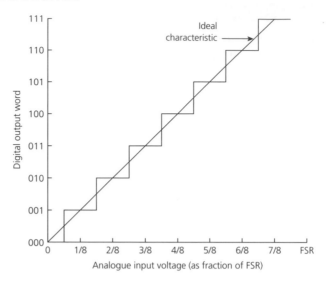

Analogue-to-digital converters

The function of an *analogue-to-digital converter* (ADC) is to convert an input analogue signal into the equivalent output digital codeword. Figure 11.20 shows the transfer function of a 3-bit ADC. A step characteristic is obtained because a range of input analogue voltages ± LSB either side of a quantization level have the same digital codeword. The ideal transfer function is the straight line drawn between the points 0, 000 and 7/8 V_{FS}, 111. The analogue input signal must be sampled by the ADC a number of times per second. The *sampling frequency* must be at least twice the highest frequency contained in the analogue signal or else signal distortion will occur because of an effect known as *aliasing*.

EXAMPLE 11.9

An 8-bit ADC has an input analogue voltage range of 0–5 V. Calculate the output digital word that represents:

(a) 2.5 V.
(b) 4 V.

Solution

2^8 = 256. Hence, there are 255 steps in the transfer function each of which represents a voltage of 5/255 = 19.608 mV.

(a) $2.5/(19.608 \times 10^{-3})$ = 127.5. This must be represented by the nearest quantization level, i.e. by 127. Therefore, the digital word is 01111111 (*Ans.*)
(b) $4/(19.608 \times 10^{-3})$ = 204. Therefore, the output digital word is 11001100 (*Ans.*)

Fig. 11.21 *Ramp ADC*

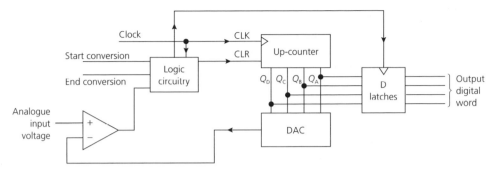

ADC circuits

Ramp ADC

The basic circuit of a ramp ADC is shown in Fig. 11.21. Before the start of a conversion the counter is reset to a count of 0. The start command, a positive pulse, initiates the conversion process. The output of the voltage comparator is either 0 or 1 depending upon which of its two input terminals is the more positive. At each clock pulse the count of the counter is incremented. The digital word held by the counter represents its current count and it is applied to both the DAC and to the output latches. The digital word is converted by the DAC into the equivalent analogue voltage and this is applied to one terminal of the voltage comparator. The output of the comparator remains HIGH as long as the analogue voltage is more positive than the output voltage of the DAC. Immediately the DAC output voltage becomes more positive than the analogue voltage the voltage comparator switches to have a LOW output. At this point the DAC output voltage is a close approximation to the analogue voltage. The LOW at the comparator output causes the logic circuitry to stop the conversion process and to generate an end-of-conversion signal that may be used elsewhere. The digital word that represents the analogue voltage is then held by the output latches. The digital output of the circuit therefore can indicate only the maximum value of the analogue input voltage attained between successive clears of the ADC. This type of ADC is relatively slow in its operation because it must be cleared and a new count started from zero for any decrease in voltage to be indicated.

EXAMPLE 11.10

A 12-bit ramp ADC is required to convert an analogue signal that occupies a bandwidth of 0–6 kHz. Calculate the minimum clock frequency. Assume that the ADC is reset to 0 at the end of each conversion.

Solution

A 6 kHz signal must be sampled at a rate of at least $2 \times 6 = 12$ kHz. Hence the conversion time is, at most, $1/(12 \times 10^3) = 83.3$ μs. A 12-bit ADC requires $2^{12} = 4096$ clock pulses to have its output voltage change from 0 to FSR. Hence, the maximum conversion time is the time occupied by 4096 clock pulses. Therefore, minimum clock frequency $= 4096 \times 12 \times 10^3 = 49.152$ MHz (*Ans.*)

[Alternatively, $83.3 \times 10^{-6} = 4096T$, where $T = 1/f$.]

EXAMPLE 11.11

An 8-bit ramp ADC has a 100 kHz clock frequency. Calculate its maximum conversion time for a full-amplitude range input analogue signal.

Solution

For an 8-bit ADC there are $2^8 = 256$ different codewords. This means that the maximum conversion time is 256 clock periods. Therefore, maximum conversion time $= 256 \times 1/10^5 = 2.56$ ms (*Ans.*)

EXAMPLE 11.12

An 8-bit ramp ADC has a maximum range input voltage of +12 V. If the clock frequency is 125 kHz, determine the time taken to convert a 6 V input signal.

Solution

Eight bits produce 256 codewords, and hence the quantization interval is $12/256 = 46.88$ mV. The count must ramp through $6000/46.88 = 128$ codewords. Therefore, conversion time $= 128 \times (1/125) \times 10^{-3} = 10.24$ ms (*Ans.*)

Tracking ADC

A considerable decrease in the conversion time can be achieved if the counter in the circuit shown in Fig. 11.21 is replaced with an up-down counter as in the circuit shown in Fig. 11.22. Clock pulses are directed to either the up-count or the down-count terminal of the counter depending upon whether the DAC output voltage is larger than, or smaller than, the analogue input voltage.

If the DAC output voltage is the smaller then the output of the voltage comparator will be HIGH. Then all the inputs to the top gate are HIGH so that its output is also HIGH. The lower gate then has one of its inputs LOW so that its output is LOW. The counter is then in its up-count mode of operation. If the DAC output voltage becomes larger than the input analogue voltage the gate outputs will switch states and this will take the counter into its count-down mode of operation.

Fig. 11.22 *Tracking ADC*

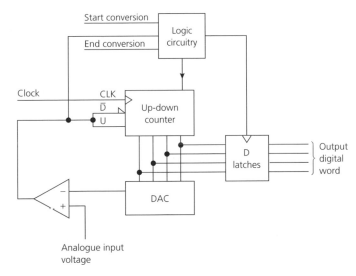

The tracking ADC is thus able to continuously track any variations in the input analogue voltage and keep updated its output digital codeword.

EXAMPLE 11.13

A tracking ADC has a maximum range input voltage of 12 V. If the clock frequency is 125 kHz, calculate the conversion time when a 6 V signal changes to 4 V.

Solution

The quantization step is 12.256 = 46.88 mV. Hence the counter must pass through (6000 − 4000)/46.88 codewords. Therefore, conversion time = 43 × (1/125) × 10^{-3} = 0.344 ms (*Ans.*)

Successive approximation ADC

The basic circuit of a successive approximation ADC is shown in Fig. 11.23. At the beginning of a conversion the shift register is reset to 0. The next clock pulse then applies the MSB of the register output to the DAC and its *half-scale* (1/2 FS) voltage is applied to the voltage comparator. If the DAC output voltage is smaller than the input analogue voltage the MSB output of the shift register is retained, but if the DAC output is the larger then the MSB output is turned OFF. The next clock pulse turns ON the next to MSB output of the shift register and this voltage causes the DAC to generate a voltage of 3/4 FS if the MSB is still ON or 1/4 FS if the MSB is OFF. The new DAC output is compared with the analogue input voltage and a decision is made either to retain the next MSB or to turn it OFF. The process of turning ON a shift register output and comparing the total DAC output with the analogue input voltage continues until the LSB is reached.

Fig. 11.23 *Successive approximation ADC*

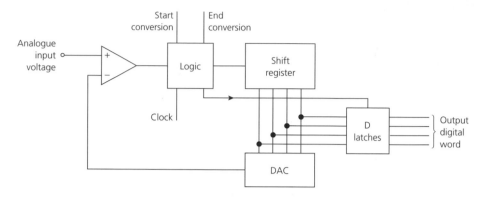

Fig. 11.24 *Action of successive approximation ADC*

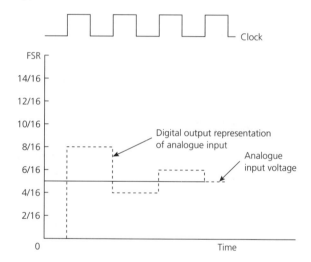

After *n* comparisons the outputs of the shift register produce the digital codeword that represents the analogue input voltage to an accuracy of ±1/2 LSB. The difference between the DAC output and the analogue input voltage is known as the *quantization error*. Each conversion takes exactly the same time regardless of the amplitude of the analogue input voltage.

The approximation process is illustrated in Fig. 11.24 for a 4-bit ADC in which the analogue input voltage is equal to $5/16 \ V_{FS}$. The first approximation is $8/16 \ V_{FS}$, which is too large. Hence, the next clock pulse turns the MSB OFF and the second approximation is tried. This is $4/16 \ V_{FS}$; this is less than the analogue voltage and hence this bit is retained. The third clock pulse turns ON the next significant shift register output and this increments the DAC output by $2/16 \ V_{FS}$ to give a total voltage of $6/16 \ V_{FS}$. Now the approximation is again too high and hence the next clock pulse turns that bit OFF and the LSB ON. The DAC output is then $4/16 \ V_{FS} + 1/16 \ V_{FS} = 5/16 \ V_{FS}$ and this is equal to the input analogue voltage. The output digital codeword is 0101.

(a) Calculate the percentage increase in conversion speed if an 8-bit successive approximation ADC is employed instead of an 8-bit ramp ADC. Assume the same maximum input voltage is applied to both circuits.

(b) Calculate the output digital word if the analogue voltage is 2.3 V and the maximum input voltage possible is 6 V.

Solution

(a) For an 8-bit ramp ADC the circuit must pass through $2^8 = 256$ states and hence the conversion time is $256T$. The successive approximation ADC will carry out the same conversion in $8 + 1 = 9$ clock periods. Therefore, the increase in conversion time is $[(256 - 9)/9] \times 100$ per cent $= 2744$ per cent (*Ans.*)

(b) Resolution $= 6/256 = 23.44$ mV. Hence, the number of steps $= 2.3/(23.44 \times 10^{-3}) \approx 98 = 1100010$ (*Ans.*)

Flash ADC

When very fast conversions are required a flash ADC can be employed. Figure 11.25 shows the circuit of an n-bit flash ADC. The circuit uses 2^n resistors connected in series

Fig. 11.25 *Flash ADC*

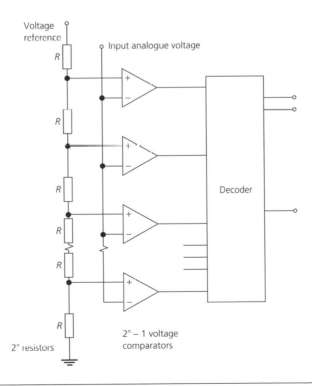

Fig. 11.26 *Single-slope integrating ADC*

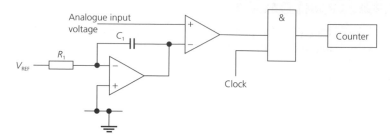

with a positive reference voltage and earth. Also, $2^n - 1$ voltage comparators are connected to the junctions of these resistors so that the comparator inputs are spaced at voltage intervals equal to the LSB. The analogue input voltage is simultaneously compared against $2^n - 1$ d.c. voltage levels. All those voltage comparators which have their + terminal at a voltage less than the analogue voltage turn ON, while all others stay OFF. The output terminals of the comparators are then decoded to give the required output digital word.

A 4-bit flash converter with a reference voltage of 6 V will require $2^4 = 16$ series resistors and 15 voltage comparators. The voltages in the resistor chain will increment in 6/16 or 0.375 V steps. If, for example, the applied analogue voltage is 2.4 V only the bottom six comparators will turn ON. The outputs of these six comparators are decoded to give the output digital word.

A flash converter gives very fast conversions but is expensive because it requires a large number of voltage comparators.

Half-flash ADC

The *half-flash* ADC operates in the same way as a flash ADC but the number of voltage comparators it uses is reduced by a factor of 8. This means that an 8-bit circuit has 32 comparators instead of 256. The half-flash ADC gives very fast conversions and is much cheaper than a flash ADC.

Integrating ADC

A number of ADC circuits operate by converting the analogue input voltage into a proportional time interval and then measuring its duration. Since the input signal is averaged over a fixed time period some reduction in noise is obtained. This type of ADC is suited to applications where high noise immunity is of prime importance.

Single-slope

The single-slope integrating ADC uses a linear ramp generator to produce a reference voltage of constant slope. Figure 11.26 shows the circuit of a single-slope integrating ADC. The op-amp is connected as an integrator so that its output is $v = -(1/CR) \int V_{REF} \, dt = (V_{REF} \, t)/CR$, which is the equation of a linear ramp waveform.

Initially, the analogue voltage is larger than the ramp voltage and the output of the voltage comparator is HIGH. The AND gate is enabled and clock pulses arrive at the counter. When the ramp voltage has increased to become slightly more positive than the analogue voltage the comparator switches to have a LOW output. The AND gate is then inhibited and no more clock pulses arrive at the counter. The count of the counter is then a measure of the analogue input voltage. At the end of each count the circuit must be reset by control logic that is not shown.

EXAMPLE 11.15

The ADC shown in Fig. 11.26 has a reference voltage of -5 V, a clock frequency of 2 MHz, and component values $R = 22$ kΩ and $C = 1$ nF. The maximum count of the counter is FF.

(a) When the counter is reset the count is 38H. Calculate the analogue input voltage.
(b) Determine the maximum input voltage that can be measured.
(c) What would be the maximum input voltage if the clock frequency were reduced to 1.5 MHz?

Solution

(a) $T = 1/(2 \times 10^6) = 0.5$ μs. $38H = 56_{10}$. $t = 0.5 \times 10^{-6} \times 56 = 28 \times 10^{-6}$
$V_{in} = [-(-5 \times 28 \times 10^{-6})]/(1 \times 10^{-9} \times 22 \times 10^3) = 6.364$ V (Ans.)

(b) $FF = 255_{10}$. $t = 0.5 \times 255 = 127.5$ μs
$V_{in} = -[-(5 \times 127.5 \times 10^{-6})]/(1 \times 10^{-9} \times 22 \times 10^3) = 28.98$ V (Ans.)

(c) $T = 1/(1.5 \times 10^6) = 666.7$ ns
$V_{in} = [(5 \times 666.7 \times 10^{-9} \times 255)]/(1 \times 10^{-9} \times 22 \times 10^3) = 38.64$ V (Ans.)

Dual-slope

The circuit of a *dual-slope integrating* ADC is shown in Fig. 11.27. The input analogue voltage is applied to the integrator (previously set to have zero output voltage) and is integrated for a fixed time T. During this time the counter will have counted N_s clock pulses. At the end of T_1 seconds the control logic switches the integrator to either the $+V_{REF}$ or the $-V_{REF}$ input, whichever is of the opposite polarity to the input voltage. This makes the integrating capacitor discharge and the voltage V_A falls linearly towards 0 V. Clock pulses are again counted while V_A falls. When V_A becomes equal to 0 V the comparator changes state and stops the counter at a count of N_R pulses. The time T_2 taken for V_A to fall to zero is directly proportional to the average value of the analogue input voltage. Hence, $V_A = (1/CR)V_{REF}T_2$ and $(V_{REF}T_2)/CR = (V_{REF}T_1)/CR$. Therefore, $T_2 = T_1 V_{in}/V_{REF}$ and $N_R = N_s V_{in}/V_{REF}$.

The digital output codeword produced by the counter (counting pulses to measure T_2) represents the ratio of the input voltage to the reference voltage. The output waveform is shown in Fig. 11.28.

Fig. 11.27 *Dual-slope integrating ADC*

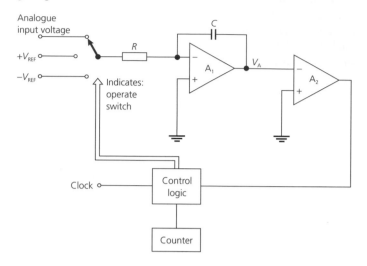

Fig. 11.28 *Operation of dual-slope integrating ADC*

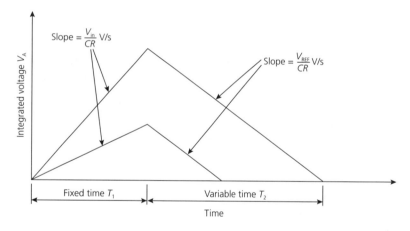

The dual-slope technique of analogue-to-digital conversion has the following advantages:

(a) The accuracy of a conversion is independent of the stability of both the clock and the integrator provided they are constant during a conversion period.
(b) The noise immunity is good.

Integrated circuit ADCs

There are a number of different types of ADC available in IC form and it is necessary to have an idea of their relative merits so that the right device can be chosen for each application.

The simplest technique that is employed for analogue-to-digital conversion is *single-slope* or *ramp conversion*. The *dual-slope* or *integrating* ADC operates using a similar principle but it has greater accuracy with a conversion time measured in milliseconds again. Both of these types of ADC are used for low-frequency data loggers and digital voltmeters. *Tracking ADCs* are faster to operate, typically less than 1 ms, but the time required before an input signal is acquired could be as much as 2^n, where n is the number of bits employed. A tracking ADC is best employed for the continuous monitoring of an analogue signal and its conversion into a digital code sequence. The conversion time of a *successive approximation ADC* may be just a few microseconds and it is independent of the amplitude of the analogue voltage to be converted. The cost of a successive approximation ADC increases rapidly with increase in the number of bits used for each conversion. Typically, the cost of a 10-bit device is about five times that of an 8-bit device, and a 14-bit device is ten, or more, times more expensive. The fastest type of converter is the *flash ADC* and its conversion time may be less than 1 μs. Flash ADCs are expensive and a cheaper alternative, namely the *half-flash ADC*, is often used since its conversion time, although slower than that of a flash ADC, is still faster than that of all other types.

In the choice of an ADC for a particular application the following points should be considered:

- The input and output requirements such as current and/or voltage ranges, impedances, logic levels and data rates.
- The required accuracy; this involves the resolution determined by the number of bits and the linearity.
- The tolerable errors such as allowable non-linearity and missing codes.
- The conversion speed required.
- Whether a microprocessor compatible device is needed.
- The digital code that is employed.

EXAMPLE 11.16

An ADC has a maximum dynamic range of 60 dB and is preceded by a Butterworth low-pass filter that has a 3 dB frequency of 1 kHz and 25 dB loss at 3 kHz. The ADC is to be used to convert an analogue signal in the frequency band 0–3000 Hz. It is required that the aliasing error is less than 3 dB. Calculate:

(a) The number of bits the ADC should have.
(b) The order of filter required.
(c) The sampling frequency.

Solution

(a) At 1 kHz the filter has 3 dB loss and therefore $63 = 6.02n - 4.24$, or $n = (63 + 4.24)/6.02 = 11.2$. Hence, a 12-bit ADC is required (*Ans.*)

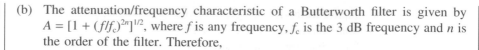

(b) The attenuation/frequency characteristic of a Butterworth filter is given by $A = [1 + (f/f_c)^{2n}]^{1/2}$, where f is any frequency, f_c is the 3 dB frequency and n is the order of the filter. Therefore,

$$25 \text{ dB} = 17.78 \text{ voltage ratio} = [1 + 3^{2n}]^{1/2}$$
$$316 = 1 + 3^{2n}, \ 2n \log_{10}3 = \log_{10}315 = 2.5. \ n = 2.5/(2 \times 0.477) = 2.6.$$

Therefore, a third-order filter is necessary (*Ans.*)

(c) For an attenuation of 63 dB or 1413 the frequency must be f'. $1413 = [1 + (f'/1)^6]^{1/2}$, $1413 \approx f'^6$, $6.3 = 6 \log_{10}f'$, and so $f' = 11.2$ kHz. The maximum signal frequency is 3 kHz $= f_s - 11.2$ kHz. Hence, minimum sampling frequency $= f_s = 14.2$ kHz (*Ans.*)

Data sheets

The layout of the data sheet of an ADC follows the usual pattern of:

(a) A brief description of the device and its main features.
(b) Its pinout and block diagram.
(c) Its electrical characteristics and absolute maximum ratings.

The data sheet continues with several pages that describe various circuit operations including methods of interfacing the device to a microprocessor. Understanding the information given in a data sheet demands a knowledge of the meanings of the several terms that are employed.

Terms used with ADCs

- Absolute accuracy. The *absolute accuracy* of an ADC is a measure of the departure from ideal of the transfer characteristic. This means that it is the difference between the theoretical and actual inputs needed to produce a given codeword.
- Relative accuracy. The relative accuracy of an ADC is a measure of the largest deviation between the transfer function and a line drawn between the points 0 V and FS. It is expressed as a percentage of the FS.
- Differential linearity and monotonicity. These terms have the same meanings as with DACs (see pp. 323–324).
- Conversion time. The conversion time is the time that elapses from a start conversion command being applied to the circuit and the digital word appearing at the output.
- Dynamic range. The dynamic range of an ADC is the ratio of the full-scale input voltage to the smallest change in the input voltage to which the circuit can respond.
- Resolution. The resolution of an ADC is the smallest change in the input analogue voltage that can be indicated at the output. The resolution is given by:

$$\text{Resolution} = (\text{full-scale voltage})/(2^n - 1) \tag{11.5}$$

(a) Calculate the resolution of a 12-bit ADC with a ±10 V analogue voltage range.

(b) Calculate the output digital word if the analogue input voltage is 2.3 V and the maximum input voltage is 6 V.

Solution

(a) Resolution = $20/(2^{12} - 1) = 20/4095 = 4.884$ mV (*Ans.*)

(b) Resolution = $6/4.095 = 1.465$ mV. Hence, number of steps = $2.3/(1.465 \times 10^{-3})$
$\approx 1570 = 0110\ 0010\ 0010$ (*Ans.*)

Short-form data for ADCs divides devices according to the number of bits used, ranging from 6 to 24, under the headings shown in Table 11.5 which give details of some popular devices.

Table 11.5 ADCs

	Linearity error ± LSB	Conversion time (1/2 LSB µs)	Power dissipation (mW)
8-bit			
0804	0.5	100	12.5
0820	1	1.0	12
10-bit			
820	1	0.05	275
12-bit			
574	1	25	265
912	1	10	95

ADC 0804

Figure 11.29 shows the pinout of a commonly used integrated circuit ADC: the 0804 which is of the successive approximation type. The IC uses the supply voltage V_{CC} as the reference voltage. Two analogue input terminals are provided so that differential operation is possible. For normal single-ended operation the V_{in}^- terminal should be connected to earth. Besides the analogue input and digital output pins there are some other pins whose functions are not so obvious:

(a) $\overline{CS}$ (conversion start; pin 1): the conversion start pin must be LOW for a conversion to take place. Then both the $\overline{RD}$ and $\overline{WR}$ pins are active.

Fig. 11.29 *Pinout of AD 0804*

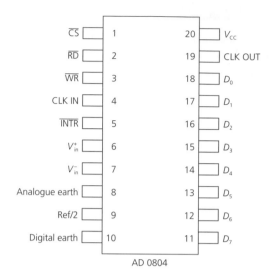

$\overline{CS}$	1	20	V_{cc}
$\overline{RD}$	2	19	CLK OUT
$\overline{WR}$	3	18	D_0
CLK IN	4	17	D_1
$\overline{INTR}$	5	16	D_2
V_{in}^+	6	15	D_3
V_{in}^-	7	14	D_4
Analogue earth	8	13	D_5
Ref/2	9	12	D_6
Digital earth	10	11	D_7

AD 0804

(b) $\overline{RD}$ (read; pin 2): the output latch is enabled only when both $\overline{RD}$ and $\overline{CS}$ are LOW. When $\overline{CS} = \overline{RD} = 0$ the outputs hold the logic levels of the last conversion to have been carried out.

(c) $\overline{WR}$ (write: pin 3): the write pin in taken LOW to reset the internal SAR and shift registers and start a new conversion. As long as both $\overline{CS}$ and $\overline{WR}$ are held LOW the ADC is in its reset state.

(d) CLK IN and CLK OUT (pins 4 and 19): the CLK IN pin is provided for an external clock to be used. CLK OUT is used when the internal clock is employed, then a resistor must be connected between the pin and CLK IN and a capacitor between their junction and earth.

(e) $\overline{INTR}$ (interrupt; pin 5): the interrupt output goes LOW when a conversion has finished and an 8-bit digital word has been transferred to the output latches. It is used to signal the end of conversion to a microprocessor and that the data is ready to be read out. If $\overline{INTR}$ is connected to $\overline{WR}$ and $\overline{CS}$ is held LOW the circuit will operate continuously.

PRACTICAL EXERCISE 11.3

Aim: to confirm the operation of an ADC.
Components and equipment: one 0804 ADC and one 0808 DAC. Function generator. CRO. Breadboard. Power supply.
Procedure:

(a) Build the circuit shown in Fig. 11.30.
(b) Set the function generator to produce a ramp waveform at about 1 kHz. Apply the appropriate logic levels to the pins of the 0804 and 0808 to get them working.

Fig. 11.30

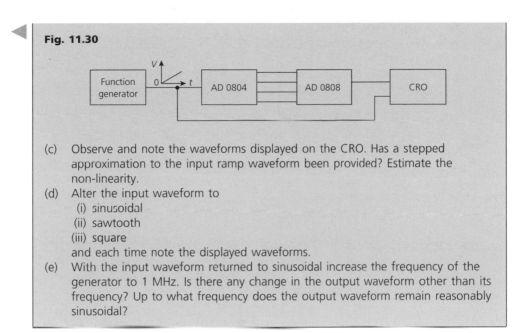

(c) Observe and note the waveforms displayed on the CRO. Has a stepped approximation to the input ramp waveform been provided? Estimate the non-linearity.

(d) Alter the input waveform to
  (i) sinusoidal
  (ii) sawtooth
  (iii) square
  and each time note the displayed waveforms.

(e) With the input waveform returned to sinusoidal increase the frequency of the generator to 1 MHz. Is there any change in the output waveform other than its frequency? Up to what frequency does the output waveform remain reasonably sinusoidal?

Sample-and-hold amplifiers

While the conversion of an analogue signal into digital form is taking place it is essential that the signal remains constant to within ±0.5 LSB. If the change in analogue signal is larger than this then a *sample-and-hold* amplifier must be connected at the input to the ADC to keep the input voltage constant during the conversion process. There is another reason why a sample-and-hold amplifier is employed in a system. An ADC places transient loads on the signal source during each bit of a conversion. This may cause transient errors in the analogue signal that produce both conversion noise and non-linearity. The effect worsens as the number of bits employed by an ADC is increased.

A sample-and-hold amplifier can sample an analogue input voltage and then hold the sampled value when it receives a hold command. The basic circuit of a sample-and-hold amplifier is shown in Fig. 11.31. When the electronic switch is closed the output voltage V_{out} will be equal to the input voltage V_{in}. When the switch is opened the charge stored in the capacitor is retained and hence the output voltage is held at the instantaneous value of the input voltage at the moment the switch opened.

Fig. 11.31 *Basic sample-and-hold circuit*

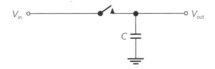

Fig. 11.32 *Sample-and-hold amplifier*

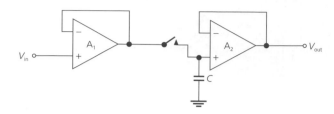

Fig. 11.33 *Sample-and-hold amplifier*

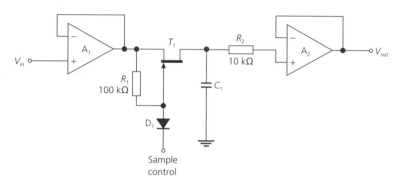

In order to prevent the storage capacitor loading the analogue source an input unity-gain amplifier is normally employed. Also, to prevent the capacitor discharging too rapidly into the load another unity-gain amplifier is connected at the output. The basic circuit of such a sample-and-hold amplifier is shown in Fig. 11.32. The input buffer amplifier is an op-amp connected as a voltage follower and so it has a very high input resistance and a low output resistance. The low output resistance keeps the time constant of the circuit small so that the voltage across the capacitor is able to follow all fluctuations in the analogue voltage. The output buffer amplifier is also an op-amp connected to act as a voltage follower.

The switch in a sample-and-hold amplifier is usually provided by an FET and Fig. 11.33 gives a typical circuit. The switch is provided by T_1 and it is turned ON and OFF by the voltage applied to the control input terminal. When this voltage is HIGH, the *hold command*, T_1 turns OFF and the sampled voltage is held by the capacitor. The charge stored in C_1 will slowly disappear because of the finite output resistance of T_1. Restore R_2 protects the input stage of the op-amp.

If the bias current of the op-amp is I_1 and the leakage current of T_1 is I_2 then the *drift rate* of the circuit is $dv/dt = (I_1 + I_2)/C_1$. The time taken for the capacitor voltage to reach the same value as the analogue input voltage depends upon the time constant of the circuit. This is $\tau = (R_{DS(on)} + R_{out})C_1$. The output resistance of the voltage follower is low and can be neglected. After a time equal to 5τ the capacitor voltage is within 1 per cent of the input voltage and within 0.1 per cent after 7τ.

EXAMPLE 11.18

If the ON resistance of T_1 in Fig. 11.33 is 30 Ω and $C_1 = 1$ μF, calculate the time required for the capacitor voltage to reach a value within 1 per cent of the input voltage.

Solution

Time constant $= 1 \times 10^{-6} \times 30 = 30$ μs. $5\tau = 150$ μs (*Ans.*)

EXAMPLE 11.19

(a) Calculate the maximum capacitor voltage that can be employed if $R_{DS(on)} = 35$ Ω and if the maximum time for the output voltage to be within 1 per cent of the input voltage is to be 10 μs.
(b) If the bias current of the op-amp is 150 pA and the FET leakage current is 200 pA, calculate the drift rate.

Solution

(a) $10 \times 10^{-6} = 5C_1 \times 35$, $C_1 = (10 \times 10^{-6})/(5 \times 35) = 0.057$ μF (*Ans.*)
(b) Drift rate $= dv/dt = I/C_1 = (350 \times 10^{-12})/(0.057 \times 10^{-6}) = 6.14$ mV/s (*Ans.*)

The maximum frequency f_{max} that may be converted without the use of a sample-and-hold amplifier can be calculated. If the input signal is sinusoidal,

$v = V_m \sin 2\pi ft = 2^n \sin 2\pi ft$

$dv/dt = $ slew rate $= 2\pi f 2^n \cos 2\pi ft$

Maximum slew rate $= dv/dt = 2\pi f_{max} 2^n$

$$dv = 2\pi f_{max} 2^n \, dt = 2\pi f_{max} 2^n T_c \qquad (11.6)$$

where T_c is the conversion time.
For 0.5 LSB error

$$f_{max} = 1/(2\pi 2^n T_c) \qquad (11.7)$$

EXAMPLE 11.20

A 10-bit, 15 μs conversion time, successive approximation ADC is used in a system. Calculate the frequency above which a sample-and-hold amplifier will have to be used.

Solution

From equation (11.7), $f_{max} = 1/(2\pi \times 1024 \times 15 \times 10^{-6}) = 10.4$ Hz (*Ans.*)

Fig. 11.34 *Action of a sample-and-hold amplifier when first a sample and then a hold command is received*

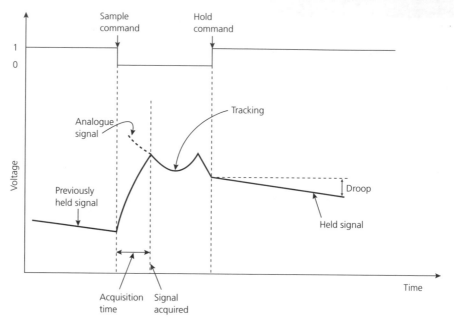

Modes of operation

A sample-and-hold amplifier can be instructed by an applied digital signal to operate in either one of two sequential modes:

(a) The *sample mode*.
(b) The *hold mode*.

In the sample mode the output of the sample-and-hold amplifier closely follows the waveform of the input analogue signal. In the hold mode the output of the sample-and-hold amplifier is held at the value it had when the command to go into the hold mode was received. Once a signal has been acquired the amplifier is switched into its hold mode and then the voltage across the sample capacitor appears at the output terminals. Figure 11.34 shows the action of a sample-and-hold amplifier when first a sample and then a hold command is received.

Terms used with sample-and-hold amplifiers

- Acquisition time. The acquisition time of a sample-and-hold amplifier is the maximum time needed for the output voltage to begin to track the input voltage after the sample command has been received (see Fig. 11.34).
- Aperture time. The aperture time of a sample-and-hold amplifier is the time that elapses between the application of the hold command and the output voltage ceasing to track the input voltage (see Fig. 11.34).
- Drift rate or droop rate. The drift (droop) rate is the maximum rate of change of the output voltage when the circuit is in its hold mode.

- Settling time. The settling time of a sample-and-hold amplifier is the time required for the output voltage to become equal to the input voltage within a specified accuracy (see Fig. 11.34).
- Hold step. This is the magnitude of the step in the output voltage produced when the circuit is switched from the sample mode to the hold mode.
- Turn-off time or aperture delay. This is the delay that occurs between the application of the hold command and the start of the hold mode.
- Full-power bandwidth. The full-power bandwidth is the highest frequency at which the sample-and-hold amplifier is able to track a sinusoidal input signal of voltage large enough to drive the circuit to its rated full-scale output voltage.
- Aperture uncertainty. This is the difference between the maximum and minimum turn-off times.

The demands made on a sample-and-hold amplifier vary according to whether the input sign is continuous or is a train of pulses. The main criterion for a sample-and-hold amplifier used to convert pulsed information is acquisition time. The acquisition time of a sample-and-hold amplifier should be less than the duration of the shortest input pulse. A continuous analogue signal is tracked continuously by a sample-and-hold amplifier and aperture delay and drift rate are the most important parameters. Some manufacturers describe an amplifier optimized for such a purpose as a *track and hold amplifier*. A track-and-hold amplifier will remain in the sample mode for relatively long periods of time and so its acquisition time is of little importance. The most important parameter is the delay that occurs when the circuit is switched from track mode to hold mode. The sampling of the analogue voltage can be carried out at regular intervals or whenever a command is received. In the latter case the most important parameter is the aperture delay; it should not be so long that the signal voltage is able to change by more than 0.5 LSB of the conversion accuracy.

If the allowable voltage change is dv and the input signal is

$$v = V_m \sin 2\pi ft, \; dv/dt = 2\pi f V_m \cos 2\pi ft$$

$$dv/dt_{(max)} = 2\pi f_{max}V, \text{ and } dv = 2\pi f_{max}V \, dt = 2\pi f_{max}VT_d \qquad (11.8)$$

EXAMPLE 11.21

Calculate the maximum frequency for a 10-bit conversion of a 10 V peak analogue waveform if ±0.5 LSB accuracy is required. The aperture delay is 24 ns.

Solution

$2^{10} = 1024$. $10/1024 = 9.766$ mV. From equation (11.8),

$$f_{max} = (9.766 \times 10^{-3})/(2\pi \times 10 \times 24 \times 10^{-9}) = 6476 \text{ Hz } (Ans.)$$

The parameters of a sample-and-hold amplifier must be matched to those of the ADC. The acquisition time of the sample-and-hold amplifier should be about 10 per cent of the converter's conversion time so that data throughput is slowed down by only about

Fig. 11.35 *Use of a sample-and-hold amplifier*

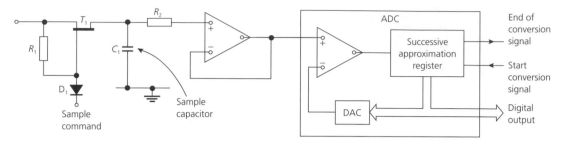

Fig. 11.36 *Timing diagram of a sample-and-hold amplifier and ADC system*

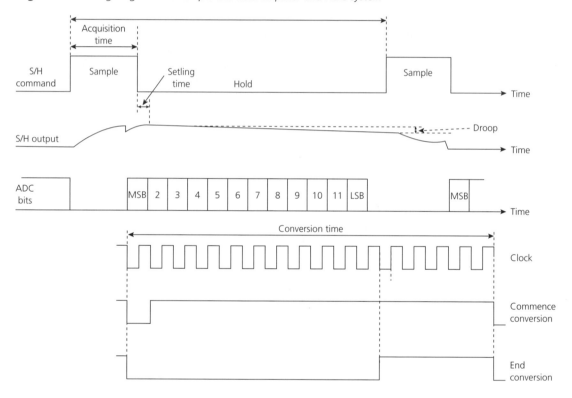

10 per cent. A typical arrangement is shown in Fig. 11.35. When the sample command input is taken HIGH FET T_1 turns ON and the analogue signal is applied to the sample capacitor C_1. This is the sample mode. When the sample command terminal is taken LOW T_1 turns OFF and the circuit enters its hold mode in which the sampled input voltage is held by C_1. R_1 is a bias component and R_2 protects the input stage of the op-amp. A 'start-conversion' command is then given to the converter and a conversion starts. At the end of the conversion an 'end-of-conversion' signal is sent to the control logic. The timing diagram for the system is given in Fig. 11.36 in which it is supposed that the conversion time is equal to the time occupied by 16 clock pulses.

Fig. 11.37 *Data acquisition system*

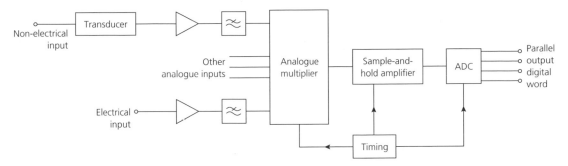

Data acquisition systems

A *data acquisition system* allows information that is represented by analogue signals to be measured. The analogue signals come from sensors, or transducers, that convert physical parameters, such as temperature, pressure, sound or flow, into voltage. The block diagram of a data acquisition system is shown in Fig. 11.37. A number of analogue signal inputs are amplified, to increase the resolution, and filtered, to remove noise and unwanted frequencies, before they are applied to an *analogue multiplexer*. This is known as *signal conditioning*. Filtering is also sometimes required to prevent *aliasing*. An *anti-aliasing filter* is a low-pass filter that has a very steep cut-off characteristic so that it almost completely removes all components at frequencies above the bandwidth of the system. The multiplexer has the same function as the digital multiplexer considered in Chapter 6 but deals with analogue signals. Each analogue signal input terminal is connected to the sample-and-hold amplifier for a short period of time and is then converted into digital form by the ADC. The analogue inputs are sampled in sequence and converted by the ADC. Some variations on this block diagram are also employed; these are:

- The analogue input signals may be connected directly to the multiplexer and the multiplexer output is amplified before it is applied to the sample-and-hold amplifier.
- There may be a sample-and-hold amplifier in each channel before the multiplexer. When sampling begins all the sample-and-hold amplifiers simultaneously go into their hold mode to freeze each input signal until it can be converted by the ADC. A simultaneous system overcomes *time skew*, an effect in which there is a time difference between each channel's reading.

EXERCISES

11.1 An ADC is to be selected for a particular application. List the main factors that should be considered in the selection of a suitable device. What is integral non-linearity and why should it be high? An ADC has an ideal differential non-linearity. How accurate is the converted output voltage?

11.2 What is meant by:
- (a) An ADC?
- (b) A DAC?
- (c) An ADC has a resolution of 16 bits. How many digital codewords does it employ?
- (d) What is meant by the range of an ADC?
- (e) List the three factors that determine the smallest change in input voltage that an ADC is able to detect.
- (f) The ideal code width of an ADC is (voltage range)/(gain $\times 2^n$), where n is the number of bits used. Calculate the ideal code width if the voltage range is 0–5 V, the gain of the circuit is 500 and n is
  - (i) 12
  - (ii) 16.

11.3 Draw the block diagram of a data acquisition system that stores information from a transducer. If the output voltage of the transducer is 0.25 V at a maximum frequency of 10 kHz, estimate the time for which the data may be continuously collected if the microprocessor has 512 bytes of directly addressable memory. Assume a 12-bit successive approximation ADC is employed.

11.4 Which specifications of a DAC determine the quality of the analogue output voltage? What differences in performance are required when the DAC is to provide:
- (a) An audio signal output.
- (b) An output that will control a heater?

11.5 A transducer is connected to a DAS.
- (a) Why should the analogue signals be amplified before being processed?
- (b) To what extent should signals be amplified?
- (c) Why is linearization required?
- (d) Why is an input filter employed?

11.6 Explain the operation of a summing amplifier type of DAC. List, and explain briefly, its sources of error. Calculate the output voltage of a 12-bit DAC with a voltage range of 12 V when the input digital word is 62H.

11.7 Calculate for a 10-bit ramp ADC with a 1 MHz clock and a voltage range of 16 V:
- (a) The voltage resolution.
- (b) The percentage resolution.
- (c) The maximum conversion time.

11.8 Explain the operation of a 4-bit successive approximation ADC. If the circuit can handle input voltages between 0 and +5 V and the clock frequency is 1 MHz, calculate:
- (a) The resolution.
- (b) The conversion time.

Table 11.6

Address	40	41	42	43	44	45	46	47	48	49	4A	4B
Data	00	11	3E	A1	D9	FB	01	05	26	5F	BD	EE

11.9 Draw the block diagram of a ramp ADC and explain its operation. Calculate the number of bits required in the DAC for the ADC to have a resolution of about 10 mV when the voltage reference is 10 V.

11.10 For the ADC in exercise 11.9 calculate the clock frequency necessary if the circuit is to perform 100 conversions per second.

11.11 An 8-bit DAC has an output voltage of 2.0 V for an input digital word of 0110 0100. Calculate the output voltage for an input digital word of:
(a) 1011 0011.
(b) 1000 1110.

11.12 The circuit shown in Fig. 11.15 is to have 0 V output for an input digital word of 000. Calculate the necessary value for R_3 if the zero offset is +0.4 V.

11.13 The $R/2R$ ladder network shown in Fig. 11.14 has series resistors of 10 kΩ and parallel resistors of 20 kΩ, and the output op-amp employs a feedback resistor of 20 kΩ. A d.c. voltage of 5 V is applied to the input terminals.
(a) Calculate the current flowing into the network.
(b) Calculate the output voltage when
  (i) only switch D_0 is in the 1 position
  (ii) only switch D_2 is in the 1 position
  (iii) all switches are in the 1 position.

11.14 Hexadecimal codes are stored in the look-up table shown in Table 11.6. They are accessed by a shift register that can be incremented/decremented between the lowest and highest addresses at time intervals of 40 μs. Once accessed, data is transferred to a DAC whose FSR is 10 V. Determine the DAC output voltage for each location in the look-up table. Draw the output waveform of the DAC.

11.15 Calculate the resolution of:
(a) An 8-bit DAC.
(b) A 12-bit DAC.
(c) A 16-bit DAC.

11.16 An 8-bit DAC has an FSR of 6 V and a full-scale error of ±0.75 per cent. Calculate the range of possible output voltages when the input digital word is 1010 1010.

11.17 An 8-bit ADC has a clock frequency of 1.5 MHz and an FSR of 10 V. Calculate:
(a) The output digital word for an input voltage of 3.6 V.
(b) The conversion time.
(c) The resolution.

11.18 A 4-bit DAC of the binary-weighted resistor type has resistor values 20, 40, 80 and 160 kΩ and the op-amp feedback resistor is 10 kΩ. Calculate:
(a) The output voltage when the input digital word is 1010.
(b) The output voltage when the 80 kΩ resistor is 15 per cent too high.

11.19 A DAC has the errors shown in Fig. 11.38. State what each error is.

Fig. 11.38

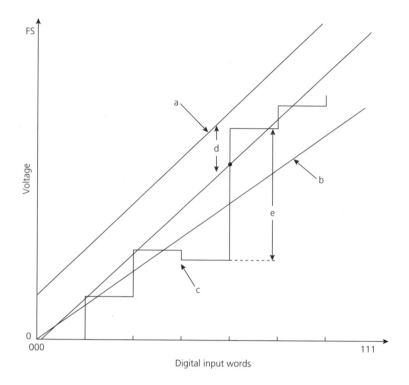

12 Digital signal transmission

After reading this chapter you should be able to:

- Understand that the connection between two ICs on a PCB may have to be treated as a transmission line.
- Recognize the different kinds of transmission line.
- Understand the need for an earthplane.
- Understand the meanings of the terms characteristic impedance and propagation delay and be able to calculate their values.
- Understand why reflections may occur on a transmission line and why they are undesirable when digital signals are transmitted.
- Calculate voltage reflection coefficient.
- Use a lattice diagram to explain and calculate reflections on a digital line.
- Use a Bergeron diagram.
- Compare the relative merits of single-ended and differential digital circuits.
- Discuss the various terminations available for digital lines.
- State the main features of the various EIA interface standards.

The ICs used in a digital system are connected together by conductors, or tracks, on a printed circuit board (PCB). The tracks have both inductance and capacitance and may therefore exhibit transmission line effects. Most PCBs use either *microstrip* or *stripline*; microstrip uses double-cladded boards and stripline employs multi-layer boards. Microstrip is easier to manufacture than stripline and hence stripline is used only when the higher packing density it gives is of importance. It is essential to assign one layer of conductor as an earthed backplane whose function is to provide a low-resistance return path for signal currents that is directly beneath the signal-carrying conductors. The only breaks in the earthplane are for necessary current paths. Because the plane is flat it provides a low-inductive path from any point on the board; hence it reduces the effects of stray capacitances by referring them to earth and so avoiding unwanted feedback paths. To reduce the length and resistance of the backplane it should have as large an area as possible and hence it usually consists of a continuous conductive plane over one surface (usually the component side) of the circuit board. The power supply line to every IC on the board should be provided with a decoupling capacitor. High-speed systems generally have the backplane constructed as a multi-layer PCB, in which two internal layers are used for earth and one layer for V_{CC}.

Lumped circuit or transmission line

At low bit rates the connection between the driving IC and the receiving IC can be regarded as a short-circuit with capacitance to the earth plane. Analysis of the behaviour of the circuit can be based upon the charging of a capacitance C, equal to the sum of the printed circuit board (PCB) track capacitance, the line capacitance and the input capacitance of the receiver IC, through the source resistance R. The risetime of the received pulse waveform is then approximately 2.2 times the time constant CR of the connection. Modern high-speed digital systems operate at high clock frequencies and this introduces problems with the interconnections between devices and sub-systems. Signals must be transmitted between devices mounted on the same printed circuit board or between different circuit boards. As the bit rate of a digital system is increased the fundamental frequency of the digital waveform increases also and the point is reached where the wavelength of the fundamental frequency, and of course all of its harmonics, is of the same order as the dimensions of the circuit board. The lengths of the conductors connecting ICs and/or different boards will then be of the same order of dimension as the signal wavelength. At these high bit rates the time taken for a signal to change from one logical state to another will be short enough for the transition to be completed before the signal has had time to reach the driven or receiving device, or circuit board. This means that the lumped-circuit approach to interconnections is no longer valid and all interconnections must be regarded as transmission lines. Unless the interconnection is *correctly terminated* some of the energy contained in a signal will be reflected by the input impedance of the receiving device and returned towards the sending device. At the sending end of the circuit the reflected signal will degrade the signal waveform. A signal propagating along a 'transmission line' is also likely to be affected by external sources of noise and interference.

Different equipments may also be interconnected. The different items of equipment may be very near to one another or they may be some considerable distance apart. Some form of cabling is generally employed as the connecting medium; this may be only about a metre in length, as is the case, for example, of a printer connected to a personal computer, or it may be tens, or even hundreds, of metres in length for data circuits. Unless the bit rate is very low these connections must always be treated as transmission lines.

If the length of a conductor is short the risetime of the input voltage transition will be greater than twice the propagation delay of the conductor. A voltage arriving at the end of the conductor will be partially reflected and the reflected wave will arrive back at the sending end while the driving circuit is still in the process of changing from one output state to another. The reflected wave is then partly over-ridden by the rising edge of the input transition and any ringing is reduced. This means that the longer the risetime of the input voltage transition the longer will be the length of line that may be employed before ringing effects must be taken into account. Any connection, no matter what its physical length, must be considered to be a transmission line if the time taken by an input voltage transition to travel over the connection and back to the sending end is longer than either the risetime or the falltime – whichever is the shorter – of the transition.

The maximum permissible length l_{max} of line that may be considered as a lumped circuit and transmission line effects ignored is given in equation (12.1):

$$l_{max} = t_r/2t_D \qquad (12.1)$$

where t_r is the risetime of the input transition and t_D is the propagation delay of the conductor per metre.

EXAMPLE 12.1

A 30 m length of cable is used to transmit:

(a) 10 kb/s signals between two points.
(b) 20 Mb/s signals between two points.

Should the connection be treated as a transmission line?

Solution

(a) Wavelength $\lambda = (3 \times 10^8)/(10 \times 10^3) = 30$ km. Since the electrical length of the cable is only 0.001λ, transmission line effects will be negligible (*Ans.*)
(b) Wavelength $\lambda - (3 \times 10^3)/(20 \times 10^6) = 15$ m. The cable is now only 0.5λ long and the connection should be treated as a transmission line (*Ans.*)

EXAMPLE 12.2

A PCB track is 10 cm in length and is to carry a signal that has a risetime of 2 ns and a falltime of 3 ns. Determine whether the track should be considered to be a transmission line.

Solution

Propagation delay $- 1/(3 \times 10^8)$ s/m. The time taken by the signal to travel up and down the track is $(2 \times 0.01)/(3 \times 10^8) = 67$ ps. This is a much shorter time than 2 ns so the track need not be treated as a transmission line (*Ans.*)

Transmission lines

Any transmission line consists of two conductors that provide a path over which current and voltage waves are able to propagate. The two conductors may consist of a twisted pair, a pair in a multi-pair cable, a coaxial pair, a PCB track and an earthplane, or even a single conductor with an earth return. In all cases the two conductors are separated from one another by a dielectric that may be air, some form of plastic, or, in the case of PCBs, Teflon or epoxy-glass. Figure 12.1 shows the twisted pair, coaxial pair, microstrip and stripline versions of transmission lines.

Fig. 12.1 *Types of transmission line: (a) twisted pair, (b) coaxial pair, (c) microstrip and (d) stripline*

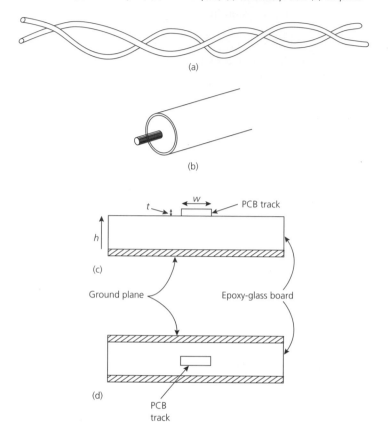

Microstrip is a PCB track that is mounted on the top surface of a dielectric material, such as Teflon or epoxy-glass, whereas stripline uses a multi-layer board with the track embedded inside the dielectric. The choice between microstrip and stripline is generally based upon the number of layers needed and the complexity of the signal path routings. Stripline is employed when the maximum packing density is wanted since increasing the number of layers reduces the lengths of the signal paths.

Any transmission line has series resistance R, series inductance L, shunt capacitance C and shunt conductance G, evenly distributed along the length of the line. Very often the two conductors have different physical dimensions and/or are made from different materials and hence their resistances are different. Most backplanes and PCB tracks have very small resistances and inductances and so their losses are negligibly small. The other types of transmission line are generally of longer length and their losses may not be negligible, particularly at higher bit rates. However, all transmission lines have negligible loss if their length is short enough and then they are known as *loss-free* lines.

The performance of a transmission line is described in terms of both its *secondary coefficients* and its capacitance. The most important secondary coefficients for a line carrying digital signals are its characteristic impedance and its propagation delay.

Fig. 12.2 *Characteristic impedance*

Characteristic impedance

The characteristic impedance Z_o of a transmission line is determined by both its phys-ical dimensions and the relative permittivity of the dielectric material. The concept of characteristic impedance is illustrated in Fig. 12.2. The characteristic impedance of a transmission line is the input impedance of a line that is terminated by its charac-teristic impedance. At all points along the line the characteristic impedance is equal to the voltage/current ratio; thus, at the sending end of the line, $Z_o = V_s/I_s$ Ω and at the receiving end, $Z_o = V_r/I_r$ Ω. A line that is terminated in its characteristic impedance is said to be *correctly terminated*. At the higher bit rates where $\omega \gg R$ and $\omega C \gg G$, Z_o is purely resistive and is given by

$$Z_o = \sqrt{(L/C)} \ \Omega \tag{12.2}$$

For a microstrip line,

$$Z_o = 87(\sqrt{\varepsilon_r} + 1.41) \log_e[(5.98h)/(0.8w + t)] \tag{12.3}$$

where h = thickness of dielectric, w = width of track, t = thickness of track and $\varepsilon_r = 5$ for epoxy-glass and 2.4 for Teflon.
   For stripline,

$$Z_o = (60/\sqrt{\varepsilon_r}) \log_e[(4b/0.67\pi)/(0.8w + t)] \tag{12.4}$$

where b = height between earth planes, w = width of track and t = thickness of track.
   Typically, the characteristic impedance of the different kinds of transmission lines are:

(a)   Twisted pair 80–125 Ω.
(b)   Multiple pair cable 62–100 Ω.
(c)   Coaxial pair 50–75 Ω.
(d)   Ribbon coaxial cable 75 or 93 Ω.
(e)   Microstrip 50–110 Ω.
(f)   Stripline 50–75 Ω.

If a PCB track is wide and near to the backplane its characteristic impedance is capacitive and nearer to 50 Ω, but if the track is narrow and further from the backplane Z_o is inductive and nearer to 100 Ω.

Propagation delay

Figure 12.3(a) shows a simple circuit in which an OR gate is connected to another OR gate by a connection that is long enough to be considered as a transmission line. The

Fig. 12.3 *Connection between two OR gates: (a) not terminated, (b) terminated and (c) equivalent circuit of (b)*

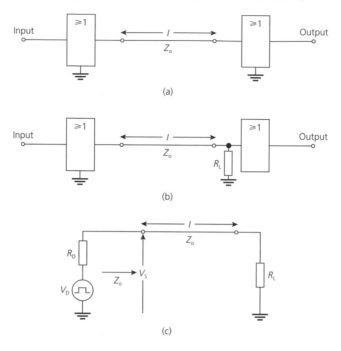

input resistance of the gate is much higher than the characteristic impedance of the line and this *mismatch* at the receiving end of the line will cause *reflections* to occur. This means that some, or all, of the incident voltage will be reflected by the receiving gate and returned back to the sending gate. To prevent reflections taking place the line must be correctly terminated. A load resistor R_L can be connected between the gate input terminal and earth as shown in Fig. 12.3(b). The value of R_L should be equal to Z_o. The input impedance of the line is then also equal to Z_o and the circuit may be re-drawn as shown in Fig. 12.3(c). When the driving gate changes state from 1 to 0, or from 0 to 1, the output voltage of the gate changes by ΔV_D and the voltage transition applied to the line is

$$V_s = \Delta V_D Z_o / (R_D + Z_o) \tag{12.5}$$

For many digital devices the output resistance R_D of the driving circuit is small compared with Z_o and then $V_s \approx \Delta V_D$. This voltage step will propagate along the line and appear across the load resistance after a time delay of T_D seconds. If the line losses are negligibly small the received voltage transition will be equal to the transmitted voltage transition. The time T_D taken for the voltage transition to travel over the line from one end to the other is known as the *propagation delay*.

The velocity with which a digital signal travels along a line is known as the *velocity of propagation* V_p and is given by

$$V_p = 1\sqrt{(LC)} \text{ m/s} \tag{12.6}$$

The propagation delay per metre is equal to the reciprocal of the velocity of propagation, i.e.

Fig. 12.4 *Line with n devices connected*

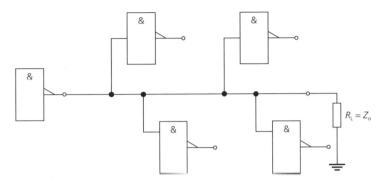

$$t_D = \sqrt{(LC)} \text{ s/m} \tag{12.7}$$

A line l metres in length has a propagation delay of $l/\sqrt{(LC)}$ seconds. For a line with air as the dielectric the velocity of propagation is equal to the velocity of light, 3×10^8 m/s. This gives a propagation delay of 3.33 ns/m. If the inter-conductor space is filled with a dielectric of relative permittivity ε_r then the propagation delay becomes $t'_D = 3.3 . \sqrt{\varepsilon_r}$ ns/m. Typical figures for propagation delay are:

(a) Twisted-pair cable 5 ns/m.
(b) Coaxial cable 5 ns/m.
(c) Stripline 7.4 ns/m.
(d) Microstrip 5.79 ns/m.

EXAMPLE 12.3

A backplane has $Z_0 = 75 \ \Omega$ and $\varepsilon_r = 4.3$. Calculate its inductance and capacitance per metre.

Solution

$$t'_D = 3.33\sqrt{4.3} = 6.91 \text{ ns/m}.$$

$$Z_0 t'_D = \sqrt{(L/C)} \times \sqrt{(LC)} = L = 75 \times 6.91 \times 10^{-9} = 0.52 \ \mu H/m \ (Ans.)$$

$$t'_D/Z_0 = C = (6.91 \times 10^{-9})/75 = 92 \text{ pF/m} \ (Ans.)$$

Device loading

Figure 12.4 shows a line of length l metres that has n devices connected to it at various points along the line. Each device has a high input resistance that does not shunt the line and an input capacitance of C_{in} pF. The total capacitance C_A added to the line is therefore $C_A = nC_{in}l$ pF/m.

The effective characteristic impedance of the line will be reduced by this added capacitance to

$$Z'_o = Z_o/\sqrt{(1 + C_A/C)} \tag{12.8}$$

and the propagation delay of the line will be increased to

$$t''_D = t'_D\sqrt{(1 + C_A/C)} \tag{12.9}$$

The loaded propagation delay must be used when deciding whether a PCB track should be considered as a transmission line.

Increasing the line capacitance has several adverse effects upon an interface:

- More current must be supplied by the driving circuit so that a higher average current is taken from the power supply with consequent increased power dissipation.
- The output impedance of the driving circuit forms a voltage divider with the characteristic impedance of the line. Increasing C decreases Z_o and this, in turn, reduces the signal amplitude at the receiver.
- The slew rate of the driving circuit will be reduced which will increase the risetime and the falltime of the voltage transitions and so limit the maximum possible data rate. The maximum allowable length of line can be determined by dividing the maximum line capacitance by the capacitance/metre of the track or cable used.

EXAMPLE 12.4

A 5 m length of line has a characteristic impedance of 68 Ω and a propagation delay of 4.8 ns/m. The line has 10 line receivers connected to it at 0.5 intervals along its length. Each receiver has an input capacitance of 7 pF. Calculate the effective value of:

(a) The characteristic impedance Z_o.
(b) The propagation delay t_D.

Solution

Total added capacitance $= 10 \times 7 = 70$ pF. Hence, added capacitance per metre $= 70/5 = 14$ pF. $Z_o = \sqrt{(L/C)} = 68$ Ω, so $L = 4624C$. $t_D = \sqrt{(4624C^2)} = 68C$, and $C = (4.8 \times 10^{-9})/68 = 70.59$ pF.

(a) $Z'_o = 68/\sqrt{(1 + 14/70.59)} = 62.2$ Ω *(Ans.)*
(b) $t'_D = 4.8\sqrt{(1 + 14/70.79)} = 5.26$ ns/m *(Ans.)*

PRACTICAL EXERCISE 12.1

Aim: to measure the characteristic impedance Z_o and the propagation delay of a transmission line.
Components and equipment: PCB with two identical parallel tracks. Pulse generator. CRO.

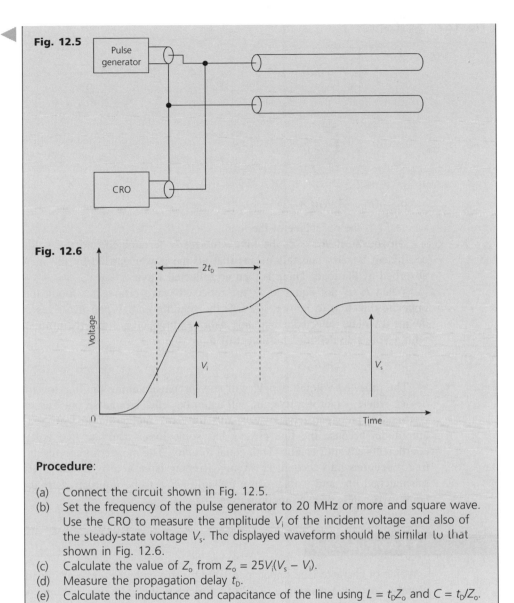

Fig. 12.5

Fig. 12.6

Procedure:

(a) Connect the circuit shown in Fig. 12.5.
(b) Set the frequency of the pulse generator to 20 MHz or more and square wave. Use the CRO to measure the amplitude V_i of the incident voltage and also of the steady-state voltage V_s. The displayed waveform should be similar to that shown in Fig. 12.6.
(c) Calculate the value of Z_o from $Z_o = 25V_i/(V_s - V_i)$.
(d) Measure the propagation delay t_D.
(e) Calculate the inductance and capacitance of the line using $L = t_D Z_o$ and $C = t_D/Z_o$.

Reflections

When a digital signal has been propagated along a transmission line and has arrived at the load presented by the driven device at the far end, it will be *reflected* and returned back along the line towards the sending end – unless the line is correctly terminated. The reflected voltage is equal to $\rho_L V_s$, where ρ_L is the *voltage reflection coefficient* at the load:

Fig. 12.7 *Digital circuit*

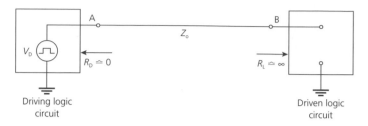

Driving logic circuit Driven logic circuit

$$\rho_L = (R_L - Z_o)/(R_L + Z_o) \tag{12.10}$$

where R_L is the resistance of the load.

Obviously, if $R_L = Z_o$ the line is correctly terminated and the voltage reflection coefficient is zero, and this means that all the power carried by the incident wave is absorbed by the load. There is then no reflected wave.

When R_L is not equal to Z_o the reflected wave propagates along the line, in the opposite direction to before, towards the sending end. When it arrives at the driving device it will be reflected again, unless the sending end is correctly terminated, this time with a voltage reflection coefficient of ρ_D

$$\rho_D = (R_D - Z_o)/(R_D + Z_o) \tag{12.11}$$

The reflected voltage $\rho_D \rho_L V_s$ will now propagate along the line towards the driven device where, when it arrives, it will again be reflected and so on. The reflected wave continually propagates backwards and forwards along the line with a diminishing amplitude, because the line always has some losses and the two voltage reflection coefficients are not (usually) both equal to unity. The most common result of not taking line reflections into account is *ringing*. Ringing is caused by multiple reflections on a mismatched line and it may cause a single pulse to be interpreted by the receiver as being two, three, or even more, pulses.

EXAMPLE 12.5

A line of characteristic impedance 100 Ω is driven by a source of impedance 50 Ω and loaded by a resistance of 120 Ω. Calculate the voltage reflection coefficient at each end of the line.

Solution

Source end: $\rho_L = (50 - 100)/(50 + 100) = -0.33$ (*Ans.*)
Load end: $\rho_D = (120 - 100)/(120 + 100) = 0.091$ (*Ans.*)

Figure 12.7 shows two digital circuits connected together by a length of line. The driving circuit has an output resistance R_D of approximately zero so that the sending end

Fig. 12.8 *Reflected waves on a digital circuit*

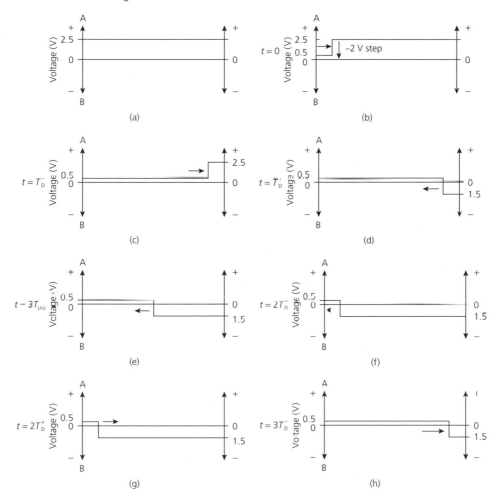

voltage reflection coefficient is −1. The driven circuit has a very high input resistance R_L so that the voltage reflection coefficient at this end is +1.

Suppose that initially there is a steady-state d.c. voltage of 2.5 V on the line as shown in Fig. 12.8(a) and that at time $t = 0$ this voltage is suddenly reduced to 0.5 V. The voltage transition is a −2 V step and this voltage step is propagated along the line towards the driven circuit (see Fig. 12.8(b)). As the −2 V step travels along the line it reduces the voltage on the line to 0.5 V. After a time slightly less than the propagation delay T_D the −2 V step has almost arrived at the open-circuit load presented by the driven circuit; this is shown in Fig. 12.8(c). At time $t = T_D$ the −2 V step reaches the open circuit and here it is completely reflected with zero change in its polarity. The total voltage at the open circuit is then equal to +0.5 − 2 = −1.5 V. The −2 V step now

Fig. 12.9 *Received voltage at end of line*

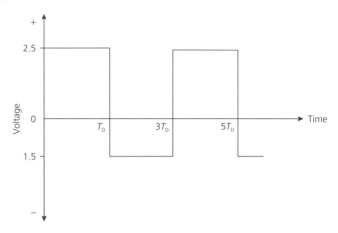

propagates back towards the driving logic circuit and this is shown in Fig. 12.8(d). At time $t = 3T_D/2$ the −2 V step has travelled half-way back to the driving circuit (Fig. 12.8(e)). Just prior to time $t = 2T_D$ (Fig. 12.8(f)), the −2 V step has very nearly arrived at the driving circuit and the voltage at all points along the line is now −1.5 V. At time $t = 2T_D$ the returned wave is completely reflected by the short-circuit load presented by the driving circuit, but this time, since $\rho_D = -1$, with a change in polarity. The resulting +2 V step causes the voltage at the sending end of the line to change abruptly from −1.5 to +0.5 V. The +2 V step now propagates along the line towards the receiving end and as it travels it increases the line voltage to its original value of +0.5 V (see Fig. 12.8(g)). At very nearly $t = 3T_D$ the +2 V step is just about to arrive at the driven circuit and the line voltage is everywhere equal to +0.5 V (see Fig. 12.8(h)). At time $t = 3T_D$ the +2 V voltage step is reflected by the open-circuit load, again with zero change in polarity, and the voltage at the input terminal of the driven circuit abruptly increases to +2.5 V. The +2 V step now travels back towards the driving circuit at the sending end of the line where it is again reflected with a change in polarity and this −2 V step reduces the line voltage to +0.5 V again and so on.

The voltage at the input to the driven circuit at the receiving end of the line oscillates continuously between +2.5 and −1.5 V, which is a voltage swing of double the original change in the voltage applied to the line. The voltage waveform at the driven circuit is shown in Fig. 12.9.

Most TTL and CMOS devices have a high input impedance that is many times greater than the characteristic impedance of the connecting 'transmission line'. Therefore, the load reflection coefficient ρ_L is usually approximately equal to 1. The output impedance R_D of the driving circuit is usually lower than the characteristic impedance of the line, giving a negative, but not unity, voltage reflection coefficient. For some devices, however, the output impedance $R_D = V_{OL}/I_{OL}$ for a high-to-low transition is much higher than Z_o, while the output impedance $R_D = V_{OH}/I_{OH}$ for a low-to-high transition is smaller than Z_o.

EXAMPLE 12.6

A 74HC20 dual 4-input NAND gate IC has $V_{OH} = 3.4$ V, $V_{OL} = 0.2$ V, $I_{OH} = 0.8$ mA and $I_{OL} = 16$ mA. Two such gates are connected together by a PCB track of characteristic impedance 70 Ω. Calculate the voltage reflection coefficient at each end of the connection.

Solution

For the driving gate $R_D = 3.4/(0.8 \times 10^{-3}) = 4250\ \Omega$ for a 1-to-0 transition and $R_D = 0.2/(16 \times 10^{-3}) = 12.5\ \Omega$ for a 0-to-1 transition. Hence,

$$\rho_D = (4250 - 70)/(4250 + 70) = 0.968\ (Ans.)$$

$$\text{or } \rho_D = (12.5 - 70)/(12.5 + 70) = -0.697\ (Ans.)$$

When the impedance of the driving device is larger than the characteristic impedance of the line the reflected voltage at the sending end of the line will be reflected with zero change in its polarity. Referring to Fig. 12.8, when the -2 V step arrives back at the driving device it is reflected without a polarity change. Figure 12.8(g) must therefore be altered as shown in Fig. 12.10(a). The -2 V voltage step travels back towards the

Fig. 12.10 *Reflections on a digital circuit*

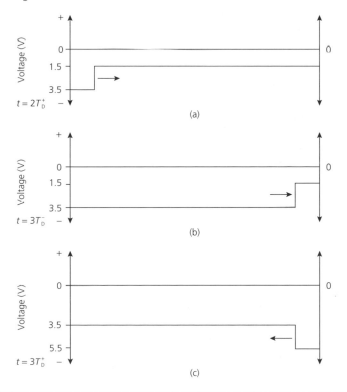

driving circuit increasing the line voltage to −3.5 V as it propagates. After time $t = 3T_D$ (Fig. 12.10(b)), the line voltage is at all points along the line equal to −3.5 V. After time $t = 3T_D$ the −2 V step is reflected, again with zero change in polarity, to increase the voltage at this end of the line to −5.5 V. The −2 V voltage step now travels along the line towards the driving circuit increasing the line voltage to −5.5 V as it travels. This is shown in Fig. 12.10(c).

Multiple reflections at both ends of the line now cause the voltage at the driving device end to 'stair-step' in −4 V steps. In practice, line losses and non-unity voltage reflection coefficients mean that the voltage steps are of progressively smaller amplitude. The signal waveform is distorted and false operation of the driven circuit may occur.

PRACTICAL EXERCISE 12.2

Aim: to investigate the action of a line that is:

(a)　Correctly terminated.
(b)　Incorrectly terminated.

Components and equipment: length of 50 Ω coaxial cable. Pulse generator. CRO.
Procedure:

(a)　Connect the output of the pulse generator to one end of the coaxial cable and set it to deliver a square waveform at 1 MHz. Connect a 50 Ω resistor between the inner and outer conductors at the other end of the cable. Connect one channel of the CRO to the output of the pulse generator and the other channel to the 50 Ω resistor.
(b)　Set the voltage of the generator to 5 V and observe the waveforms displayed on the CRO. Increase the frequency of the pulse generator in suitable steps up to 30 MHz. Comment on the output waveform of the cable.
(c)　Remove the 50 Ω resistor from the output terminals of the coaxial cable and connect the CRO across the inner and outer conductors. Reduce the frequency of the pulse generator to 1 MHz and observe the output waveform. Repeat at other frequencies up to 30 MHz.
(d)　Short-circuit the output terminals of the coaxial cable and repeat procedure (c).

The lattice diagram

Consideration of the effects of multiple reflections on a digital transmission line is made easier if a *lattice diagram* is employed. A lattice diagram has axes of line length in the horizontal direction and of time in the vertical downwards direction. It enables the progress of a voltage transition over a mismatched line to be indicated by a zigzag plot in which the gradient of the zigzag path represents the velocity of propagation on the line.

An example of a lattice diagram is shown in Fig. 12.11. The initial steady-state d.c. voltage on the line is V. When a voltage transition ΔV_D occurs at the sending end of the line the voltage applied to the line is

Fig. 12.11 *Lattice diagram*

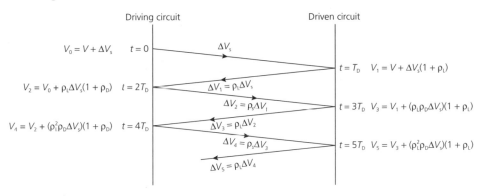

$$V_o = V + \Delta V_D Z_o/(R_D + Z_o) = V + \Delta V_s$$

The voltage transition ΔV_s propagates along the line and arrives at the other end after a time $t = T_D$. Some of the incident voltage is now reflected so that the total voltage at this end of the line is $V_1 = V + \Delta V_s(1 + \rho_L)$. If R_L is larger than Z_o, ρ_L is positive and voltage waves arriving at the load are reflected without a change in phase. The voltage at the receiving terminals of the line is then larger than the incident voltage. If, on the other hand, R_L is smaller than Z_o, ρ_L is negative and the total voltage at the receiving end is reduced. The reflected voltage $\Delta V_1 = \rho_L \Delta V_s$ propagates along the line towards the sending end where some of it is reflected. The total voltage at the sending end is now $V_2 = V_0 + \rho_L \Delta V_s(1 + \rho_D)$. If $R_D < Z_o$ then ρ_D is negative.

A fraction of this voltage, $\Delta V_2 = \rho_D \Delta V_1$, now propagates towards the driven circuit where it will be partially reflected and so on. The amplitudes of the voltages at each end of the line are given in the lattice diagram. The multiple reflections continue to propagate back and forth on the line until their amplitudes become negligibly small.

When ρ_D and ρ_L are of opposite signs the load voltage has a dampened oscillatory waveform whose period is equal to four propagation delays ($4t_D$). This oscillation is generally known as *ringing*. A typical waveform is shown in Fig. 12.12.

The final value of the voltage applied to the driven device is equal to the initial voltage plus the sum to infinity of the incident and reflected voltages at the receiving end of the line. This is a geometric series whose sum to infinity is given by

Sum to infinity = (initial term)/(1 − common ratio)   (12.12)

In the source-to-load direction the initial term is $\Delta V_s = \Delta V_D Z_o/(Z_o + R_D)$ and the common ratio is $\rho_L \rho_D$. Hence, $V'_L = \Delta V_D(R_L + Z_o)/2(R_L + R_D)$.

In the load-to-source direction the initial value is $\Delta V_I = \rho_L \Delta V_s$ and the common ratio is again $\rho_L \rho_D$. Hence, $V''_L = \Delta V_D(R_L - Z_o)/2(R_L + R_D)$.

The total voltage applied to the driven circuit is the sum of V_I, V'_L and V''_L. Therefore,

$$V_L = V_I + \Delta V_D R_L/(R_L + R_D)$$ (12.13)

Consider the system shown in Fig. 12.8 and apply the equations given in the lattice diagram for three different cases.

Fig. 12.12 *Ringing*

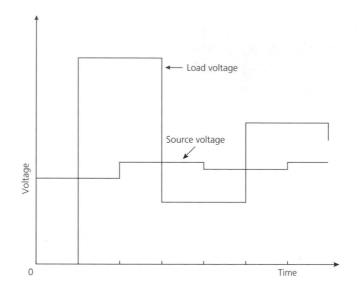

(a) $\rho_L = +1$, $\rho_D = -1$

$V_I = 2.5$ V $\quad \Delta V_s = -2$ V

$V_0 = 2.5 - 2 = 0.5$ V

$V_1 = 2.5 - 2(1 + 1) = -1.5$ V

$V_2 = 0.5 + (1 \times -2)(1 - 1) = 0.5$ V

$V_3 = -1.5 + (1 \times -1 \times -2)(1 + 1) = 2.5$ V

$V_4 = 0.5 + (1^2 \times -1 \times -2)(1 - 1) = 0.5$ V

$V_5 = 2.5 + (1^2 \times (-1)^2 \times -2)(1 + 1) = -1.5$ V etc.

as was obtained before in an alternative manner.

(b) $\rho_L = \rho_D = +1$

Assume that a -2 V transition is applied to the line.

$V_I = 2.5$ V $\quad \Delta V_s = -2$ V

$V_0 = 2.5 - 2 = 0.5$ V

$V_1 = 2.5 - 2(1 + 1) = -1.5$ V

$V_2 = 0.5 + (1 \times -2)(1 + 1) = -3.5$ V

$V_3 = -1.5 + (1 \times 1 \times -2)(1 + 1) = -5.5$ V

$V_4 = -3.5 + (1^2 \times 1 \times -2)(1 + 1) = -7.5$ V

$V_5 = -5.5 + (1^2 \times 1^2 \times -2)(1 + 1) = -9.5$ V etc.

(c) $\rho_L = +1$, $\rho_D = -0.3$

Since R_D is no longer approximately equal to 0 Ω not all of the −2 V transition will be applied to the line; $(R_D - Z_o)/(R_D + Z_o) = -0.3$ and hence $R_D = 0.54Z_o$. Now

$$\Delta V_s = -2Z_o/1.54Z_o = -1.3 \text{ V}$$

$$V_0 = 2.5 - 1.3 = 1.2 \text{ V}$$

$$V_1 = 2.5 - 1.3(1 + 1) = -0.1 \text{ V}$$

$$V_2 = 1.2 + (1 \times -1.3)(1 - 0.3) = 0.29 \text{ V}$$

$$V_3 = -0.1 + (1 \times -0.3 \times -1.3)(1 + 1) = 0.68 \text{ V}$$

$$V_4 = 0.29 + (1^2 \times -0.3 \times -1.3)(1 - 0.3) = 0.563 \text{ V}$$

$$V_5 = 0.68 + (1^2 \times (-0.3)^2 \times -1.3)(1 + 1) = 0.446 \text{ V etc.}$$

The Bergeron diagram

A lattice diagram is able to deal only with systems in which both the driver and receiver have linear output/input resistances but, in practice, these resistances vary with current and voltage. The *Bergeron diagram* provides an alternative method for predicting the behaviour of a line that can take account of non-linear resistances at both the driver and the receiver. The diagram consists of a graph of voltage plotted against current on which are drawn the current–voltage characteristics of both the driving device and the receiving device.

Consider a driver of e.m.f. V_D and resistance R_D that is connected via a transmission line of characteristic impedance Z_o to a receiver whose input resistance is R_R. On a graph of voltage plotted against current draw a straight line joining the points $V = V_D$, $I = 0$ to $V = 0$, $I = V_D/R_D$. This line is the current–voltage characteristic of the driver (see Fig. 12.13(a)).

In the steady state, when $t \to \infty$, the voltage at the receiver is $V_R = V_D R_R/(R_D + R_R)$. Draw another straight line with a slope of R_R that starts from the origin of the graph. The point where this line intersects with the first line gives the steady-state voltage and is labelled as the quiescent point. This is shown in Fig. 12.13(b).

Fig. 12.13 *Bergeron diagram: (a) driver characteristic and (b) receiver characteristic*

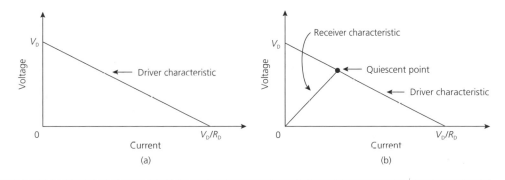

Fig. 12.14 *Illustration of how (a) the incident voltage and (b) the reflected voltage is shown on a Bergeron diagram*

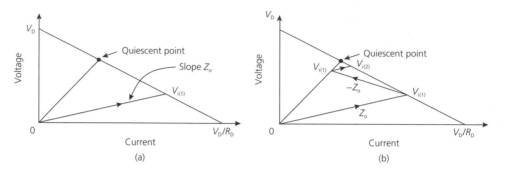

Fig. 12.15 *Using a Bergeron diagram to obtain waveforms of the voltages at each end of a line. (Courtesy of Texas Instruments)*

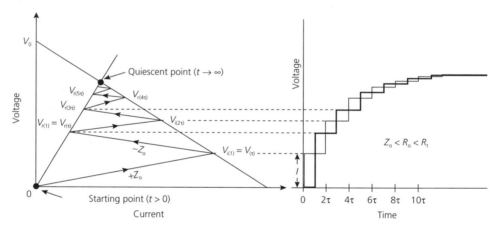

At time $t = 0$ the current and voltage on the line are zero, so the starting point of the Bergeron diagram is the origin of the graph. The incident voltage is represented by a straight line drawn from the origin with a slope of Z_o. This is shown in Fig. 12.14(a). The intersection of this line with the driver characteristic gives the value of the incident voltage at the receiver.

The first reflected voltage from the mismatched load is represented by yet another straight line, this time drawn with a slope of $-Z_o$. The intersection of this line with the receiver characteristic gives the value of the first reflected voltage to reach the driver (see Fig. 12.14(b)).

When the reflected voltage arrives at the driver it is again reflected and travels back to the receiving end of the line. At the receiver it is again reflected and returned towards the driver. The voltages travelling in both directions along the line are shown in the diagram by a series of lines drawn with slopes of $\pm Z_o$ to give a zigzag path that ends at the quiescent point. Figure 12.15 shows the line reflections for a line in which

Fig. 12.16 *Bergeron diagram when $R_L \gg Z_o$*

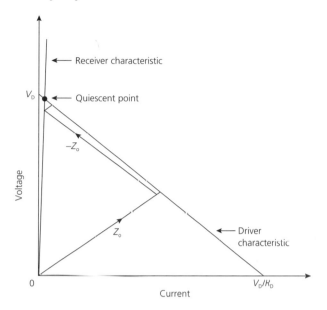

$Z_o < R_D < R_R$, and how the voltage waveform at the receiver is obtained by projecting from the Bergeron diagram.

If the receiver has a very high input resistance (nearly an open circuit), the driver characteristic will be a vertical line more or less co-incident with the voltage axis. Figure 12.16 shows the Bergeron diagram for this case.

EXAMPLE 12.7

A line driver has an output resistance of 100 Ω and delivers a 5 V pulse into a line of characteristic impedance 50 Ω. The line is terminated at the other end by a line receiver of input resistance 220 Ω. Draw the Bergeron diagram and from it plot the waveforms at the receiving end of the line.

Solution

$V_D/R_D = 5/100 = 50$ mA. A straight line is drawn between the points $V = 5$ V, $I = 0$ and $V = 0$, $I = 50$ mA to give the driver characteristic. The steady-state voltage is $(5 \times 22)/(100 + 220) = 3.44$ V; drawing a line from this value on the driver characteristic to the origin gives the receiver characteristic. On this graph straight lines with a slope of ± 50 Ω have been drawn (see Fig. 12.17(a)). From the intersects of the zigzag path with the driver and receiver characteristics, values for the voltages at the far end of the line have been projected to give the waveform shown in Fig. 12.17(b). (*Ans.*)

Fig. 12.17

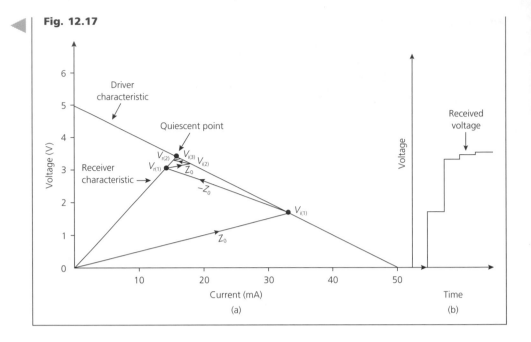

Fig. 12.18 *Overshoot*

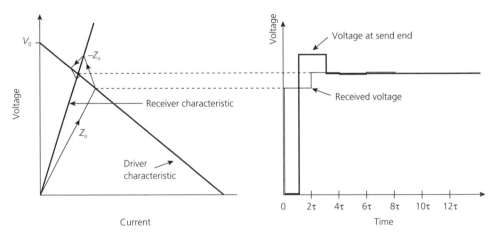

Overshoot

If the resistance of the driver is smaller than the characteristic resistance of the line the initial value of the voltage at the driver will be larger than the steady-state voltage. This extra voltage, known as *overshoot*, is shown in the Bergeron diagram in Fig. 12.18.

Fig. 12.19 *Input and output characteristics of a digital IC. (Courtesy of Texas Instruments)*

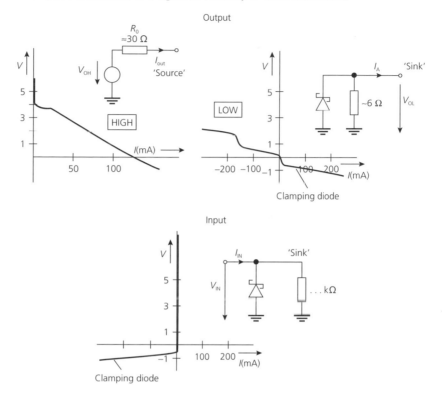

Non-linear driver and receiver resistances

The output resistance of a driver and the input resistance of a receiver are not linear quantities but, instead, vary with voltage and current. Typical variations are shown in Fig. 12.19; the current–voltage characteristic differs according to whether the device is sinking or sourcing current. Since when a driver is working it will be switching between its output HIGH and LOW states, the Bergeron diagram requires that both output characteristics and the input characteristic are drawn on the voltage–current graph. The point at which the input characteristic intersects with an output characteristic gives the steady-state voltage; there will be two quiescent points, one for output HIGH and one for output LOW. These are the starting points for the straight lines that are drawn with slopes of $\pm R_o$.

Line terminations

Ideally, to minimize ringing and/or stair-stepping on a digital line, the line should be correctly terminated at the receiving end, or at both ends if it is a bi-directional or party line.

Fig. 12.20 *Series termination*

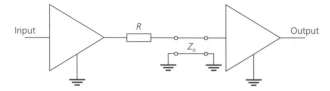

Fig. 12.21 *Parallel termination*

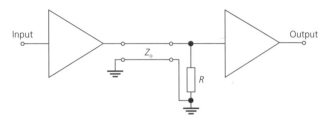

Single-ended series termination

A resistor can be connected in series with the output of the driver as shown in Fig. 12.20. There is no termination resistor at the receiver. The resistance value should be equal to the difference between the characteristic impedance of the line and the output resistance R_D of the driving device, i.e. $R = Z_o - R_D$. The method can be used only when $R_D < Z_o$. The voltage transition applied to the line will be equal to one-half of the voltage change developed by the driving device only when it changes state. If the driven device has a high input impedance the voltage transition at its input terminals will, because of a zero-polarity change reflection, be equal to the generated transition. This is then the same voltage as would be obtained if the line had been correctly terminated at the far end. Any reflections reaching the driving device will be absorbed without further reflection because the line and the effective driver resistance are equal to one another. The series termination will cause a small increase, typically 2 ns, in the propagation delay over the line. The series termination is preferred when the load is lumped at the end of the line since it gives a good noise margin, but it is not suitable for a line that is supplying several distributed loads.

Single-ended operation offers the advantages of low cost and simplicity but it is prone to radiate *radio-frequency interference* (RFI), it is prone to crosstalk from adjacent circuits, and the maximum line length is limited. A better performance can be obtained if coaxial cable is employed, but at increased cost.

Single-ended parallel termination

The parallel termination resistor method is shown in Fig. 12.21. The parallel resistor should have the same value as the characteristic impedance of the line. The resistor is not always connected to earth as shown in Fig. 12.21; it may instead be connected to either $+V_{CC}$ or to +3 V. This method of terminating a line does not increase the propagation delay of the line as much as does the series resistor method, but the increased d.c. power dissipation may present a problem.

Fig. 12.22 *Series termination of a differential circuit*

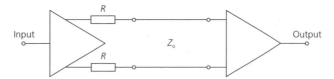

Fig. 12.23 *Differential methods of terminating a differential circuit*

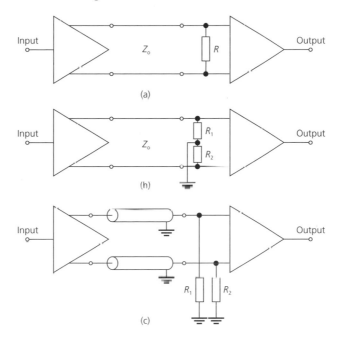

Differential transmission

It is often necessary to transmit digital signals via a high-noise environment and in such cases it is better to convert the output from the driver circuit into a differential signal, and then transmit this signal over a balanced twisted-pair line. At the receiver any induced noise will appear equally in both conductors and a differential receiver will be able to reject the noise. When differential drivers and receivers are employed it may be necessary to connect a resistor in series with both conductors to minimize reflections, as shown in Fig. 12.22. Differential transmission has the advantages of:

(a) Good noise immunity.
(b) Less RFI than single-ended.
(c) Ability to operate at higher bit rates over longer lengths of line.

A differential line driver/receiver circuit may connect the terminating resistor between the two input terminals of the receiver, as shown in Fig. 12.23(a). This circuit has a poor noise immunity and better circuits are given in Fig. 12.23(b) and (c). Each resistor should have a resistance equal to $Z_{o/2}$.

Fig. 12.24 *Thevenin termination*

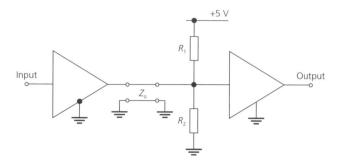

Fig. 12.25 *RC termination*

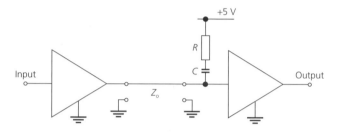

Table 12.1		
Type	*Advantages*	*Disadvantages*
Single-ended	Simple Low cost	Radiates RFI Poor noise immunity Limited line length
Differential	Good noise immunity Reduced RFI Increased speed Longer line lengths	More costly Requires twisted-pair cable

The Thevenin circuit shown in Fig. 12.24 uses two resistors, one connected to earth and the other connected to $+V_{CC}$. The mid-point d.c. voltage should not be allowed to be anywhere in between the HIGH and LOW voltage levels since this would reduce the noise margin of the circuit. Typically, $R_1 = 5Z_0/3$ and $R_2 = 5Z_0/2$, when the effective terminating resistance is Z_0 and the mid-point d.c. voltage is +3 V. This form of termination gives only a small increase in the propagation delay of the circuit but some power is dissipated.

The power dissipation problem can be reduced by the use of the *CR* termination shown in Fig. 12.25. The resistor value should be equal to Z_0 and the capacitor value should be smaller than $1/(6Z_0f_c)$, where f_c is the clock frequency. The power dissipated in R is then $V_{CC}^2 C f_c$.

The relative merits of the single-ended and differential methods of transmission are listed in Table 12.1.

Fig. 12.26 *EIA 232E connection*

Line driver   Line receiver

Fig. 12.27 *EIA 232E/ITU-T V24 voltage ranges*

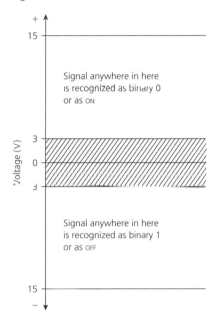

Interface standards

EIA 232E

The EIA 232 standard was first introduced in 1969 and has since gone through several revisions, with the latest, dating from 1991, being EIA 232E. EIA 232E defines a single-ended interface for the connection of a DTE to a DCE for serial transmission (see Fig. 12.26). Usually, the line is not terminated at the receiving end. The standard is also commonly employed for low-cost interfaces between a DTE and a peripheral equipment. The standard specifies that a positive voltage between 3 and 15 V should be used to indicate logic 0 for a data circuit, and ON for a control circuit; and a negative voltage of between −3 and −15 V should indicate logic 1, or OFF. The EIA 232E voltage ranges are shown in Fig. 12.27. All voltages are with reference to the signal earth line. The maximum speed of transmission is set as

$$\text{(time to change from } -3 \text{ to } 3 \text{ V)/(time duration of a bit)} \tag{12.14}$$

Fig. 12.28 *EIA 422B connection*

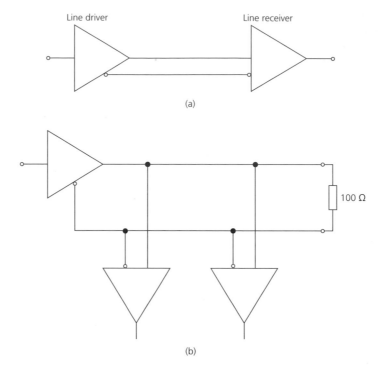

Line driver

Line receiver

(a)

100 Ω

(b)

The EIA 232E standard specifies that for baud speeds up to 8000 this ratio should be 4 μs, and for higher baud speeds it should be less than 5 μs. Other parameters specified by EIA 232E are:

(a)   The sensitivity of the receiver is ±3 V.
(b)   The output resistance of the driver must be between 3000 and 7000 Ω.
(c)   The driver must have a maximum slew rate of 30 V/μs.

As the length of a line increases, or the voltages used are larger, the time taken for a signal at the end of the line to change from 1 to 0, or from 0 to 1, increases also. The maximum bit rate is limited by both the capacitance of the line and the maximum output current of the driving device. EIA 232E quotes a maximum line capacitance of 2500 pF.

At relatively low bit rates transmission line effects may be ignored and the resistance and capacitance of the line regarded as lumped parameters. Another disadvantage of EIA 232E is that it has a poor noise immunity. The performance of the EIA 232E standard is perfectly adequate for many applications and it is widely employed. EIA 232E is equivalent to the ITU-T recommendations V24 and V28.

EIA 422B

The EIA 422B standard, equivalent to the ITU-T V11 recommendations, defines the electrical characteristics of a balanced or differential interface (see Fig. 12.28). It specifies that line drivers and line receivers are employed for bit rates of up to 10 Mb/s over

Fig. 12.29 *EIA 422B voltage ranges*

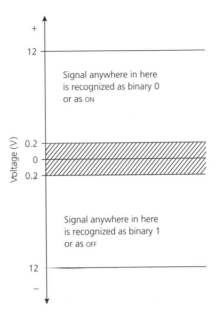

Signal anywhere in here
is recognized as binary 0
or as ON

Signal anywhere in here
is recognized as binary 1
or as OFF

Fig. 12.30 *EIA 423B connection*

distances of up to 12 m, or for bit rates of up to 100 kb/s over distances of up to 1200 m. EIA 422B caters for one line driver and up to ten receivers. The voltage ranges used are shown in Fig. 12.29. It is an electrical only standard that specifies an unbalanced uni-directional point-to-point interface. It is similar to EIA 232E.

EIA 423B

The EIA 423B standard is shown in Fig. 12.30 and is equivalent to the ITU-T V10 recommendations. It defines the electrical characteristics of an unbalanced digital interface circuit. The specification can give longer line lengths than the EIA 232E standard because it uses maximum driver voltages of only ±5.4 V and a differential receiver sensitivity of ±200 mV. The transmitted pulses are pre-shaped to reduce the amplitudes of the high-frequency components in the waveform. An EIA 423B interface can provide bit rates of up to 100 kb/s over distances of up to 10 m, and 1 kb/s for distances of up to 1200 m.

Fig. 12.31 *EIA 485 connection*

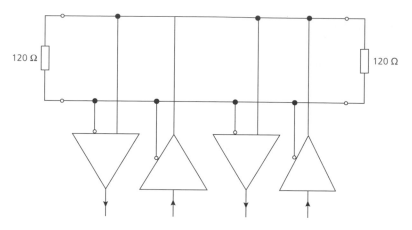

EIA 485

The EIA 485 (ISO 8482) standard has been based upon EIA 422 but it has been modified to allow a multi-point interface to be employed. Figure 12.31 shows its basic arrangement. The standard can handle up to 32 line drivers and line receivers connected to a common line. The maximum line length is not defined but it is generally taken to be 1200 m with a maximum data rate of 10 Mb/s.

EIA 530

The EIA 530 is a standard that is intended to allow interfaces to be operated in the 20 kb/s–2 Mb/s region. It is electrically similar to EIA 422.

EIA 562

This is a standard that has been developed for single-ended interfaces operating at bit rates of up to 64 kb/s. It is an electrical only standard that specifies an unbalanced uni-directional point-to-point interface. It is similar to EIA 232E.

EIA 694

The EIA 694 interface standard specifies that the line drivers and receivers are able to operate at bit rates of up to 512 kb/s. The driver is similar to that specified in EIA 423B and the receiver is similar to the one specified in EIA 232E. Both the noise margin and the maximum line capacitance are reduced from the figures quoted in EIA 232E. The standard gives a simple high-speed unbalanced interface.

Table 12.2

	EIA 232E	EIA 422A	EIA 423B	EIA 485	EIA 562	EIA 694
Operation	Single-ended	Differential	Single-ended	Differential	Single-ended	Single-ended
Line length	2500 pF	1.22 km	1.22 km	1.22 km	2500 pF	1100 pF
Data rate (b/s)	20 k	10 M	100 k	10 M	64 k	512 k
Voltage range (V)	$\pm5-\pm15$	$\pm2-\pm5$	$\pm3.6-\pm5.4$	$\pm1.5-\pm5$	$\pm3.7-\pm13.2$	$\pm3-\pm6$
Receiver sensitivity (V)	±0.2	±0.2	±0.2	±3		±0.2

Comparison of EIA standards

Table 12.2 gives the main parameters of the different EIA interface standards.

Line drivers and receivers

The function of a line driver is to convert TTL or CMOS signals to the voltage levels that are specified by the interface standard that is being used. An EIA 232E line driver, for example, will convert the +5 V logic 1 level of a TTL and some CMOS systems into the −5 to −15 V required by the EIA 232E specification.

Data sheets

The data sheet of a line driver or receiver follows a similar format to the data sheets of logic devices. A general description and feature section is followed by the pinout, logic symbol, and a functional diagram. Next, the absolute maximum ratings, the electrical characteristics and the switching characteristics are found; the latter two are generally split into two sections, the driver section and the receiver section. The last part of a data sheet then gives information on the measurement of parameters and possible applications, as well as typical performance characteristics. A volume of data sheets divides the devices into groups according to the EIA specification for which they have been designed. Some devices can be used as either drivers or receivers and these are known as transceivers. Four examples are listed below.

185 EIA 232E: three drivers and five receivers. The logic symbol of this device is shown in Fig. 12.32.
196 EIA 232E: five drivers and three receivers.
941 EIA 422: dual differential drivers and receivers.
1177 EIA 485: dual drivers and receivers.

Fig. 12.32 *Logic symbol of a line driver/receiver*

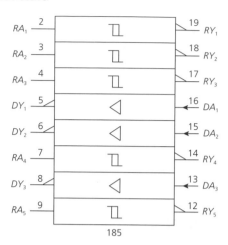

12.1 The output of a digital circuit is connected by a transmission line of characteristic impedance 50 Ω and velocity of propagation 2×10^8 m/s to its load. If the line is 1 m in length and the load impedance is 100 Ω sketch the load voltage when the signal applied to the line is a 1 V positive step.

12.2 A 50 Ω transmission line is open-circuited at one end and driven by a digital circuit at the other. The propagation delay of the line is 100 ns. In the steady state the line voltage is zero. At time $t = 0$ a voltage transition of +2 V is applied to the line. Determine the voltage at the open circuit after 110 ns.

12.3 An RC termination for a digital line of characteristic impedance 80 Ω has a 12 MHz clock waveform applied to it.
(a) Determine a suitable value for the capacitance C.
(b) Assuming the receiving device uses a 5 V power supply calculate the power dissipated in the termination.

12.4 A microstrip line is formed by a 1 mm thick, 0.5 cm wide track on a PCB that is 0.75 cm thick and of relative permitivity 4.9. Calculate the characteristic impedance of the line.

12.5 The connection between two gates having $V_{OH} = 0.25$ V and $I_{OH} = 0.45$ mA is made by a line of characteristic impedance 80 Ω. The steady-state voltage on the line is 2 V and a negative 2 V transition is applied to the sending end. Draw the lattice diagram of the system.

12.6 Explain, with the aid of sketches, the various ways in which a digital line may be terminated. List, and discuss, the relative merits of the methods shown.

12.7 A train of rectangular pulses has an amplitude of 1.5 V, a pulse width of 100 μs and a pulse repetition frequency of 1 kHz. What is the approximate bandwidth needed to transmit this signal and retain a more or less rectangular shape?

12.8 A bus driver is 6 m long and has 25 receivers connected to it, each of which has an input capacitance of 6 pF. The line has a propagation delay of 5 ns/m and a characteristic impedance of 70 Ω. Calculate the bus delay and the characteristic impedance when the 25 receivers are connected at regular intervals along the line.

12.9 Calculate the minimum length of:
(a) twisted-pair line
(b) bus line
that must be regarded as a transmission line if the risetime of the signal is 2 ns. For twisted-pair $t_D = 5$ ns/m and for bus line $t_D = 20$ ns/m.

12.10 A 5 V voltage source of 75 Ω resistance is connected to a line of characteristic impedance 50 Ω. The line is terminated by a non-linear resistance whose current–voltage characteristic is $I = 2V^2$ mA. Use a Bergeron diagram to determine:
(a) The steady-state voltage on the line.
(b) The incident voltage at the load.

12.11 Draw Bergeron diagrams for:
(a) $R_D < R_o$ and $R_R =$ open circuit.
(b) $R_R < R_D < R_o$.
(c) $R_D = R_o < R_R$.
(d) $R_R = R_o > R_R$.
(e) $R_D = R_o = R_R$.

12.12 A driver provides a voltage pulse of 4 V to a line.
(a) Draw waveforms to show the voltages at each end of the line a short time after the first reflection from the load if
(i) $R_D = R_o = 3R_R$
(ii) $R_D = R_o = R_R/3$.
(b) For how long does each step last?
(c) Calculate the steady-state voltage on the line for each case.

12.13 A digital circuit is to send signals over the pair of conductors that link two circuit boards. The propagation delay is 4.92 ns/m and the minimum transition time is 1.6 ns. Calculate the maximum line length that may be used without reflections being taken into account.

12.14 A digital line driver has an output resistance of 100 Ω and delivers a 5 V pulse to a line of characteristic impedance 50 Ω. The line is terminated by a line receiver of input resistance 22 Ω. Draw the Bergeron diagram of the system and use it to obtain the waveforms at both ends of the circuit.

13 Digital system design

After reading this chapter you should be able to:

- Understand the meaning and use of specifications in electronic engineering.
- Design digital circuits, select suitable components for the design, and then implement and test the circuit.

A digital electronic system consists of a number of ICs that are connected together to perform a particular function that is defined by the equipment specification. To simplify the design process the initial definition of the project is used to produce a block diagram of the required system. Each of the blocks shown in the diagram will represent a separate circuit, e.g. a counter or an analogue-to-digital converter, or a sub-system. The operation of each block must be defined in some way. As each circuit, or sub-system, is designed separately its operation should be simulated using an electronic design package, such as *Electronics Workbench*. The simulated circuit can be tested to check that it works correctly. Once the operation of the simulated circuit is satisfactory, the circuit should be breadboarded and again tested and, if found necessary, modifications made. Once a satisfactory breadboard circuit has been made a prototype circuit can be built and tested to check that its operation is as expected. If it is not the circuit design will need to be modified and then re-tested until it does work correctly. This method of design is not suitable for VLSI circuits because of their sheer complexity and the need to implement them directly in silicon. Such circuits are therefore designed using a *computer-aided design* software package in which software tools are employed to model and evaluate designs and layouts.

Specifications

A specification is a detailed description of the required characteristics of a material, component, circuit, system or item of equipment. The specification explains in detail what the item concerned should be able to do in response to all kinds of inputs, and under what environmental conditions this performance is to be guaranteed. Any restrictions, such as physical size and/or weight, power consumption, etc., are stated as well. The usual format for a specification is:

- A description of what the equipment is to do and its intended applications.
- Electrical data such as voltages and frequencies.
- Power requirements, e.g. mains and/or battery operated.
- Environmental data, such as the working temperature range.
- Mechanical data such as the physical dimensions and weight.

The specification ought to state clearly exactly what the system is required to do; it should concentrate on the functions to be performed rather than how those objects are accomplished.

A typical specification for a digital system could be one for a drinks vending machine. The machine should provide a choice between eight different drinks some of which, such as tea and coffee, are to be made on demand. Hot drinks are to be delivered at a specified temperature after cold water has been heated in the machine. Any combinations of coins are to be accepted up to a total value that can be reset as, and when, it becomes necessary to increase prices. No change is to be given.

When the correct sum, or more, has been inserted into the machine all the drink-choice buttons should light up and become active. When a particular button is pressed: (i) all the inserted coins should fall into the machine, (ii) all the drink-choice buttons should become inactive, (iii) a cup should be made to fall into the correct position ready to receive the chosen drink, and (iv) the requested drink should be delivered in the quantity that very nearly fills the cup. After a short time delay the machine should be ready for the next customer.

A specification like this could be met in a number of different ways and it would demand both electronic and mechanical design. Factors such as cost and reliability are taken into account in the choice between competing technologies. Essentially, the choice is between SSI/MSI devices, PLDs or a microprocessor-based system. The use of SSI/MSI devices is relatively easy and it requires no programming equipment or expertise. The use of PLDs reduces the number of ICs used in the system, and hence increases the reliability, but some programming is necessary. A PLD design will be physically smaller but it may dissipate more power and it may not be as fast. Microprocessor control is usually best for any system that requires a function to be controlled via an input/output port, but again software is necessary. A microprocessor solution to a design problem will be slower than a hardware solution, because of the time taken to execute the instructions, and it will prove to be more expensive for a small system. On the other hand, microprocessor systems are more versatile, since any change in the system's requirements will need a change only in the programming. Microprocessor systems are outside the scope of this book.

Besides the relative technical merits of the alternative technologies, the costs of each technology and the availability of devices are important considerations. 74LS devices, for example, have been used for several years but are presently being superseded by devices in the various 74HC CMOS logic families, for reasons discussed in Chapter 3, but 74LS still offers by far the widest range of devices. Texas Instruments have promised that LS devices will remain available for the foreseeable future. Another important consideration is reliability; this is improved by reducing the number of devices used in a design and is a pointer towards using PLDs.

The specification of a piece of equipment may be used by a potential buyer to evaluate and compare different equipments from different manufacturers in order to decide which one is best suited, on both technical and economic grounds, to his/her application. Sometimes an equipment specification is derived as the result of discussions between the manufacturer and the customer. The customer may request that certain materials/components are used and perhaps decree exactly what the equipment must be able to do. In many cases British Standards are quoted in parts of a specification.

A *test specification* is written by the system designer once the system has passed the prototype stage. It details the measurements and tests that should be carried out on a piece of equipment. It will give details of the test points to be used and the measured values that ought to be obtained at those points. A test specification is used by the manufacturer's test department to test each piece of equipment as it is made. It often also specifies the test instruments that ought to be used, and when this is so it is important that the specified instruments are used, otherwise different instrument loading may invalidate the test results. The testing instructions given in the handbook for the equipment, when produced, may be the same as the test specification or at least be based upon it.

The test specification should contain detailed instructions on how to set the system up for test, how to perform the tests, and the expected results if the system is working correctly. Often included is a checklist so that items may be ticked off as each test is completed. It is best, and faster in the end, to test each block as it is designed/manufactured rather than wait until all the blocks have been assembled to produce the complete system.

Design of a digital system

The design of a digital system is based upon the specification that has been obtained from a request by a customer, from market research, or from a circuit designer. The designer must analyse the specification to determine whether it is possible to make a circuit/system that can perform the specified function(s). If the specified requirements seem to be reasonable then the various ways in which it may be possible to implement the circuit/system will need to be evaluated to select the method that appears to be the best.

The essential steps to be taken in the design of a digital electronic system are:

- Define the problem based on the specification. There may be a full specification to be satisfied or the design may have to be based upon a vague idea, or very often some combination of both.
- Draw a block diagram of a possible system that will meet the specification. The block diagram should identify all the inputs and output(s) to/from the system. How many blocks are shown depends upon the complexity of the required system.
- For each block decide on which seems to be the most suitable technology and/or circuit to employ. The block, or modular, approach to system design allows complex systems to be assembled from relatively simple modules. Each module can be separately designed, built and tested before being assembled into the complete system. Modules can be built on their own PCBs and this will allow the PCB boards to be designed and manufactured separately.

- Should SSI/MSI ICs, PLDs or a microprocessor be used? Or some appropriate mixture of these technologies? The choice will be based upon a knowledge of the characteristics of the various technologies and the system requirements. Important factors are:
  - the speed of operation
  - power dissipation
  - cost
  - availability of components and devices.

  Once one of these options has been decided upon it will then be necessary to decide on the particular circuit(s) to be employed and to select the components.
- Design the hardware and, if a PLD or microprocessor is to be used, design the required software.
- The chosen circuit can then be simulated using a software package on a PC and tested. If the circuit does not work as expected it can first be checked to ensure that no wiring errors have been made. If not it will be necessary to re-design a part, or all, of the circuit or even, in some cases, scrap the circuit and try another approach. Once the simulated circuit works correctly a circuit should be built on a breadboard and tested.
- Once the breadboarded circuit works correctly a prototype can be built. When the prototype has been tested the test results should be evaluated against the design criteria. If necessary, the design can be modified to improve the performance of the circuit and then be re-tested. Sometimes it may prove necessary to revise the specification. This, of course, should be done only with the full agreement of all concerned.
- When all the prototypes are satisfactory the various modules can be connected together to form the complete system. This must then be tested to check its performance against the specification.
- Design a suitable test procedure and specify suitable test points and test equipment. Then measure the voltages etc. at each test point and record the values obtained.
- An essential part of any designed circuit/system is documentation that informs people how the system works, gives constructional details, gives test procedures which should include expected waveforms at the test points when specified test signals are applied to specified input terminals, etc. This information should be written down before it is lost and even the designer is no longer certain of how he/she arrived at the finished design. British Standard symbols and terms should be employed to ensure that any reader will understand the printed information. The documentation should include such items as the circuit/system specification, a schematic diagram, a component layout diagram, a parts list and the test procedure. The parts list gives details of all the components used in the system. Components of the same type are often listed together to make the list easier to follow.

Design exercises

In this concluding section of the chapter a number of design exercises are given. These do not include random logic designs (see Chapter 3), but designs that involve the use of MSI devices. Once a design has been completed it should be simulated and/or breadboarded and tested. Once a designed circuit is working correctly it could then be implemented using either a PAL or a GAL (see Chapter 10), if the necessary programming facilities are available. Exercises 13.1 through to 13.15 have been provided with a suggested solution in Chapter 14, but access to these solutions requires the reader to have Electronics Workbench software and to access the web site ftp://ftp.awl.co.uk/pub/awl-he/engineering/green. Each of these exercises refers to the appropriate Electronics Workbench file.

If possible, a design should be followed through from the initial concept and block diagram, through simulation and breadboarding, to construction and, lastly, testing.

EXERCISE 13.1

A factory process requires that two tanks are each filled with different liquids that are then pumped into a mixer tank; here they are mixed thoroughly before being heated to a pre-determined temperature. The heated mixture is then pumped out of the mixer tank for use in some further process. Design a suitable system.

Suggested solution

The two tanks must each be fitted with sensors that can detect the level of liquid in the tank. Two sensors are needed for each tank: one to give a LOW output when the tank is nearly empty and the other to give a HIGH output when the tank is nearly full. The signals from these sensors can then be used to control the opening and closing of valves that let liquid into, or out of, a tank. When each tank is full, signals are required to operate two pumps to pump the liquids from their separate tanks and into the mixer tank. When the liquids are in the mixer tank a signal is required to operate the mixer for a fixed length of time and then another signal must turn on the heater. A temperature sensor is needed in the mixer tank to monitor the temperature of the liquid. When the correct temperature has been reached the sensor should produce a HIGH that will control both the turning off of the heater and the operation of another pump to remove the mixed and heated liquids from the mixer tank.

The block diagram of one possible solution is shown in Fig. 13.1. Using MSI devices design the system. Use LEDs in series with 270 Ω resistors to simulate the signals to the two valves, the three pumps, the mixer, and the heater. Simulate the liquid level and temperature sensors with two-pole switches connected to a HIGH and a LOW. Build the circuit and test its operation.

Suggested MSI devices are two 74HC112A dual J-K flip-flops, one 74HC138 3-to-8 line decoder and one 74HC151 8-to-1 line multiplexer.

A solution is given in Electronics Workbench file ade 13.1.ewb.

EXERCISE 13.2

Design a system that will monitor four doors and indicate those doors that are open.

Suggested solution

A possible block diagram of the required system is given in Fig. 13.2. Each of the four doors should be fitted with a switch that will generate a HIGH when the door is open and

Fig. 13.1

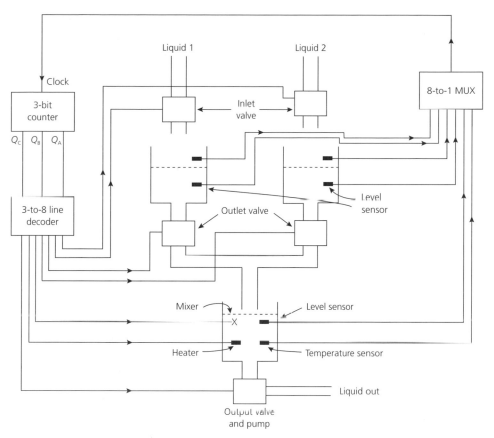

Fig. 13.2

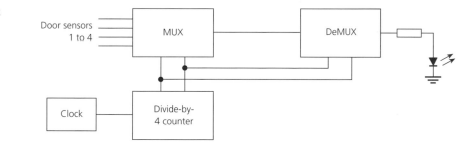

a LOW when it is shut. The Q_A and Q_B outputs of the counter select the same numbered inputs of the multiplexer and outputs of the demultiplexer at any instant in time. The clock applied to the counter causes the four inputs and outputs to be selected in sequence. A selected output will be LOW only if the corresponding input is HIGH, i.e. that door is open. Design the system and build the circuit. Test that it works correctly.

Suggested MSI devices are the 74HC112A dual J-K flip-flops, the 74HC138 3-to-8 line decoder, and the 74HC153 dual 4-to-1 multiplexer.

A possible solution to the problem is given in Electronics Workbench file ade 13.2.ewb.

EXERCISE 13.3

Design a system that will indicate which of three contestants in a quiz game presses their answer button first. The first button pressed is to be indicated by a lamp turning ON and this must inhibit the other two contestants. The quiz compère is to have a reset button that he/she can use to reset the system.

Suggested solution

The Boolean equations for the required circuit are:

$$a = AbcH + aH, \ b = aBcH + bH \text{ and } c = abCH + cH$$

where a capital letter denotes a contestant's button and a lower case letter denotes an LED or lamp. The equations can be implemented on a PAL or a GAL if one is available along with a programmer. Otherwise SSI logic can be used. Construct a circuit to implement the system equations, and then test the circuit to ensure correct operation.

Suggested SSI devices are two 74HC21 dual 4-input AND gates, one 74HC08 quad 2-input AND gate, one 74HC32A quad 2-input OR gate, and one 74HC04 hex inverter.

An SSI logic solution is given in Electronics Workbench file ade 13.3.ewb.

EXERCISE 13.4

Design an egg timer. The egg timer should start when a button is pressed and a buzzer should sound when a selected time has elapsed. Design the timer to give a required delay in 1 minute steps.

Suggested solution

The basic block diagram of one solution to the design problem is shown in Fig. 13.3. The 300 Hz voltage source can be obtained in various ways but one of the easiest is to use a 555 timer IC. The three counter circuits can be made using flip-flops and/or MSI devices. Possible devices that could be used include the 74HC112A dual J-K flip-flop IC, the 74HC174 hex D flip-flop IC and the 74HC161 counter IC.

Fig. 13.3

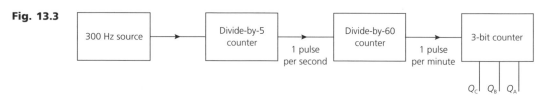

(a) Design and build the 555 timer pulse generator and get it running at approximately 300 Hz. A precise frequency is not necessary since the boiling of an egg need not be accurate to fractions of minutes.

(b) Design and build the divide-by-5 circuit. Connect its input to the output of the 555 timer circuit and confirm its correct operation.

(c) Design and build the divide-by-60 circuit. Connect the output of the divide-by-5 counter to its input terminal and confirm its correct operation. The output of this counter should then be approximately one pulse per minute.

(d) Design and build the 3-bit counter. Decide what time delays can be obtained by taking the output of the system from
   (i) the Q_A output
   (ii) the Q_B output
   (iii) the Q_C output.
   Check that the anticipated time delays are obtained.

(e) Add a switching arrangement to the output of the circuit so that any one of the three outputs can be selected.

(f) Determine a way in which some other timings can also be obtained. State the proposed method but do not implement it.

(g) The timer is required to always start from zero time. Does it? If it does not then it is necessary to modify the circuit so that all stages can be cleared before a new timing begins when the start button is pressed.

EXERCISE 13.5

Design and build a counter that counts from 0 to 999.

Suggested solution

A divide-by-1000 counter can best be made by cascading three decade counters. The block diagram of the suggested arrangement is shown in Fig. 13.4.

Decade counters are not available in the HC logic family and hence the counters could be constructed using either 74HC112A dual J-K flip-flop ICs or 74HC174 hex D flip-flop ICs. A rather complex circuit will then result, however, and perhaps a better alternative is to use LS devices. Then three 74LS90 decade counters can be used. The 74HC404 12-bit non-synchronous counter is also available in the HC logic family. If

Fig. 13.4

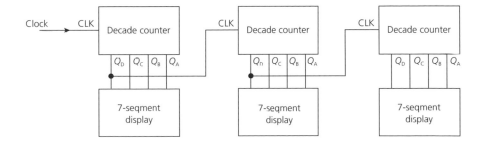

the Q outputs of each stage are connected to seven-segment decoder/drivers a hexadeci-mal count is obtained from a very simple circuit. Investigate whether the counter can be modified to provide the required decade count.

Having decided which devices to use design the divide-by-1000 counter. Connect the Q_A, Q_B, Q_C and Q_D outputs of each counter to the inputs of a seven-segment decoder/driver IC. Connect a clock to the input terminal. Set the clock frequency to a low value and then switch ON. Observe the visual display and check that the circuit works correctly.

A suggested solution is given in Electronics Workbench file ade 13.5.ewb.

EXERCISE 13.6

Design a toyshop window display in which three lights, RED, GREEN and BLUE, go ON and OFF following a preset sequence at 10 s, and multiples of 10 s, intervals.

Suggested solution

The block diagram of one possible circuit is shown in Fig. 13.5. A 1 Hz pulse source is required. A 1 Hz generator could be built or a higher frequency source could be divided down to 1 Hz. The 1 Hz pulse train must be divided by 10 to produce one pulse every 10 s at the input to the shift register. The Q outputs of the shift register are connected to LEDs, or, via relays, lamps, of different colour which will glow visibly when that Q output goes HIGH.

Start with the simplest sequence to implement, i.e. RED; RED and GREEN; RED, GREEN and BLUE; RED etc. When the third state, RED, GREEN and BLUE, is reached the circuit must reset to begin another sequence. Hence, connect the Q outputs of the shift register to a logic circuit that will generate a clear signal that can then be applied to the counter.

Design the circuit using LEDs in series with resistors to act as the lights. Suggested devices are one 74HC163 synchronous counter, one 74HC174 shift register, one 74HC00 quad 2-input NAND gate and one 74HC20 dual 4-input AND gate. Test the circuit and check that it gives the specified lighting sequence.

Fig. 13.5

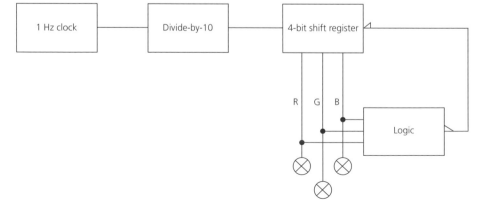

Re-design the circuit to give the lighting sequence RED, RED and GREEN, GREEN, GREEN and BLUE, BLUE, BLUE and RED, RED, and so on. Modify the previous circuit to give this new lighting sequence and check that it works correctly.

Suggested solution

A possible solution to this problem is given in Electronics Workbench file ade 13.6.cwb.

EXERCISE 13.7

Design a circuit to monitor the fluid level in each of eight tanks. If any level reaches the maximum allowable a signal must be generated that can be applied to a microprocessor (or microcontroller), which has been programmed to take corrective action. To reduce the number of microprocessor ports that are occupied by this task use an encoder to reduce the number of signals to be monitored by the microprocessor.

Suggested solution

The block diagram of a possible design is shown in Fig. 13.6. The eight fluid level sensors are connected to the eight inputs of an encoder and the three encoder outputs are connected to a microprocessor. The microprocessor should receive an interrupt signal when an encoder output becomes active so that it can start corrective action to reduce the over-high liquid level.

If the encoder used has active-HIGH inputs the level sensors employed must generate a HIGH voltage when operated; conversely, if the encoder has active-LOW inputs the sensors must also be active-LOW devices. The suggested device for the design is the 74HC148 8-to-3 priority encoder (the only one in the HC logic family).

Build the circuit using LEDs with series resistors to indicate the logical state of the encoder outputs. [A more complex project involving both digital electronics and microprocessors is clearly possible here.] Unless eight liquid containers and eight level sensors are available use switches connected to 0 and 5 V to simulate the sensor output voltages.

Check the correct operation of the circuit and determine its truth table. The GS output terminal of the encoder can be used to provide the interrupt signal for the microprocessor; is it active-HIGH or active-LOW?

Fig. 13.6

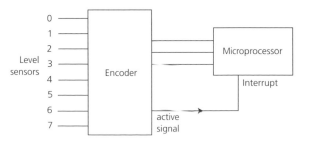

The IC used is a priority encoder and not an encoder. What does priority mean in this context? Does it affect the operation of the circuit in any way?

A possible solution is given in Electronics Workbench file ade 13.7.ewb.

EXERCISE 13.8

Design a simple burglar alarm system. The system is to monitor the three windows and two doors in a room. When the system is turned on the opening of any door or window is to produce an audible alarm. The block diagram of the system is shown in Fig. 13.7.

Suggested solution

Build the circuit and test its operation. The suggested devices are two 74HC32 quad 2-input OR gate ICs and one 74HC00 quad 2-input NAND gate IC.

The circuit can be extended to give an indication of which door/window is open. Two alternatives are:

(a) Use a separated LED to monitor each door; the LEDs could then be numbered 1 through to 5.
(b) Use a seven-segment display.

If the latter choice is made it will be necessary to employ a decimal-to-binary (or decimal-to-BCD) converter to convert a high input into a 3-bit signal for the seven-segment display. There is no suitable device in the 74HC logic family but there is in the 74LS family. Select a suitable device and modify the circuit.

For one solution to this problem see Electronics Workbench file ade 13.8.ewb.

EXERCISE 13.9

A car park that contains 16 spaces has one entrance and one exit. A running count is to be kept of the number of cars entering and leaving the car park so that the number parked inside is known. When the car park is full a FULL display is to be turned ON and the entrance barrier should be kept down to prevent any further access. Design the system.

Fig. 13.7

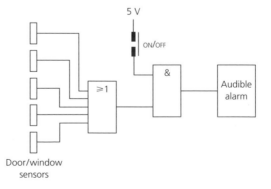

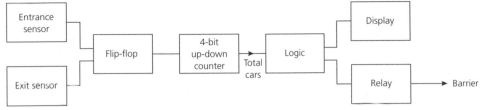

Fig. 13.8

Suggested solution

Sensors are required at both the entrance and the exit to monitor the passage of cars into and out of the car park. Infra-red sensors can be employed that generate a positive pulse whenever a beam is broken. These pulses can then be applied to an up-down counter which can keep a count of the number of cars inside the car park. The basic block diagram of the system is shown in Fig. 13.8.

In the 74HC family there are two devices, the '191 and the '193. Look at the data sheets of the two devices and decide which to use.

A possible solution to this design problem is given in Electronics Workbench file ade 13.9.ewb.

EXERCISE 13.10

The count of a 4-bit counter is to be displayed on a seven-segment display but after each digit has been shown the display is to reset to 0 before the next digit appears. Design the circuit.

Suggested solution

A solution is given in Electronics Workbench file ade 13.10.ewb.

EXERCISE 13.11

A stall at a garden fête wants to offer a car racing game in which six toy cars are simultaneously released at the high end of a wooden ramp with six lanes. A prize is offered to the 'owner' of the car that reached the finishing line first. Electronic timing of the race is required to deter the vicar throwing cakes around when his car loses. Design, build, and test the circuit.

Suggested solution

When the 'cars' are simultaneously released at the top of the ramp a microswitch closes to start six counters. When a car reaches the finishing line another microswitch is operated (one in each lane) that stops the counter associated with that lane. The count can be used to determine the time taken by each car to reach the finish. The suggested device to be used is the 74HC163 synchronous 4-bit counter whose ENP and ENT terminals can be used to start/stop the count of each counter when required. The precision of the timing depends upon the counting speed and this, in turn, depends upon the clock frequency. Depending upon the clock frequency chosen and the time it takes the 'cars' to cover the distance it may be necessary to use more than one counter connected in cascade.

A solution to this exercise is simulated by Electronics Workbench file ade 13.11.ewb.

EXERCISE 13.12

The number of vehicles travelling along a road in one direction is to be counted. Each time a vehicle passes over a pressure sensor fixed to the surface of the road a HIGH voltage is generated. When 255 vehicles have passed the point a pulse is to be applied to a divide-by-15 counter. The count of the counter is to be displayed by a seven-segment display. Design, build and test the circuit.

Suggested solution

Suggested devices to be used are the 74HC164 8-bit parallel-out shift register, the 74HC21 dual 4-input AND gate, and the 74HC163 4-bit synchronous counter. Calculate the maximum count of the circuit.

Extend the maximum count by arranging that when the display is F the counter is cleared and a pulse is inputted to another 4-bit counter whose outputs, in turn, are connected to another seven-segment display. What is now the maximum count of the system? One solution to the problem is given in Electronics Workbench file ade 13.12.ewb.

EXERCISE 13.13

Design a circuit that will operate a buzzer and turn ON a red light when a fault occurs in the monitored equipment. Correct operation is to be indicated by a green light. A switch is to be provided that can turn off the buzzer but not the red light while the fault is being repaired. A reset button is required to reset the monitoring circuit once the fault has been cleared. A test switch is also to be provided to check the operation of the red light and the buzzer.

Suggested solution

See Electronics Workbench file ade 13.13.ewb.

Fig. 13.9

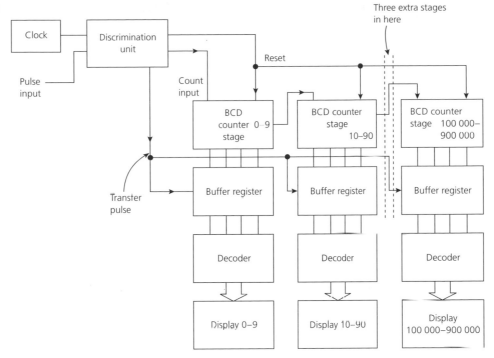

Fig. 13.10 *(a) Pulse input, (b) clock, (c) count input, (d) transfer pulse and (e) reset*

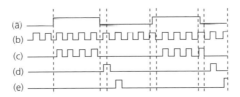

EXERCISE 13.14

Figure 13.9 shows a digital pulse length measurement system that is used to measure pulses to within ±5 μs using a 200 kHz clock. Design the logic for the discrimination unit which is to provide the pulses shown in Fig. 13.10.

Suggested solution

See Electronics Workbench file ade 13.14.ewb.

EXERCISE 13.15

Figure 13.11 shows a hall with four doors through which people are able to both enter and leave the hall. The doors are wide enough for only one person to pass at a time.

Fig. 13.11

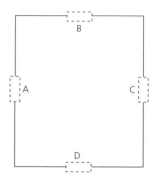

Fig. 13.12

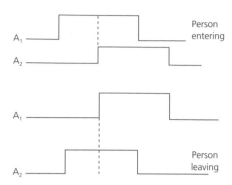

Each time a person goes through a door a light beam is broken generating the pulse shown in Fig. 13.12. Design the circuitry that will maintain a count of the number of people in the hall at any time.

Suggested solution

See Electronics Workbench file ade 13.15.ewb.

EXERCISE 13.16

A circuit is required that will output a 1 μs pulse when an input pulse of any greater width is applied to its input terminal. Design the circuit.

Suggested solution

See Electronics Workbench file ade 13.16.ewb for one possible solution to this design problem.

EXERCISES

[No solutions given.]

13.1 Four spring-loaded push-buttons A, B, C and D operate four similarly labelled lights. Initially only light A is ON. It is required that the lights are operated in the sequence A, B, C, D, C, B, A, etc. with only one light ON at any time. Whenever a particular light is ON it should not be possible to turn any other light ON other than the next light in the given sequence. Design the digital circuitry that will meet this requirement.

13.2 Design a system that will indicate the direction in which rectangular boxes pass by a given point on a conveyor belt. Two light-sensitive detectors A and B are used that each give a HIGH output when an incident light beam is interrupted by a box. Assume that the length of each box is less than the distance between the detectors.

Fig. 13.13

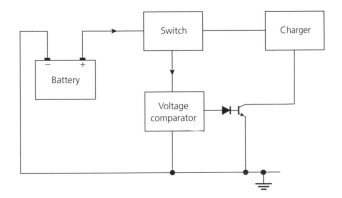

13.3 Figure 13.13 shows the essential parts of a circuit that monitors the output voltage of a 12 V battery. Design the circuit so that the battery is connected to the battery charger when its voltage falls below 10.5 V, and is disconnected when its voltage has risen to 13.5 V.

13.4 Design a circuit that is able to monitor 100 lamps and give an indication which lamps, if any, are faulty. The lamps should be continually tested in sequence.

13.5 A pressure sensor is used to monitor the pressure in a chamber. The pressure is to be displayed by a seven-segment display. Design the system.

13.6 Design a temperature monitoring system that will give an indication whether the temperature in a container is within one of five temperature ranges. The actual temperature range at any instance is to be indicated by a seven-segment display. The display must be updated every 2 minutes.

13.7 Design a system that could be used in a large building to automatically turn the lights ON and OFF when a person enters the room and then leaves the room. If more than one person enters the room the lights must stay ON until the last person has left the room.

13.8 In a television panel game the audience is invited to vote on the performances of the four panel members. Each person can vote for one panel member or can decide to withhold his/her vote. Design a system that will indicate the votes given to each member when the chair asks for a vote. The system is to be cleared by the chair pressing a control button.

13.9 Design a digital bathroom weighing scale.

14 Electronics Workbench

After reading this chapter you should be able to:

- Use Electronics Workbench to simulate digital circuits.
- Use Electronics Workbench to check the operation of a digital circuit that appears in the circuit window when a file has been opened.
- Check the correct operation of many of the practical exercises given throughout the book.
- Test the operation of the circuit you designed after reading Chapter 13.

EXERCISE EWB 2.1. PARITY ERROR DETECTOR

Aim: to carry out practical exercise 2.1 using Electronics Workbench.
Procedure:

(a) Open file ade 2.1.ewb and activate the circuit.
(b) Ensure the switch A is in the position such that the inputs A to the transmitting and receiving halves of the system are connected together. Then all four inputs, A, B, C and D, are linked together and hence there is zero error in transmitting data from the transmitter to the receiver.
(b) Apply different combinations of logic 1 and 0 to the four inputs and check that the correct polarity bit for an odd parity system is generated each time. The parity indicator LED will glow visibly whenever the parity bit is 1. Write down a truth table showing the performance of the sending part of the circuit.
(c) Repeat procedure (a) and this time note the logical states of the error indicator LED. Since there is zero error in data transmission the indicator ought to stay dim at all times.
(d) Now operate switch A so that the A input to the receiving end of the system is always at logical 1. Apply different inputs to A, B, C and D and check whether the error-indicating LED glows visibly whenever there is an error. This, of course, will be every time input A to the sending end of the system is at logical 0. Give a truth table to explain the circuit operation.

Table 14.1

C	0	1	0	1	0	1	0	1
B	0	0	1	1	0	0	1	1
A	0	0	0	0	1	1	1	1
F	0	0	1	1	1	0	0	1

EXERCISE EWB 2.2. SOP AND POS EQUATIONS FOR A GIVEN CIRCUIT

Aim: to carry out practical exercise 2.2 using Electronics Workbench.
Procedure:

(a) Open file ade 2.2.ewb and activate the circuit.
(b) The truth table given in Table 2.8 represents a digital circuit whose operation can be described by the equations $F = \bar{A}B\bar{C} + AB\bar{C} + A B\bar{C} + ABC$ and $F = (A + B + C)$ $(\bar{A} + B + C)(A + B + \bar{C})(A + \bar{B} + \bar{C})$ Open file ade 2.2.
(c) The logic converter can be used to derive the required SOP equation. Note, however, that it takes A as the most significant bit and C as the least significant. Hence, the truth table must be re-written as shown in Table 14.1.
From this the SOP equation is $F = C'BA' + CBA' + C'B'A + ABC$, and the POS equation is $F = (C + B + A)(C' + B + A)(C' + B + A')(C + B' + A')$.
(d) Double click the logic converter icon to gain access to its control buttons. Then, at the bottom of the display, enter the SOP equation. Click the Boolean to Circuit button and the AND/OR gate circuit that implements the equation will be displayed on the circuit window. Click X to shrink the converter and then connect the input terminals C, B and A to three switches that are, in turn, connected to the logic 1 and 0 voltage levels. Also connect the logic probe to the output of the circuit.
(e) Double click on each of the switches and give them values A, B and C, respectively. The three input switches are then controlled by keys A, B and C, respectively, on the PC keyboard. Use these keys to apply the input variable combinations in the table and thence obtain the truth table for the circuit.
(f) Double click the logic converter icon to gain access to its control buttons. Now enter the POS equation at the bottom of the converter display. Click the Boolean to Circuit button to get the AND/OR gate implementation of this expression displayed on the circuit window. The new circuit will be mixed up with the original and it may be rather difficult to distinguish between the two circuits. If this is so there are three possibilities:
  (i) remove the first circuit before displaying the second circuit
  (ii) move the gates and/or connections around until the two circuits can be easily distinguished, or
  (iii) leave the circuits as they appear and be solely concerned with the inputs and outputs of the two circuits.

(g) Connect the second circuit to the input variables C, B and A by means of the switches and connect the output to a logic probe.

(h) Repeat procedure (d). Compare the two truth tables obtained from the circuits. State whether or not the two different circuits gave the same results.

EXERCISE EWB 2.3. BCD DETECTOR

Aim: to implement, using Electronics Workbench, the circuit in practical exercise 2.4, i.e. a circuit whose output goes HIGH only when its BCD input is decimal 7, 8 or 9.
Procedure:

(a) Open file ade 2.3.ewb and activate the circuit.

(b) Apply, in turn, each of the input variable combinations shown in Table 2.9. Each time note the logical state of the logic probe. Compare the results with the truth table. Did the circuit operate correctly?

(c) Close the file and then click NEW.

(d) Construct in the circuit window a circuit that will implement the simplified equation $F = D + ABC$.

(e) Repeat procedure (b).

(f) The circuit is now to be modified so that its output will also go HIGH when the decimal input 4 is applied. Determine the Boolean expression for the modified circuit and implement it on the circuit window. Check that your circuit works correctly.

EXERCISE EWB 3.1. STATIC HAZARDS

Aim: to investigate the presence of a static hazard in a combinational circuit.
Procedure:

(a) Open file ade 3.1.ewb and activate the circuit. This is the circuit used in practical exercise 3.1.

(b) Apply each of the possible combinations of 1 and 0 to the inputs and each time note the logical state of the logic probe. Hence, obtain the truth table for the circuit. Derive the simplified Boolean equation that describes the operation of the circuit.

(c) Modify the circuit by disconnecting the voltage sources A, B and C and replacing them by the logic converter. Note that this has A as the most significant bit and C as the least significant bit. Hence, connect the left-hand output terminal of the converter to both pin 1A of the 74HC04 and pin 2B of the 74HC08, the next terminal to pin 2A of the '08, and the third terminal to pin 1B of the '04. Double click on the logic converter icon to access its controls and then click on the circuit to truth table button. Check that the truth table displayed agrees with that previously obtained – A and C are interchanged. Click the truth table to Boolean equation button to get $F = \bar{A}C + AB$ displayed; interchanging A and C gives $F = A\bar{C} + BC$, as before.

(d) Having confirmed that the circuit works correctly in generating the required Boolean function, check for static hazards. Connect the clock, set to 1000 Hz square wave to the C input (delete the switch and voltage source). Connect the output of the circuit to the CRO and then activate the circuit. Note the presence of a glitch. This is a static error. Note that the logic probe remains lit all the time so that it fails to indicate the presence of the glitch.

(e) Draw the Karnaugh map of the equation $F = A\overline{C} + BC$ and note that the addition of the term AB ought to remove the hazard. Modify the circuit to include this term and then repeat procedure (d). Note that the static error has now disappeared.

EWB EXERCISE 3.2. STATIC HAZARDS

Aim: to investigate further a static hazard and to use

(a) The word generator.
(b) The logic analyser.

Procedure:

(a) Open file ade 3.1a.ewb. This is the same circuit as in file ade 3.1. The least significant input C is connected to the right-hand input terminal of the word generator and the most significant input A is connected to the third input terminal. The Boolean equation describing the logical operation of the circuit is $F = AB + \overline{A}C$.

(b) Double click the word generator icon to access its controls. Ensure that the edit field is 0000. Place the cursor into the final field and alter it to 0007.

(c) (i) Place the cursor on the second line of the scrollable hex field and click. Then move the cursor to the binary field and alter the right-hand bit from 0 to 1. The second word in the hex field will then change to 0001.

 (ii) Place the cursor in the third line of the hex files and click, then alter the second 0 in the binary field to 1. The third line in the hex field will then change to 0002.

 (iii) Select the fourth line of the hex field and click. Move the cursor to the binary field and change the two right-hand 0s into 1s. The fourth hex word will now be 0003.

 (iv) Carry on in this way to set the next hex words to 0004, 0005, 0006 and, finally, 0007. Click the step button and then exit the word generator.

(d) Bring the logic analyser to the circuit window and connect it to the output of the circuit and to all three inputs. Double click on the logic analyser icon to display its screen. There will be no displayed waveform and each of the four circles on the left will contain 0.

(e) Click the pause/resume button once to input 100 to the logic analyser. One input pulse will then be displayed and the left-hand circles will contain, downwards, 1001. Click repeatedly on the pause/resume button to obtain the other input signal combinations and each time note the input and output waveforms. The waveforms will be displayed in the same colours as the wiring to the inputs/output.

(f) Leave the logic analyser and double click the word generator icon. Set its frequency to 10 Hz and click the cycle icon. Observe that the numbers in the hex field cycle from 0000 to 1111 repeatedly. Leave the word generator and double click the logic analyser icon. Click the pause/resume button and watch the input and output waveforms appear on the screen.

(g) Vary the clocks per division control and note the effect on the displayed waveforms. Also note the effect of changing the internal clock rate.

(h) Why is a glitch not visible? Connect the CRO to the output of the circuit and use it to observe the output waveform. Is a glitch visible now? Explain.

EWB EXERCISE 4.1. MULTIPLEXER USED TO GENERATE A LOGIC FUNCTION

Aim: to become familiar with the use of a multiplexer to generate logic functions.
Procedure:

(a) Open file ade 4.1.ewb and activate the circuit. The circuit shown implements the circuit given in Fig. 4.21 using the 74HC00 quad 2-input NAND gate, the 74HC04 hex inverter and the 74HC153 dual 4-to-1 multiplexer.

(b) The data inputs A and B can be varied between 1 and 0 by pressing keys A and B, respectively. Select inputs C and D can be varied by altering the value of the voltage source. To do this double click the icon and type either 0 or 5. Vary the values of A, B, C and D and each time note the logical state of the LED probe. Hence, obtain the truth table for the circuit.

(c) From the truth table obtain the Boolean equation that describes the operation of the circuit. Compare it with the equation given in Example 4.10.

(d) Close file ade 4.1 and click NEW. Click the digital IC button and drag the 741XX series icon to the circuit window. Select 74151 and double click the IC. Select CMOS and then HC.

(e) Construct the circuit that will implement the Boolean equation using this IC. Once the circuit has been wired check that it gives the same truth table as in (b).

EWB EXERCISE 5.1. TO USE TWO 74HC138 3-TO-8 LINE DECODERS AS A 2-BIT MULTIPLIER

Aim: this exercise simulates the problem given in Example 5.3.
Procedure:

(a) Open file ade 5.1.ewb and activate the circuit.

(b) One of the two 2-bit numbers to be multiplied together is represented by the switches labelled A and B, and the other 2-bit number is represented by switches C and D. These switches are controlled by A, B, C and D, respectively, on the keyboard. Set $A = B = 1$ and $C = D = 0$ and note the logical state of the logical probes $P1$ and $P2$ that simulate the outputs of the circuit.

(c) Set the input switches to give all the possible combinations of the two 2-bit numbers and each time note the logical states of outputs $P1$ and $P2$.

(d) Use your results to draw up a truth table for the circuit. Use the truth table to confirm whether the circuit has acted as a 2-bit multiplier.
(e) Discuss the practicality of using this method to multiply together
 (i) two 3-bit numbers
 (ii) two 4-bit numbers.
Suggest another way of achieving (i) and (ii).

EWB EXERCISE 5.2. MULTIPLE INPUT NAND GATE

Aim: the 74CMOS logic families do not include any NAND gates with more than four inputs. Yet sometimes there is a need for a NAND gate with five or more inputs.
Procedure:

(a) Open file ade 5.1.ewb and activate the circuit.
(b) Vary the voltages of the five voltage sources A through to E to obtain, in turn, all possible combinations of 0s and 1s. Each time note the logical state of the logic probe. Hence, determine the logical function carried out by the circuit.
(c) Modify the circuit so that it provides the logical function of a 6-input NAND gate.

EWB EXERCISE 5.3. 74HC148 PRIORITY ENCODER

Aim: to determine the operation of a 74HC148 priority encoder.
Procedure:

(a) Open file ade 5.3.cwb and activate the circuit.
(b) Set all the inputs 0 through to 7 to 0 V and note the logical state of the output logic probe.
(c) Vary the input voltages applied to the inputs by operating keys 0 through to 7 on the keyboard and each time note the logical state of the logic probe.
(d) Write down the truth table of the circuit and then confirm with the function table given in Fig. 5.12. Did the circuit work correctly?
(e) Mention some possible applications for a priority encoder.

EWB EXERCISE 6.1. DIVIDE-BY-256 COUNTER

Aim: to determine the operation of a divide-by-256 binary counter formed by connecting two 74HC163 4-bit counters in cascade.
Procedure:

(a) Open file ade 6.1.ewb and activate the circuit.
(b) Get the circuit going and confirm that it acts as a divide-by-256 counter.
(c) Modify the circuit to operate as a divide-by-153 counter circuit by connecting the Q_A and Q_B outputs of both ICs to the four inputs of a 74HC20 dual 4-input NAND gate IC. Connect the output of the 74HC20 gate to the $\overline{CLR}$ pin of both 74HC163 ICs. The required HIGH required at the $\overline{CLR}$ pins to prevent the circuit

clearing is provided by the output of the NAND gate which will be HIGH until $Q_{A1}Q_{D1}Q_{A2}Q_{D2} = 1$.

(d) The required circuit is available in file number ade 6.1a.ewb. Load this file and then use the logic analyser to investigate and confirm its correct operation.

EWB EXERCISE 6.2. J-K FLIP-FLOP

Aim: to determine the operation of a J-K flip-flop connected as a modulo 2 counter.
Procedure:

(a) Open file ade 6.2.ewb and activate the circuit. It has its J and K input terminals connected to 1 and so it ought to act as a divide-by-2 circuit.
(b) Observe the logic probes and check if the one connected to the output flashes at one-half the rate of the input logic probe.
(c) Bring the CRO icon down on to the circuit window and use it to observe the input and output waveforms of the flip-flop.
(d) Use the CRO to measure the propagation delay and the division ratio of the flip-flop.

EWB EXERCISE 6.3. J-K FLIP-FLOP SET-UP TIME

Aim: to determine the importance of set-up time in a J-K flip-flop. To measure propagation delay.
Procedure:

(a) Open file ade 6.3.ewb and activate the circuit. With the links in the green connections missing the circuit is the same as that given in Fig. 6.9.
(b) Try different clock frequencies starting from 1 kHz to confirm that the flip-flop will not toggle with a set-up time of 0 ns.
(c) Link the connections in the green line to include the shift register in the circuit. The clock signal to the flip-flop is now delayed relative to the J signal by an amount equal to the total propagation delay of the shift register between the B input and the Q_{H} output.
(d) Restore the frequency of the pulse generator to 1 kHz and confirm that the flip-flop now operates correctly.
(e) Use the CRO to measure
  (i) the division ratio
  (ii) the propagation delay
  of the flip-flop.
(f) Repeat procedure (e) for frequencies of
  (i) 10 MHz
  (ii) 25 MHz
  (iii) 30 MHz.
(g) Determine the number of shift register stages that are required to provide the minimum set-up time for the flip-flop. Alter the circuit to give this delay, or as near to it as possible, and again check the operation of the circuit.

Aim: to repeat practical exercise 6.3 using Electronics Workbench.
Procedure:

(a) Open file ade 6.4.ewb and activate the circuit.
(b) Put the switch labelled D/A to the left (by pressing key D), and the Q switch up (key Q), so that the D flip-flop is not in the circuit.
(b) Repeated operation of the space switch (press space bar to operate) will apply random 1s and 0s to the *1A* input of the AND gate. Use the CRO to observe the input and output waveforms of the AND gate. Note that the width of the output pulse is erratic.
(c) Press keys D and Q to connect the *1Q* output of the AND gate to the *D* input of the D flip-flip. Repeatedly operate the space switch and observe the waveforms displayed by the CRO.
(d) Note that the pulses appearing at the output of the AND gate are now of the same width as the input pulses, no matter how erratically the space switch is operated.
(e) Comment on the practical importance of the results.

Aim: to simulate practical exercise 6.4.
Procedure:

(a) Open file ade 6.5.ewb and activate the circuit. It consists of a 74HC4040 12-bit non-synchronous counter that provides the input to a 74HC42 BCD-to-decimal decoder.
(b) Double click on the logic analyser icon to access its controls and make its screen visible. Set the logic analyser to 16 clocks per division. The logic analyser has been connected to five different inputs and outputs in the circuit. Note the displayed waveforms.
(c) Connect the logic analyser to other parts of the circuit and observe the displayed waveforms.
(d) Experiment with the clocks per division and reset controls to determine what effect(s) they have on the displayed waveforms. Click on the set button and try out the various trigger patterns that are available.

Aim: to repeat practical exercise 6.5, i.e. to simulate the operation of a divide-by-153 counter obtained by modifying a divide-by-256 counter.
Procedure:

(a) Open file ade 6.6.ewb and activate the circuit.
(b) With the links in the purple connections missing the circuit is just a single 74HC163 4-bit synchronous counter. Check that the circuit repeatedly counts correctly from 0 to 15.

(c) Restore the missing links in the purple connections. This will connect a second 74HC163 counter in cascade with the first. Check that this circuit counts correctly from 1 to 255 and then back to 0.

(d) Increase the clock frequency to, in turn,
   (i) 17 MHz
   (ii) 20 MHz
   (iii) 25 MHz.
   Each time note the effect, if any, on the circuit.

(e) The circuit is to be modified to operate as a divide-by-153 counter. Devise a method of doing this; construct the circuit and check its operation.

(f) One way in which the divide-by-153 counter can be obtained is shown in file ade 6.6a.ewb. Load this file and activate the circuit. The data word 0110 0111 can be loaded into the circuit by pressing key L, which operates both switches and takes the $\overline{\text{LOAD}}$ terminals LOW. Once the data word has been loaded press key L again to take the $\overline{\text{LOAD}}$ terminals HIGH and so make them inactive.

(g) The circuit will now start to count from its initial state of 0110 0111. Note the count reached when all the stages are set ($Q = 1$). What happens next?

(h) When the maximum count is reached the circuit is required to go to the initial state and not to the all 0s state. Think of a method of achieving this and try to implement it.

(i) Open file ade 6.6b.ewb and activate the circuit. The previous circuit has now been modified to make it restore to 0110 0111 after reaching the all 1s maximum state. Note the details of the circuit and check its operation.

(j) If time allows suggest another method of implementing a divide-by-153 counter and implement it in the circuit window.

EWB EXERCISE 6.7. 74HC191 UP-DOWN COUNTER

Aim: to repeat practical exercise 6.6 using Electronics Workbench.
Procedure:

(a) Open file ade 6.7.ewb and activate the circuit.

(b) The $\overline{\text{LOAD}}$ terminal is held LOW so that the data at inputs A, B, C, and D is loaded into the counter to give its initial count. Determine the input data word. Check that the logic probes indicate the same number. Press key L to take the $\overline{\text{LOAD}}$ terminal HIGH and hence make it inactive.

(c) Note the logical states of the logic probes as the counter goes through its counting sequence. Does the circuit count up or down? [Note that the IC icon has the $\overline{\text{U}}$/D pin wrongly labelled as U/$\overline{\text{D}}$.] What counting sequence does it follow?

(d) State when
   (i) $\overline{\text{RCO}}$
   (ii) MAX/MIN go HIGH.
   What is the function of these two terminals?

(e) Determine what happens if terminal $\overline{\text{CTEN}}$ is taken HIGH.

(f) Note what happens when the count reaches 0000. It is required that after 0000 the count next goes into its initial state of 0111 and then repeats the counting sequence. Does the circuit do this?

(g) If it does not say why and then modify the circuit so that it counts repeatedly through the sequence 7, 6, 5, 4, 3, 2, 1, 0, 7, etc.

Aim: to check the operation of an up-down counter and to modify its counting sequence.
Procedure:

(a) Open file ade 6.8.ewb and activate the circuit.
(b) The circuit shows the 74HC191 up-down counter. With the $\overline{\text{LOAD}}$ terminal switch in its earth position the data word 0111 is loaded into the counter. This gives the starting point for the counting sequence. Press key L on the keyboard and the circuit will start to count.
(c) Note the count sequence indicated by the logic probes. Does it agree with the labelling of the IC icon's terminal 5?
(d) Operate the other switch to change the function of the circuit from down-counter to up-counter and again check its operation.
(e) The circuit is required to follow the counting sequence 7, 8, 9, 10, 15, 0, 1, 7, 8, etc. Modify the circuit to obtain this counting sequence. Confirm the correct operation of the modified circuit.
(f) Alter the up/down switch so that the modified circuit counts down and note the counting sequence then followed.

Aim: to repeat practical exercise 7.1 using Electronics Workbench.
Procedure:

(a) Open file ade 7.1.ewb and activate the circuit.
(b) Press key B on the keyboard to operate switch B and enter data into the shift register. Note the logical states of the logic probes and hence confirm the correct operation of the circuit. Operate the B switch quickly so that only 1 bit is entered.
(c) Double click the logic analyser icon to bring up its screen and then observe the circuit waveforms as switch B is operated. Enter bits first singly, and then in twos, and then in threes.
(d) Connect the generator labelled as 'input' to the B input terminal and again note the circuit waveforms on the logic analyser.
(e) See what happens when the frequency of the input to terminal B is taken higher than the clock frequency.
(f) Determine the maximum clock frequency of the 74HC164 from its data sheet (p. 191) and operate the circuit at
  (i) the maximum clock frequency
  (ii) a frequency higher than the maximum clock frequency.
  Use the logic analyser to see what happens.

Aim: to repeat practical exercise 7.2, i.e. to investigate a ring counter.
Procedure:

(a) Open file ade 7.2.ewb and activate the circuit. It is of a 74HC164 shift register connected as a ring counter.
(b) It will be found that the circuit will not start its counting sequence. Connect the $\overline{\text{CLR}}$ terminal to earth by pressing key L on the keyboard and then press again to take $\overline{\text{CLR}}$ HIGH.
(c) Press key B on the keyboard to connect input B to +5 V long enough for stage A to be set. Then press key B again so that input B is connected to output Q_H. The circuit will now operate as a ring counter. Observe the logical states of the LED probes and hence determine the counting sequence of the circuit.
(d) Repeat procedure (b) but this time press key B for a time sufficient to allow
 (i) two bits
 (ii) three bits
 to circulate around the counter.
(e) The circuit is now to be modified to make it self-starting. The method shown in Fig. 7.11 is to be employed. To reduce the amount of modification to the circuit required and to avoid the circuit window becoming too cluttered only the first four stages will be used. Remove the connection between output Q_H and input B. Connect Q_A, Q_B and Q_C to the inputs of a 74HC27 triple 3-input NOR gate. Connect the output of the gate to the B input. The circuit will now automatically start its counting sequence as soon as the circuit is activated.
(f) Confirm that the first four stages act as a ring counter and note the counting sequence. What happens to the four most significant stages?
(g) Open file ade 7.2a.ewb and activate the circuit. The circuit is that of the self-starting ring counter of part (f). Confirm the correct operation of the circuit.
(h) Determine what happens if the inputs to the NOR gate are Q_B, Q_C and Q_D and explain why.

Aim: to simulate practical exercise 7.5 using Electronics Workbench.
Procedure:

(a) Open file ade 7.3.ewb and activate the circuit.
(b) The circuit shown is a 74HC174 hex D flip-flop connected as a divide-by-5 ring counter whose $5Q$ output is connected to the CLK input of a 74HC112A J-K flip-flop. The J and K terminals of this flip-flop are both connected to the logic 1 voltage level and hence the flip-flop acts as a divide-by-2 circuit. The overall division ratio is $5 \times 2 = 10$.
(c) The circuit is not self-starting so input a single pulse by pressing key S on the keyboard.
(d) Note the logical states of the logic probes and hence determine the count of the circuit.

(e) Modify the circuit to make it self-starting and then check the modified circuit.
(f) Further modify the circuit so that it starts in the other of the two possible counting sequences. Observe the logic probes when the modified circuit operates and hence determine the new counting sequence.

Aim: to observe the operation of a ring counter.
Procedure:

(a) Open file ade 7.4.ewb and activate the circuit.
(b) With the S switch connected to the *1Y* output of the hex inverter and the pulse generator connected to the CLK input observe the operation of the circuit.
(c) Press key S to operate switch S and apply +5 V to the *B* input terminal. How does the circuit work now?
(d) Operate switch C by pressing key C on the keyboard so that the CLK input is connected to the switch labelled IN. Press key I repeatedly to provide a very low frequency clock signal and determine the counting sequence of (b) and (c).

Aim: to implement practical exercise 8.1 using Electronics Workbench.
Procedure:

(a) Open file ade 8.1.ewb and activate the circuit.
(b) Determine the count of the circuit and write down its counting sequence.
(c) Open file ade 8.1a.ewb and activate the circuit.
(d) Determine the count and the counting sequence of this new circuit. How does it differ from the first circuit?
(e) Connect the CRO to the input and output terminals of the circuit. Double click on the CRO icon to access its controls and observe the input and output waveforms. Explain the screen display.
(f) Remove the CRO from the circuit and connect the logic analyser instead. Use it to observe the waveforms at each of the *Q* outputs.
(g) Increase the frequency of the clock generator to
   (i) 1 MHz below the maximum clock frequency
   (ii) the maximum clock frequency
   (iii) 1 MHz above the maximum clock frequency for the flip-flop.
   Observe the effect in each case on the waveforms displayed on the logic analyser.

Aim: to build a counter that does not follow a strict binary sequence using

(a) D flip-flops
(b) J-K flip-flops.

Procedure:

(a) Open file ade 8.2.ewb and activate the circuit. It shows a counter made using D flip-flops. Draw the schematic diagram of the circuit.

(b) Determine
   (i) the count
   (ii) the counting sequence
   of the circuit.

(c) Use
   (i) the CRO
   (ii) the logic analyser
   to observe the waveforms at the input and at all three Q outputs of the counter.

(d) Increase the frequency of operation first to 10 MHz, and then 25 MHz, and note any differences in the behaviour of the circuit that are observed.

(e) The use of D flip-flops results in rather complex circuit for trivial results. Use the design equations given in Example 8.3, i.e. $J_A = Q_B Q_C$, $K_A = Q_B + Q_C$, $J_B = 1$, $K_B = \bar{Q}_A$, $J_C = \bar{Q}_A Q_B$ and $K_C = Q_A + Q_B$, to set up the J-K version of the counter.

(f) Check the operation of your circuit and confirm that it has the same count and counting sequence as the D flip-flop version.

(g) If required, the J-K flip-flop version of the counter can be obtained by loading file ade 8.2a.ewb.

EWB EXERCISE 8.3. DECADE COUNTER

Aim: to determine the error in a decade counter and then to correct the error.
Procedure:

(a) Open file ade 8.3.ewb and activate the circuit.

(b) State what kinds of flip-flops are employed.

(c) Determine the count and the counting sequence of the counter.

(d) The circuit is supposed to be that of a decade counter that follows a straight binary sequence, but both the count and the counting sequence are incorrect. Check the circuit and discover the error in the connections.

(e) When the error has been found, modify the circuit and again check its operation. If need be the correct circuit can be found by loading file ade 8.3a.ewb.

EWB EXERCISE 9.1. SYNCHRONOUS SEQUENTIAL CIRCUIT

Aim: to simulate the operation of a synchronous sequential circuit.
Procedure:

(a) Open file ade 9.1.ewb and activate the circuit.

(b) The circuit is that of a synchronous sequential circuit with a single input x. The output of the circuit is to go HIGH only when a particular bit sequence is entered at input x. The logic analyser is connected to both the input x to the circuit and to the output terminal. The logical state of the output is indicated by the logic probe. Set the clock frequency to 1 kHz and then double click the logic analyser icon.

(c) The input x to the circuit is under the control of the key X on the keyboard. Pressing this key changes x from 0 to 1, or from 1 to 0. The blue line that is connected to the upper switch is the x input, while the green line gives the complement of x. Use the X key to input various combinations of 1 and 0 to the circuit. The input waveform is shown by the top trace on the logic analyser display. Note when the output of the circuit goes HIGH – it is indicated by the logic probe glowing red and by a 1 pulse appearing on the lower trace of the logic analyser's display.

(d) Determine the input bit sequence that is detected by the circuit.

EWB EXERCISE 9.2. SYNCHRONOUS BIT SEQUENCE DETECTOR

Aim: to simulate the circuit given in Fig. 9.11 and check its operation.
Procedure:

(a) Open file ade 9.2.ewb and activate the circuit.
(b) Operate the input switch ON and OFF once by pressing key I on the keyboard to enter one 1 bit into the circuit. Note whether the logic probe glows red.
(c) Press the key 1 several times rapidly and note if the logic probe glows red.
(d) Press the I key once to input a HIGH and press again after a short delay. When the input switch has been closed for a time period equal to three 1 bits the logic probe will glow red. It will stay red until such time as a 0 bit is entered into the circuit.
(e) Double click on the logic analyser icon to access its screen controls and set the clocks per division to 8. Input different 1 and 0 combinations by pressing the I key and note the displayed waveforms. The lower purple trace shows the input bit stream and the upper brown trace shows the output waveform.
(f) Input the bit stream 1011 1001 1110 0111 and when two output pulses appear on the upper trace click 'pause' to freeze the display. Note the input and output waveforms of the sequence detector.
(g) Increase the clock frequency to
    (i) 1 MHz
    (ii) 20 MHz
    (iii) 30 MHz
    and each time observe the input and output waveforms. Comment.

EWB EXERCISE 9.3. SEQUENCE DETECTOR

Aim: to determine the input code detected by a sequence detector.
Procedure:

(a) Open file ade 9.3.ewb and activate the circuit.
(b) The circuit shown is of a sequence detector that is able to detect the presence of a 3-bit word in a bit stream. Operate key N on the keyboard to input a bit stream. Note when the logic probe goes red.

(c) Double click on the logic analyser icon to access its screen. Observe the input and output waveforms as switch N is operated. Note the input bit sequence that causes the lower pale blue trace to show a pulse. When does the output pulse appear? Is it at the leading edge, or the trailing edge, of an input pulse or somewhere in between? Explain.

Aim: to determine the action of a simple non-synchronous sequential circuit.
Procedure:

(a) Open file ade 9.5.ewb and activate the circuit.
(b) The circuit is of a simple sequential circuit in which an input signal is applied to one input of a 2-input OR gate and the other input is connected to the output of the gate. Apply a HIGH voltage to input *1A* by pressing the space bar and note the logical state of the logic probe. What happens when the HIGH voltage is replaced by a LOW? Explain the action of the circuit.
(c) Open file ade 9.5a.ewb and activate the circuit.
(d) This is another simple non-synchronous sequential circuit. Apply each of the four possible combinations of 1 and 0 to the two inputs *A* and *B* of the circuit by pressing keys A and B on the keyboard. For each combination note the logical state of the logic probe. Hence determine the Boolean expression that describes the operation of the circuit.
(e) Open file ade 9.5b.ewb and activate the circuit.

Aim: to determine the transfer function of a DAC.
Procedure:

(a) Open file ade 11.1.ewb and activate the circuit.
(b) Connect all eight digital inputs to 0 V and then measure the analogue output voltage. Note its value. Then connect, in turn, inputs 0 through to 7 to +5 V and each time note the output voltage.
(c) Use the values obtained in (b) to plot the transfer function of the DAC. Is any non-linearity or other defect noticeable?
(d) (i) measure the resolution of the circuit
     (ii) determine the FSR of the circuit and the number of bits employed in its conversion and calculate the resolution
     (iii) compare the two values thus obtained.
(e) Measure the analogue output voltages obtained for input digital words of
     (i) 1010 0011
     (ii) 0101 1100
     (iii) 1100 0010.

EWB EXERCISE 11.2. OPERATION OF A DAC

Aim: to determine the operation of a DAC.
Procedure:

(a) Open file ade 11.2.ewb and activate the circuit.
(b) Determine the count of the 74HC163 counter.
(c) Note whether the selected output of the 3-to-8 line decoder goes HIGH or LOW.
(d) Double click on the CRO icon to access its controls and its screen and adjust the timebase until the output waveform of the DAC is displayed.
(e) (i) determine how many bits the DAC uses
     (ii) measure the maximum output voltage
     (iii) measure the resolution of the circuit.
(f) Bring two 74HC04 hex inverter ICs onto the circuit window and use them to invert all the decoded outputs before they are applied to the DAC. Note the waveform displayed on the CRO screen.
(g) Increase the clock frequency to
     (i) 1 MHz
     (ii) 20 MHz
     (iii) 30 MHz
     and each time repeat procedure (f).
(h) Discuss how the generic DAC differs from a practical device.

EWB EXERCISE 11.3. ADC TRANSFER FUNCTION

Aim: to determine the transfer function of an ADC.
Procedure:

(a) Open file ade 11.3.ewb and activate the circuit.
(b) With CAPS LOCK *not* ON pressing key R on the keyboard will increase the voltage applied to the input terminal of the ADC. With CAPS LOCK *on* pressing key R will reduce the input voltage. Press key R until the DVM indicates the smallest possible value. State what this value is. The reference voltage is 10 V.
(c) Take pin SOC (start conversion) HIGH by pressing key S. Note that all the logic probes are OFF except the one connected to the EOC (end of conversion) pin. The EOC logic probe will always come ON to indicate when a conversion has ended. One press on the R key moves the potentiometer by 5 per cent to give $V_{in} = 500$ mV. Press key S and note which logic probes turn ON. Note the digital word produced. Press key R again to move the potentiometer another 5 per cent and note both the analogue input voltage and the output digital word.
(d) Repeat procedure (c) for each step of the potentiometer from 0 to 100 per cent and use the results to plot the transfer function of the ADC.
(e) Change the reference voltage to 5 V and repeat procedure (d). Hence, explain the effect on the circuit of changing the reference voltage.

Aim: to observe the operation of an ADC connected to a DAC.
Procedure:

(a) Open file ade 11.4.ewb and activate the system.
(b) Vary the value of the variable resistor R by pressing key R on the keyboard. Note the logical states of the logic probes. From the transfer function obtained in EWB Exercise 11.3 determine the analogue input voltages for each percentage value of R (quoted as R/10 kΩ x per cent). For each input voltage note the indication of the DVM.
(c) Compare the values of analogue voltage into the ADC and the analogue voltage out of the DAC.
(d) Replace the d.c. voltage source that has supplied the input voltage to the ADC with the function generator. Double click the function generator icon to access its controls and set it to sinusoidal at 1 kHz. Discover what happens if this sinusoidal voltage is applied just to the ADC and explain.
(e) Modify the circuit so that it is able to convert the sinusoidal voltage and test the operation of your circuit. Draw waveforms to show the effect of different clock frequencies.

Aim: to measure the characteristic impedance and propagation delay of a transmission line.
Procedure:

(a) Open file ade 12.1.ewb and activate the circuit.
(b) Set the function generator to 500 kHz, 5 V square wave. Double click on the lossy line icon and click default. Click edit and make
  (i) $R = 0$ so that the line has zero loss
  (ii) $L = 50$ mH and $C = 20$ pF so that $Z_o = 50$ Ω.
  Note the default length of line.
(c) Observe the waveform displayed on the CRO which ought to be of the form shown in Fig. 12.6. Measure the values of V_i and V_s and calculate $Z_o = 12V_i(V_s - V_i)$.
(d) Measure the propagation time t_D and use it to calculate the inductance and capacitance of the line using $L = t_D Z_o$ and $C = t_D/Z_o$. Compare these values with those set in part (b).
(e) Comment on the method employed to measure the line parameters and suggest at least one alternative method.

Aim: to observe the waveforms on both correctly terminated and incorrectly terminated lines when a square wave is applied to the sending end.

Procedure:

(a) Open file ade 12.2.ewb and activate the circuit.
(b) Set the function generator to 1 MHz, 5 V and square wave. Double click on the lossy line icon and then click default. Click edit and make
   (i) $R = 0$ so the line is lossless
   (ii) $L = 50$ nH and $C = 20$ pF so that $Z_o = 50 \, \Omega$.
(c) Use the CRO to note the input and output waveforms of the correctly terminated line
   (i) when the source resistance is zero
   (ii) when the source resistance is 50 Ω.
      Switch between either (i) or (ii) by pressing the space bar on the keyboard.
(d) Disconnect the 50 Ω resistor from the output terminals of the line so that the line is open-circuited. Repeat procedure (c). Try to measure the ringing that is seen.
(e) Short-circuit the output terminals of the line and then repeat procedure (c).

EWB EXERCISE 13.1. DESIGN PROBLEM 13.1

Procedure:

(a) Open file ade 13.1.ewb and activate the circuit.
(b) The circuit shows a 3 bit ripple counter. Check that the counter works correctly. The counter is to be used in the simulation of the system given in Fig. 13.1.
(c) Open file ade 13.1a.ewb and activate the circuit.
(d) The counter used in file ade 13.1.ewb now has its Q outputs connected to both a 74HC138 3-to 8 line decoder and a 74HC151 8-to-1 line multiplexer. Check the operation of the circuit under the control of the 1 kHz clock. Confirm that the inputs/ outputs of the two ICs are selected in sequence. State how the circuit thus far works.
(e) The circuit is required to step sequentially through the various inputs and outputs under the control of the logical states of the inputs to the multiplexer. This means that the clock generator must be removed from the counter and the clock signal obtained from the output of the multiplexer. This has been done in the next file.
(f) Open file ade 13.1b.ewb and activate the circuit.
(g) Initially, it will be found that the circuit will not start until input *D0* and output *Y0* have been selected. Press the space bar on the keyboard momentarily to set the Q_A output of the counter HIGH. Note that output *Y1* and input *D1* have now been selected. Press key 1 and note whether the logical state of the output *Y* changes from 0 to 1 (and output *W* from 1 to 0). It will depend upon the logical state of the selected input, i.e. *D0*, and this can be altered by pressing key 0. This transition increments the counter so that now *Y2* and *D2* are selected. Depending on the logical state of an input *D*, pressing the corresponding key may step the system on to its next state.
(h) Vary the logical states of the signals applied to the multiplexer inputs and check the correct operation of the circuit.
(i) Read the required sequence of events for the system being simulated and then input the signals to the multiplexer accordingly. Hence check that the system operates in the required manner. [Note: the logic probes simulate the signals used to control devices and the switch inputs simulate the sensor outputs.]

EWB EXERCISE 13.2. DESIGN PROBLEM 13.2

Procedure:

(a) Load file ade 13.2.ewb and activate the circuit.

(b) Initially, all the doors are closed so the sensor inputs are LOW. Note the logical states of the logic probes connected to the decoder outputs.

(c) Press key 0 on the keyboard to simulate door 0 only being open. Note how this is indicated by the logic probe.

(d) Press key 1, so that doors 0 and 1 are open. How is this indicated by the logic probes?

(e) Press different keys and note which probes light each time. What happens if all the doors are open?

(f) Modify the circuit so that the door open condition is indicated by the associated logic probe turning OFF instead of flashing.

EWB EXERCISE 13.3. DESIGN PROBLEM 13.3

Procedure:

(a) Open file ade 13.3.ewb and activate the circuit.

(b) The circuit in the circuit window gives an SSI implementation of a system to detect which one of three contestants in a quiz presses his/her answer button first. The circuit equations are given in Exercise 13.3 on p. 394. Generic gates have been used in this simulation because the circuit is rather complex and using generic devices makes it seem somewhat less complex.

(c) Reset the circuit by pressing key H on the keyboard when all three logic probes will be OFF. Ensure that all three switches A, B and C are in their LOW positions.

(d) Press key H again. Note which, if any, logic probes glow red. Then press any of the input keys and note that only the corresponding logic probe turns ON. What happens if a second key is then pressed?

(e) Reset the circuit and see what happens when any two of keys A, B and C are pressed simultaneously.

(f) It might be desirable in a practical case for the circuit to be inhibited for a short time after the quiz master resets the system to avoid a crafty early press of a button by a contestant. Modify the circuit to achieve this.

EWB EXERCISE 13.4. DESIGN PROBLEM 13.5

Procedure:

(a) Open file ade 13.5.ewb and activate the circuit.

(b) The circuit is of a divide-by-1000 counter. Start the circuit and watch it count from 000 to 999.

(c) Counters available in the 74HC logic family and provided by Electronics Workbench include the 74HC161/3 4-bit synchronous counters, and the 74HC4040 12-bit non-synchronous counter. Pull down each counter to the circuit window and investigate the possibility of using one of the devices to implement the divide-by-1000 counter.

(d) Open file ade 13.5a.ewb and activate the circuit.
(e) It shows a 74HC4040 counter connected to a seven-segment display and operating as a hexadecimal counter. Is there some way of modifying the circuit so that it acts as a divide-by-1000 counter? Implement any ideas and check the operation of the modified circuit.

EWB EXERCISE 13.5. DESIGN PROBLEM 13.6

Procedure:

(a) Open file ade 13.6.ewb and activate the circuit. This is the circuit of a flashing light display for the window of a toyshop.
(b) The 74HC163 counter is connected as a divide-by-10 counter (using the 74HC00). Add logic probes to the Q outputs of the counter and use them to confirm that the counter does count continuously from 0 to 9. The initial state of the counter is set by pressing key L on the keyboard.
(c) Start the circuit and carefully note the count sequence and the timing with which the logic probes turn ON. Why does the logic probe connected to the Q_D output remain OFF?
(d) Modify the circuit to give the lighting sequence RED, GREEN, BLUE, RED, etc.
(e) Modify the circuit again, this time to give the lighting sequence RED, RED and GREEN, GREEN, GREEN and BLUE, BLUE, BLUE and RED, RED, etc.
(f) Lastly, modify the circuit so that the fourth logic probe, connected to the Q_D output, takes part in a lighting sequence of your choice.

EWB EXERCISE 13.6. DESIGN PROBLEM 13.7

Procedure:

(a) Open file ade 13.7.ewb and activate the circuit.
(b) Simulate the action of the liquid-level sensors when the liquid level is high by operating one of the switches. This is done by pressing keys 0 through to 7 on the keyboard. Set all switches to their inactive position, i.e. connected to +5 V, and note the logical states of the logic probes. In turn, have just one input LOW and each time note the logical state of the associated logic probe. Hence, determine the truth table of the circuit.
(c) Press two keys (0 through to 7) simultaneously and note what happens. Explain.
(d) If the outputs of the encoder are to be applied to a microprocessor that will control the flow of liquids into the eight tanks explain how they will be connected. What would be the function of the GS output?
(e) Modify the circuit so that a seven-segment display is used to give a decimal readout of the tank in which the liquid level is HIGH. If all the encoder outputs are connected directly to the inputs of the display it will be found that a reading of 0

will be given both when sensor 0 is active-LOW *and* when no sensor is active. How can this be overcome?

(f) Open file ade 13.7a.ewb and activate the circuit.

(g) It is the same circuit as before with the addition of an AND gate and a seven-segment display. Operate the input switches to determine the operation of the circuit. Note that now the 'all sensors active' state is indicated by a display of 8, and a display of 0 occurs only when sensor 0 is active-LOW.

(h) Further modify the circuit so that the visual indication of 'all sensors active' is changed from 8 to A.

EWB EXERCISE 13.7. DESIGN PROBLEM 13.8

Procedure:

(a) Open file ade 13.8.ewb and activate the circuit.

(b) The circuit shown is of a simple burglar alarm system in which sensors monitor the doors and windows in a room. When a door/window is open a HIGH voltage is generated by a sensor. If the ON/OFF control is turned ON the opening of a door or window ought to sound an alarm. In this simulation the door/window sensors are represented by the keyboard-operated switches and the audible alarm by the logic probe.

(c) Open any switch by pressing the key of the same label and note if the logic probe comes ON. Note the effect of the ON/OFF switch (key 0).

(d) It might be thought desirable to have an indication of the actual door or window that is open. This can be acheived by simply connecting an LED probe to the outputs of the switches, but suppose that a decimal display is required. This means that a decimal-to-binary, or a decimal-to-BCD, converter is necessary to convert a HIGH voltage into the signals required by a seven-segment display. Either design a suitable converter or use an MSI device. There is no 74HC device available but there is more than one in the 74LS logic family. Click the digital icon and find a suitable device in the drop-down list. Take this device onto the circuit window and using it modify the circuit to give a decimal indication of which door/window is open. What happens if two, or more, doors/windows are open at the same time?

(e) Open file ade 13.8a.ewb and activate the circuit.

(f) This circuit uses a 74HC148 8-to-3 line priority encoder. With all inputs HIGH the $A0$, $A1$ and $A2$ outputs are also HIGH, and so is the GS output. Operate, in turn, each of the keys 0 through to 4, on the keyboard, and observe the logical states of the logic probes. Hence, determine the truth table for the circuit. The inputs 5 through to 7 are not required in the circuit and so they are held HIGH all the time.

(g) The truth table shows that if the highest number $E = 4$ is LOW the output of the encoder is $A2 = $ LOW, $A1 = A0 = $ HIGH, no matter what the logical states of the other inputs may be. If $E = 4$ is HIGH, then $D = 3 = $ LOW gives $A2 = $ HIGH, $A1 = A0 = $ LOW and so on. Connect the input switches directly to the encoder and incorrect numbers will be displayed. This is, of course, because of the active-LOW outputs of the encoder. Use a 74HC04 hex inverter to invert each encoder output before it is applied to the seven-segment display. Test the circuit to check its correct operation.

(h) Open file ade 13.8b.ewb and activate the circuit.

(i) This circuit implements the suggestion made in part (g). Check its operation and note that the correct door numbers are displayed *except* that 0 is indicated for both door 0 open *and* for all doors closed.

(j) Modify the circuit so that when all doors and windows are closed the display is A.

EWB EXERCISE 13.8. DESIGN PROBLEM 13.9

Procedure:

(a) Open file ade 13.9.ewb and activate the circuit.

(b) The entrance and exit sensors are simulated by two switches that are labelled as I (IN) and O (OUT). The blue logic probe replaces the FULL display and the red logic probes would not be used in a practical design but are used here to aid in understanding the operation of the circuit. The initial state of the circuit should be set to zero by pressing key L on the keyboard to apply a LOW to the LOAD input.

(c) Simulate cars entering the car park by pressing the I key a number of times. Note that the count shown by the red probes increments every other time the key I is pressed. Why? When a count of 15 is reached – indicating that all 16 spaces in the car park are occupied – the blue logic probe turns ON. This acts as the FULL display. The HIGH signal would also be used to prevent the barrier at the entrance to the car park lifting to allow another car to enter.

(d) Now press the OUT (O) key a number of times. The count indicated by the red logic probes will now decrement each time and the FULL indicator will turn OFF.

(e) Step the count of the circuit to 15 and then press key I once more. Note that the count changes to 0 and the FULL display is OFF. Clearly, this ought not to happen. Design a method of preventing the count exceeding 15 and then modify the circuit accordingly. Check that the modified circuit works correctly.

(f) State how the design can be extended to cater for a 600 space car park

EWB EXERCISE 13.9. DESIGN PROBLEM 13.10

Procedure:

(a) Open file ade 13.10.ewb and activate the circuit.

(b) The circuit is a simulation of the problem posed in EWB Exercise 13.10. Observe that the count displayed on the right-hand seven-segment display resets to 0 after each digit while the left-hand display counts directly from 0 through to F.

EWB EXERCISE 13.10. DESIGN PROBLEM 13.11

Procedure:

(a) Open file ade 13.11.ewb and activate the circuit. Only the 'cars' in two lanes are timed in order to reduce the complexity of the circuit but the system operation for six lanes is merely an extension of this circuit.

(b) Ensure that the switches S (START) and F/G (FINISH for two different cars) are in the positions that take the ENP/ENT terminals LOW. Clear the counters to zero by pressing key L on the keyboard to take the LOAD terminals LOW and hence load the logical states of the A, B, C and D inputs into the four stages.

(c) Press keys F and G taking the two ENT terminals HIGH. It will be found that the circuit will not start.

(d) Now press key S. This action simulates the start microswitches being tripped and takes the two ENP terminals HIGH and the counters start their counting sequences. Note that the seven-segment displays indicate the time for which each 'car' has been running.

(e) Press first key F and then key G (or the other way around) to simulate microswitches operating as first one car and then the other reaching the finishing line. Note that the two counters stop and the displays are stationary.

(f) Note the clock frequency and calculate the time taken by each 'car' to travel over the race track. The clock frequency can be altered to give different timings and it may well be necessary to use two counters connected in cascade to obtain a higher count.

(g) Insert another counter into the circuit and use a second seven-segment display to increase the count and hence get a more precise timing of the 'cars'.

(h) Modify the circuit so that a light (or LED) illuminates to indicate which lane contained the winning 'car'.

EWB EXERCISE 13.11. DESIGN PROBLEM 13.12

Procedure:

(a) Open file ade 13.12.ewb and activate the circuit.

(b) The passing of a car over a pressure sensor on the surface of the road is simulated by the switch V (vehicle), which is operated by key V on the keyboard. Operating key V applies a HIGH voltage to the B serial input terminal of the 74HC164 shift register. Press key V so that a LOW is applied to the B terminal; this simulates zero cars passing the monitored point. Note that nothing happens in the circuit and the display indicates 0.

(c) Press key V to input a HIGH voltage to terminal B of the shift register. Note that the display starts to count from 0 towards F (15). Each step in the count corresponds to 255 cars passing over the pressure sensor. [There is a practical difficulty here; what is it and how may it be overcome?]

(d) Confirm the correct operation of the circuit. Then modify it so that when the displayed count reaches F a pulse is inputted to another counter. This counter can also have its Q outputs connected to a seven-segment display. Check the operation of the modified circuit and state what the maximum count of the circuit is now.

(e) Open file ade 13.12a.ewb and activate the circuit.

(f) When the count of the previous circuit reaches F the output of the 4-input AND gate goes HIGH and an input pulse is delivered to the second counter. As a result the right-hand display indicates 1. The circuit will repeat the counting sequence until, eventually, the right-hand display indicates F. What is then the count of the circuit?

(g) Modify the circuit so that it resets to zero when the right-hand display indicates F. Check that the modified circuit works correctly.

EWB EXERCISE 13.12. DESIGN PROBLEM 13.13

Procedure:

(a) Open file ade 13.13.ewb and activate the circuit.
(b) A fault in the circuit is indicated by the test point going LOW. When the circuit is working correctly the test point is HIGH. The fault status of the monitored circuit is simulated by the switch marked as I (IN). When this switch is in the HIGH position the green logic probe is ON to indicate the correct operation of the monitored circuit. When a fault occurs the IN terminal goes LOW; simulate this by pressing key I on the keyboard. The green logic probe now turns OFF and both the buzzer and the red logic probe turn ON. The buzzer can be turned OFF by pressing key B.
(c) When the fault has been cleared and the input to the circuit is HIGH (press key I to simulate this), the red logic probe stays ON and the green logic probe stays OFF. Press the R key to reset the flip-flop and reset the circuit. The green logic probe is now ON and the red logic probe is OFF.
(d) The correct operation of the red logic probe can be tested by pressing key T (TEST).

EWB EXERCISE 13.13. DESIGN PROBLEM 13.14

Procedure:

(a) Open file ade 13.14.ewb and activate the circuit.
(b) Double click on the logic analyser icon to access its controls and display. Observe the waveforms at the specified points in the circuit. Note the frequency to which the pulse input waveform is set. Test the circuit at a few other frequencies.
(c) To reduce the complexity of the circuit the two flip-flops used are generic devices. Click on the flip-flop icon on the toolbar and select a practical J-K flip-flop from the drop-down menu. Select a flip-flop and drag it on to the circuit window. Modify the circuit to include this device instead of the two generic devices. Test the modified circuit to confirm that it still works correctly.

EWB EXERCISE 13.14. DESIGN PROBLEM 13.15

Procedure:

(a) Open file ade 13.15.ewb and activate the circuit.
(b) To keep the circuit easy to see only the circuitry monitoring one door is shown. The logic for the entry light beams is $A1\overline{A2}$ and $\overline{A1}A2$ for exit; this is simulated by the switches labelled as I (IN) and O (OUT) which are operated by keys I and O, respectively.

(c) Operate the I and O switches to simulate people entering or leaving the hall by this one door and see if the circuit keeps a count of up to 15 (F).

(d) Extend the design, firstly to increase the count to 255 and, secondly, to monitor all four doors. The first is straightforward and merely requires two counters to be connected in cascade, but there are several different ways in which the circuit given can be extended, or modified, to cover all four doors. Try connecting the outputs of four IN and four OUT switches to OR gates and then to the rest of the circuit shown.

EWB EXERCISE 13.15. DESIGN PROBLEM 13.16

Procedure:

(a) Open file ade 13.16 and activate the circuit. The circuit should output a 1 μs pulse when an input pulse is applied to its PRE terminal.

(b) Double click on the CRO icon to access its controls and set the timebase to 5 μs/div. Press the space bar momentarily to input an (approximate) 1 ms input pulse to the circuit. Observe the input and output waveforms displayed by the CRO. When the input (blue) is LOW the output (red) is HIGH. When the input is taken HIGH momentarily by pressing the space bar a 1 μs negative-going pulse is generated.

(c) Modify the circuit so that a positive-going output pulse is generated.

(d) Remove the space bar switch from the circuit and instead connect the clock generator to the 1PRE terminal. Observe the input and output waveforms of the circuit.

Answers to exercises

Chapter 1

1.2 (a) 4.6 kHz. (b) 8 μs, 6 μs. (c) $t_r = 2$ μs and $t_f = 3$ μs. $\Delta v/\Delta t = (2 \times 10^{-6})/(0.8 \times 5) = 500$ ns/V.

1.3 (a) 5 ns. (b) (i) 5 ns (ii) 3.5 ns.

1.4 1 G, 2 H, 3 F, 4 D, 5 B, 6 A, 7 C, 8 H.

1.5 (a) $F_1 = AB$, $F_2 = AB\bar{C}$. (b) Yes. A 74HC can sink 4 mA and two 74LS inputs require $2 \times 0.4 = 0.8$ mA.

1.6 $R = 5/(8 \times 10^{-3} - 8 \times 10^{-6}) = 626$ Ω.

1.8 $C_{PD} = 22$ pF/gate from the data sheet. (a) $P_D = 4[(8 \times 10^{-12} + 22 \times 10^{-12}) \times 25 \times 10^6] = 3$ mW. $P_S = 4[5 \times 16 \times 10^{-6}] = 320$ μW. Total power dissipation $= 3.32$ mW. (b) $t_D = t'_d + 0.5R_oC_L$: $18 \times 10^{-9} = t'_D + 0.5 \times 5/(17 \times 10^3) \times 50 \times 10^{-12}$. $t'_D = 10.65$ ns. (c) Actual load capacitance $= 8$ pF. Hence, $t_D = 10.65 + 0.5 \times 5/(17 \times 10^{-3}) \times 8 \times 10^{-12} = 11.18$ ns.

1.10 Total load capacitance $= 100$ pF. $t_D = 80$ ns.

1.11 (b) 80 MHz if $C_L = 15$ pF, 65 MHz if $C_L = 50$ pF. (c) $P = V_{CC}(I_{CC} + n\Delta I_{CC}) + N_o(C_{PD}V_{CC}^2 f_i + C_L V_{CC}^2 f_o)$. On no-load $N_o = 0$, so $P - 5(20 \times 10^{-6} + 2 \times 1.5 \times 10^{-3}) = 15.1$ mW. (d) 2 V, 0.8 V.

1.12 (a) 1.4 V. (b) $8 \times 10^{-9} \times 10^{-3} = 38.4$ pJ. (d) $18/2 = 9$ ns.

1.13 $8/0.2 = 40$.

1.14 (a) B, its average propagation delay is less. (b) Equal at 60 μW. (c) B, its noise margin is $2.46 - 2.1 = 0.36$ V, as opposed to 0.46 V.

Chapter 2

2.2 (a) $F = (A + BC) \cdot \overline{AB} \cdot \overline{BC} = A\bar{B} + \bar{A}BC$. (b) $F = (A + \bar{C})(B + C)$.

2.3 (a) $F = (\bar{A}C + D) + \overline{\bar{A} + C} + \overline{\bar{B}D} = 1$. (b) $F = AC + B + CD$.

2.4 (a) $F = \overline{\overline{A}\overline{BC} \cdot \overline{A}\overline{BD} \cdot \overline{B}\overline{CD} \cdot \overline{B}\overline{C}} = (\overline{A} + B + \overline{C})(\overline{A} + B + D)(\overline{B} + \overline{C} + \overline{D})(\overline{B} + C) = (B + \overline{A} + \overline{C}D)(\overline{B} + \overline{C} + \overline{D})(\overline{B} + C) = (B\overline{C} + B\overline{D} + \overline{A}\overline{C} + \overline{A}D + \overline{A}B + \overline{C}D)(\overline{B} + C) = \overline{A}B + BC\overline{D} + \overline{A}C\overline{D} + \overline{B}\overline{C}D = \overline{A}B + BC\overline{D} + \overline{B}\overline{C}D$. (b) $F = \overline{A}\overline{D}(\overline{B} + C) + B(\overline{C} + D) = \overline{A}\overline{D}(\overline{B} + C) \cdot \overline{B(\overline{C} + D)} = \overline{A}\overline{D} + \overline{B} + C \cdot \overline{B} + \overline{C} + D = (\overline{A} + D + B\overline{C})(B + C\overline{D}) = \overline{A}B + \overline{A}C\overline{D} + BD + B\overline{C}$.

2.5 (a) $F = \overline{\overline{A}B(\overline{A}\overline{C})} = \overline{A}B\overline{C}$. (b) $F = \overline{\overline{A}\overline{B} \cdot \overline{B}\overline{C} \cdot \overline{A}\overline{C} \cdot \overline{A}\overline{B}\overline{C}} = (A + \overline{B})(B + \overline{C})(\overline{A} + C)(\overline{A} + B + \overline{C}) = (AB + A\overline{C} + \overline{B}\overline{C})(\overline{A} + C)(\overline{A} + B + \overline{C}) = (ABC + \overline{A}\overline{B}\overline{C})(\overline{A} + B + \overline{C}) = ABC + \overline{A}\overline{B}\overline{C}$. (c) $F = \overline{(A + \overline{B}C)(\overline{A}\overline{B}C)} = \overline{(A + \overline{B}C)} + (\overline{\overline{A}\overline{B}C}) = (\overline{A} + \overline{B} + \overline{C}) + A + \overline{B} + \overline{C} = A + \overline{B} + \overline{C}$. (d) $F = A(\overline{C} + \overline{D}) + C + (\overline{A + \overline{B}}) = A + \overline{C} \cdot D + C + \overline{A}B = A + B + C + D$.

2.6 (a) $F = (\overline{A} + \overline{B}) + (\overline{C} + \overline{AB}) = \overline{A} + \overline{B} + C \cdot AB = ABC + \overline{A} + \overline{B}$. (b) $F = \overline{A}\overline{B} + A\overline{B}\overline{C} + \overline{A}\overline{B}\overline{C} = \overline{A}(\overline{B} + B\overline{C}) + A\overline{B}\overline{C} = \overline{A}\overline{B} + \overline{A}\overline{C} + A\overline{B}\overline{C} = \overline{B}(\overline{A} + A\overline{C}) + \overline{A}\overline{C} = \overline{A}\overline{B} + \overline{B}\overline{C} + \overline{A}\overline{C}$.

2.7 (a) $F = \overline{(\overline{ABCD})(A\overline{B})} \cdot \overline{B}C = (\overline{ABCD} + A\overline{B})(\overline{B} + \overline{C}) = (\overline{A} + \overline{B} + \overline{C} + \overline{D} + \overline{A} + B)(\overline{B} + \overline{C}) = (\overline{A} + B + \overline{B} + \overline{C} + \overline{D})(\overline{B} + \overline{C}) = (1 + \overline{A} + \overline{C} + \overline{D})(\overline{B} + \overline{C}) = \overline{B} + \overline{C}$. (b) $F = \overline{A}\overline{B}C + A\overline{B}D + \overline{A}\overline{B}\overline{C} + ABCD = \overline{A}\overline{B} + A\overline{B}D + ABCD$. (c) $F = \overline{(A + B)(C + D)} = \overline{(A + B)} + \overline{(C + D)} = \overline{A}\overline{B} + \overline{C}\overline{D}$.

2.8 $F = \overline{(A\overline{B}C)(\overline{A}\overline{B})}(\overline{B}C) = (\overline{A}\overline{B}C + A\overline{B})(\overline{B}C) = (\overline{A} + B + \overline{C} + A\overline{B})(\overline{B} + \overline{C}) = \overline{A}\overline{B} + \overline{C}(B + \overline{B} + \overline{A} + A\overline{B}) + A\overline{B} = \overline{B} + \overline{C}$.

2.9 $F = (A + B)(A + \overline{B})C + \overline{A}B + ABC + \overline{A(\overline{B} + \overline{C})} = (A + A\overline{B} + AB)C + \overline{A}B + ABC + \overline{A} + \overline{B} + \overline{C} = AC + \overline{A}B + ABC + \overline{A} + BC = \overline{A} + C$.

2.10 (Method 1) $(A + B + C)(A + B + \overline{C})(A + \overline{B} + C)(\overline{A} + B + C)(\overline{A} + \overline{B} + C) = (A + BC)(A + \overline{B} + C)(\ )(\ ) = (A + BC)(\overline{A} + B + C)(\ ) = AB + AC + \overline{A}BC + BC + BC)(\overline{A} + \overline{B} + C) = (AB + AC + BC)(\overline{A} + \overline{B} + C) = \overline{A}BC + A\overline{B}C + AC + \overline{A}BC + BC = AC + BC$. (Method 2) $\Pi(CBA) = \Pi(0, 1, 2, 3, 4)$. SOP $= \Sigma(5, 6, 7) = A\overline{B}C + \overline{A}BC + ABC = A\overline{B}C + \overline{A}BC + ABC + ABC = AC(B + \overline{B}) + BC(A + \overline{A}) = AC + BC$.

2.11 (a) $\overline{F} = \overline{\overline{BC}(A + \overline{CD})} = BC + \overline{A} + \overline{CD} = BC + \overline{A}(\overline{C} + \overline{D})$.
(b) $F' = \overline{B} + \overline{C} + A(\overline{C} + D) = \overline{B}\overline{C} + A(\overline{C} + D)$. So, $\overline{F} = BC + \overline{A}(C + D)$.

2.12 $A = ID + (H + S)D + HE$.

2.13 Add redundant terms $\overline{A}B\overline{C}$ and ABC. Then, $F = \overline{A}\overline{B}\overline{C} + A\overline{B}\overline{C} + \overline{A}B\overline{C} + \overline{A}BC + \overline{A}BC + ABC + \overline{A}B\overline{C} + ABC = BC(A + \overline{A}) + \overline{B}\overline{C}(A + \overline{A}) + \overline{A}\overline{B}(C + \overline{C}) + \overline{A}B(C + \overline{C}) = BC + \overline{B}\overline{C} + \overline{A}(B + \overline{B}) = BC + \overline{B}\overline{C} + \overline{A}$.

2.14 (a) The truth table is

A	0	1	0	1	0	1	0	1	0	1	0	1	0	1	0	1
B	0	0	1	1	0	0	1	1	0	0	1	1	0	0	1	1
C	0	0	0	0	1	1	1	1	0	0	0	0	1	1	1	1
D	0	0	0	0	0	0	0	0	1	1	1	1	1	1	1	1
F_1	0	0	0	0	0	0	0	1	0	0	0	1	0	1	1	1
F_2	0	0	0	1	0	1	1	0	0	1	1	0	1	0	0	0

(b) $F_1 = ABC\overline{D} + AB\overline{C}D + A\overline{B}CD + \overline{A}BCD + ABCD$. $F_2 = AB\overline{C}\overline{D} + A\overline{B}\overline{C}D + \overline{A}B\overline{C}D + A\overline{B}C\overline{D} + \overline{A}BC\overline{D} + \overline{A}\overline{B}C\overline{D}$. (c) $F_1 = ABC(D + \overline{D}) + AB\overline{C}D + A\overline{B}CD + \overline{A}BCD$. To reduce further add three redundant $ABCD$ terms; then $F = ABC + ABD + ACD + BCD$. F_2 cannot be simplified.

2.15 The truth table is

A	0	1	0	1	0	1	0	1	0	1	0	1	0	1	0	1
B	0	0	1	1	0	0	1	1	0	0	1	1	0	0	1	1
C	0	0	0	0	1	1	1	1	0	0	0	0	1	1	1	1
D	0	0	0	0	0	0	0	0	1	1	1	1	1	1	1	1
F	0	0	0	0	1	1	1	1	1	1	1	0	0	0	0	0

From the truth table, $F = \bar{A}\bar{B}C\bar{D} + A\bar{B}C\bar{D} + \bar{A}BC\bar{D} + ABC\bar{D} + \bar{A}\bar{B}CD + A\bar{B}CD + \bar{A}B\bar{C}D$
$= \bar{B}C\bar{D} + BC\bar{D} + \bar{B}CD + \bar{A}B\bar{C}D = C\bar{D} + \bar{B}CD + \bar{A}B\bar{C}D$. Implementation requires one 74HC21 dual 4-input AND gate, one 74HC08 quad 2-input AND gate and one 74HC32A quad 2-input OR gate.

2.16 $F = A\bar{B} + \bar{A}B$. Implement with one 74HC84 quad 2-input exclusive-OR gate.

2.17 (a) $F = ABC + BC + A = A + BC$. (b) $F = A\bar{B}C + B + \bar{B}\bar{C} = AC + B + \bar{C}$. (c) $F = (AB + AC + \bar{B}C)\bar{A} = \bar{A}\bar{B}C$. (d) $F = (ABC + BC) + \bar{A}B + CD = BC + \bar{A}B + CD$.

2.18 (a) $F = \bar{B}C + AB\bar{C} + BC = C + AB\bar{C} = AB + C$. (b) $F = AC + \bar{A}B + BC = AC + \bar{A}B$. (c) $F = A + ABC + \bar{A}BC + \bar{A}B + AD + A\bar{D} = A + \bar{A}BC + \bar{A}B = \bar{A} + AB = \bar{A} + B$. (d) $F = (AC + CD + BC + ABC)(ACD + BD) = (AC + CD + BC)(ACD + BD) = ACD + ABCD + ACD + BCD + ABCD + ABCD + BCD = ACD + BCD$.

2.19 (a) (i) $F = ABC + AB$. $F' = (A + B + C)(A + B) = A + B$. $\bar{F} = \bar{A} + \bar{B}$. (ii) $\bar{F} = \overline{ABC + AB} = \overline{ABC} \cdot \overline{AB} = (\bar{A} + \bar{B} + \bar{C})(\bar{A} + \bar{B}) = \bar{A} + \bar{B}$. (b) (i) $F = (AB + C)(ABD + BC) = (A + B)C + (A + B + D)(B + C) = B + AC + CD$. $\bar{F} = \bar{B} + \bar{A}\bar{C} + \bar{C}\bar{D}$. (ii) $\bar{F} = \overline{(AB + C)(ABD + BC)} = \overline{(AB + C)} + \overline{(ABD + BC)} = \overline{AB} \cdot \bar{C} + \overline{ABD} \cdot \overline{BC} = (\bar{A} + \bar{B})\bar{C} + (\bar{A} + \bar{B} + \bar{D})(\bar{B} + \bar{C}) = \bar{B} + \bar{A}\bar{C} + \bar{C}\bar{D}$.

2.20 (a) (i) $\Sigma(0, 2, 3, 4, 6, 7, 9, 10, 11, 12, 13)$. (ii) $\Sigma(1, 3, 4, 5)$. (b) (i) $\Pi(0, 6, 7)$. (ii) $\Pi(0, 1, 3, 6, 7, 8, 9, 10, 11, 13, 15)$.

2.21 (a) (i) $F = \bar{A}\bar{B}CD + A\bar{B}CD + AB\bar{C}D + ABC = \bar{A}\bar{B}CD + A\bar{B}CD + AB\bar{C}D + ABCD + ABC\bar{D} = \Sigma(3, 5, 7, 8, 15)$. (ii) $F = A\bar{B}C + \bar{A}B + B\bar{C} = A\bar{B}C + A\bar{B}\bar{C} + \bar{A}B\bar{C} + AB\bar{C} + \bar{A}BC = \Sigma(0, 2, 3, 4, 5)$. (b) (i) $F = (A + B + C + D)(\bar{A} + \bar{B} + \bar{C} + \bar{D})(\bar{A} + B + \bar{C} + D)(A + \bar{B} + C + D)(A + \bar{B} + C + \bar{D}) = \Pi(0, 2, 5, 10, 15)$. (ii) $F = (A + \bar{B} + C)(\bar{A} + \bar{B} + \bar{C})(\bar{A} + B + \bar{C})(\bar{A} + B + C)(A + \bar{B} + \bar{C}) = \Pi(1, 2, 5, 6, 7)$.

Chapter 3

3.1

	A	B	C	D
1	$\bar{A}B\bar{C}\bar{D}$	0100 ✓	010-	10--
2	$A\bar{B}\bar{C}\bar{D}$	1000 ✓	100-	-01-
3	$\bar{A}\bar{B}C\bar{D}$	0010 ✓	10-0 ✓	
4	$\bar{A}\bar{B}CD$	0011 ✓	001- ✓	
5	$A\bar{B}\bar{C}D$	1001 ✓	-011	
6	$\bar{A}B\bar{C}D$	0101 ✓	0-11	
7	$A\bar{B}C\bar{D}$	1010 ✓	10-1 ✓	
8	$A\bar{B}CD$	1011 ✓	01-1	
9	$\bar{A}BCD$	0111 ✓	101- ✓	

The prime implicants are: $\bar{A}B\bar{C} + AB\bar{C} + \bar{B}CD + \bar{A}CD + \bar{A}BD + A\bar{B} + \bar{B}C$.

$A\bar{B}--$		$\bar{A}B\bar{C}-$		$-\bar{B}CD$		$\bar{A}-CD$		$\bar{A}B-D$		$-\bar{B}C-$		$A\bar{B}\bar{C}-$	
1000	1	0100	2	0011	12	0011	12	0101	10	0010	4	1000	1
1001	9	0101	10	1011	13	0111	14	0111	14	1010	5	1001	9
1010	5									0011	12		
1011	13									1011	13		

Prime implicant table

	1	2	3	4	5	6	7	8	9	10	11	12	13	14	15
$A\bar{B}$	✓				✓				✓				✓		
$\bar{A}B\bar{C}$		✓								✓					
$\bar{B}CD$												✓	✓		
$\bar{A}CD$												✓		✓	
$\bar{B}C$				✓	✓							✓	✓		
$\bar{A}BD$										✓				✓	
$A\bar{B}\bar{C}$	✓								✓						

The essential prime implicants are $A\bar{B}$, $\bar{A}B\bar{C}$, $\bar{B}C$ and either $\bar{A}CD$ or $\bar{A}BD$.

3.2 AB is covered by both of the other prime implicants. So, $F = \bar{A}\bar{B}\bar{C} + BCD$.

3.3 (a)

$F = \bar{A} + \bar{B}$

(b)

$F = \bar{B}\bar{C} + BC + AB$

(c)

$\bar{A} + \bar{B}$

$\bar{A} + B$

$A + B$

$F = \bar{A}B$

F

(d)

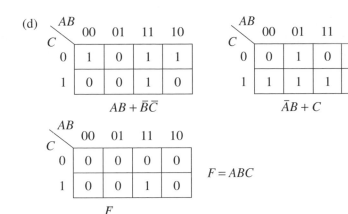

C \ AB	00	01	11	10
0	1	0	1	1
1	0	0	1	0

$$AB + \bar{B}\bar{C}$$

C \ AB	00	01	11	10
0	0	1	0	0
1	1	1	1	1

$$\bar{A}B + C$$

C \ AB	00	01	11	10
0	0	0	0	0
1	0	0	1	0

$F = ABC$

F

3.4 The map is

CD \ AB	00	01	11	10
00	1	1	0	1
01	1	1	0	0
11	0	0	1	0
10	1	0	1	1

From the map, $F = \bar{A}\bar{C} + \bar{B}\bar{D} + ABC$.
A static hazard exists between ABC and $\bar{B}\bar{D}$. This can be eliminated by adding the term $AC\bar{D}$.

3.5 (a) $F = AB + \bar{A}C + BC = AB + \bar{A}C$.
The map is

C \ AB	00	01	11	10
0	0	0	1	0
1	1	1	1	0

From the map, $F = AB + \bar{A}C$.

(b) $F = \bar{A}\bar{B}\bar{D} + \bar{A}BC + A\bar{B}C\bar{D} + \bar{A}\bar{B}D = \bar{A}\bar{B}(D + \bar{D}) + \bar{A}BC + A\bar{B}C\bar{D} = \bar{A}\bar{B} + \bar{A}BC + A\bar{B}C\bar{D} = \bar{A}\bar{B} + \bar{A}C + A\bar{B}C\bar{D} = \bar{A}\bar{B} + \bar{A}C + \bar{A}\bar{B} + A\bar{B}C\bar{D} = \bar{A}\bar{B} + \bar{A}C + \bar{B}(\bar{A} + AC\bar{D})$
$= \bar{A}\bar{B} + \bar{A}C + \bar{A}\bar{B} + \bar{B}C\bar{D} = \bar{A}\bar{B} + \bar{A}C + \bar{B}C\bar{D}$.

The map is

CD \ AB	00	01	11	10
00	1	0	0	0
01	1	0	0	0
11	1	1	0	0
10	1	1	0	1

From the map, $F = \bar{A}\bar{B} + \bar{A}C + \bar{B}C\bar{D}$.

(c) $F = \bar{A}B\bar{C}D + AB\bar{C}D + \bar{A}\bar{B}C\bar{D} + \bar{A}C = \bar{A}C(1 + \bar{B}\bar{D}) + \bar{A}B\bar{C}D + AB\bar{C}D = \bar{A}C +$
$\bar{A}B\bar{C}D + AB\bar{C}D = \bar{A}C(1 + BD) + \bar{A}B\bar{C}D + \bar{A}C(1 + \bar{B}D) + AB\bar{C}D = \bar{A}C +$
$\bar{A}BD (C + \bar{C}) + \bar{A}C + \bar{B}CD(A + \bar{A}) = \bar{A}C + \bar{A}BD + \bar{B}CD.$

The map is

CD \ AB	00	01	11	10
00	0	0	0	0
01	0	1	0	0
11	1	1	0	0
10	1	1	0	1

From the map, $F = \bar{A}C + \bar{A}BD + \bar{B}CD.$

3.6 The map is

CD \ AB	00	01	11	10
00	0	0	1	0
01	1	0	1	0
11	1	1	1	0
10	0	0	0	0

(a) From the map, $F = BCD + \bar{A}\bar{B}D + AB\bar{C}\bar{D}$; (b) $\bar{F} = B\bar{D} + \bar{A}D + C\bar{D} + \bar{A}B\bar{C}.$

3.7 (a)

CD \ AB	00	01	11	10
00	1	0	0	1
01	1	0	1	1
11	1	0	1	0
10	0	0	0	0

$F = \bar{B}\bar{C} + \bar{A}\bar{B}D + ABD.$

(b)

CD \ AB	00	01	11	10
00	1	0	0	0
01	0	0	0	0
11	1	0	0	1
10	1	1	1	1 '

$F = C\bar{D} + \bar{B}C + \bar{A}\bar{B}\bar{D}.$

(c)

CD \ AB	00	01	11	10
00	1	0	1	1
01	1	1	0	1
11	0	0	0	0
10	1	1	1	1

$$F = C\bar{D} + \bar{B}\bar{D} + \bar{A}CD + \bar{B}\bar{C}.$$

3.8

$\bar{E}$

CD \ AB	00	01	11	10
00	0	0	0	1
01	0	0	0	0
11	1	1	1	1
10	0	0	0	1

$$F = CD + A\bar{B}\bar{D}\bar{E} + \bar{A}\bar{B}\bar{D}E.$$

E

CD \ AB	00	01	11	10
00	1	0	0	0
01	0	0	0	0
11	1	1	1	1
10	1	0	0	0

3.9 Put into canonical form: $F = \bar{A}\bar{B}CDE + \bar{A}\bar{B}CD\bar{E} + ABCDE + \bar{A}BCDE + A\bar{B}\bar{C}\bar{D}\bar{E} + ABCD\bar{E} + \bar{A}BCD\bar{E} + A\bar{B}CD\bar{E} + \bar{A}\bar{B}\bar{C}DE + A\bar{B}CDE + A\bar{B}CD\bar{E} + \bar{A}\bar{B}CD\bar{E}.$

A	B	C	D	E
$A\bar{B}\bar{C}\bar{D}\bar{E}$	10000 ✓	10-00	-011- ✓	--11-
$\bar{A}\bar{B}\bar{C}\bar{D}E$	00001 ✓	00-01	--110 ✓	--11-
$\bar{A}\bar{B}CD\bar{E}$	00110 ✓	0011- ✓	0-11- ✓	--11-
$A\bar{B}C\bar{D}\bar{E}$	10100 ✓	0-110 ✓	-011- ✓	--11-
$\bar{A}\bar{B}C\bar{D}E$	00101 ✓	-0110 ✓	-011- ✓	
$\bar{A}\bar{B}CDE$	00111 ✓	001-1	--111 ✓	
$\bar{A}BCD\bar{E}$	01110 ✓	0-111 ✓	-111- ✓	
$A\bar{B}CD\bar{E}$	10110 ✓	-0111 ✓	1-11- ✓	
$\bar{A}BCDE$	01111 ✓	-1110 ✓		
$ABCD\bar{E}$	11110 ✓	1-110 ✓		
$A\bar{B}CDE$	10111 ✓	1011- ✓		
$ABCDE$	11111 ✓	-1111 ✓		
		1111- ✓		
		1-111 ✓		

Hence, $F = A\bar{B}\bar{D}\bar{E} + \bar{A}\bar{B}\bar{D}E + \bar{A}BCE + CD.$

$A\bar{B}\text{-}\bar{D}\bar{E}$		$\bar{A}\bar{B}\text{-}\bar{D}E$		$\bar{A}\bar{B}C\text{-}E$		$\text{-}\,\text{-}\,\text{-}\,CD$	
10000	1	00001	16	00101	20	00011	24
10100	5	00101	20	00111	28	10011	25
						01011	26
						11011	27
						00111	28
						10111	29
						01111	30
						11111	31

The prime implicant table shows that $\bar{A}\bar{B}CE$ is not required since it is covered by the other (essential) prime implicants.

3.10 Map 0s in cells $A\bar{B}\bar{C}\bar{D}$, $\bar{A}\bar{B}\bar{C}\bar{D}$, $ABC\bar{D}$, $\bar{A}B\bar{C}\bar{D}$, $A\bar{B}C\bar{D}$, $\bar{A}B\bar{C}\bar{D}$, $AB\bar{C}\bar{D}$ and $\bar{A}BC\bar{D}$.

CD \ AB	00	01	11	10
00	0	0	0	0
01	1	1	1	1
11	1	1	1	1
10	0	0	0	0

From the map, $F = D$ and $\bar{F} = \bar{D}$.

3.11 (a) $F = (A + B + C + D)(\bar{A} + B + C + D)(A + \bar{B} + C + D)(\bar{A} + \bar{B} + C + D)(\bar{A} + B + \bar{C} + D)(A + \bar{B} + \bar{C} + D)(\bar{A} + \bar{B} + \bar{C} + D)(A + B + C + \bar{D})(\bar{A} + B + C + \bar{D})$.

(b) Mapping 0s in cells $\bar{A}\bar{B}\bar{C}\bar{D}$, $A\bar{B}\bar{C}\bar{D}$, $\bar{A}B\bar{C}\bar{D}$, $AB\bar{C}\bar{D}$, $A\bar{B}C\bar{D}$, $\bar{A}\bar{B}C\bar{D}$, $ABC\bar{D}$, $\bar{A}\bar{B}\bar{C}D$ and $A\bar{B}\bar{C}D$ gives

CD \ AB	00	01	11	10
00	0	0	0	0
01	0	1	1	0
11	1	1	1	1
10	1	0	0	0

From the looped 1 cells, $F = BD + \bar{A}\bar{B}C + CD$.

3.12 (a) $F = AB + BCD + \bar{B}\bar{C}\bar{D}$. The map is

CD \ AB	00	01	11	10
00	0	0	1	0
01	0	0	1	0
11	0	1	1	0
10	1	0	1	1

From the 0 cells, $\bar{F} = \bar{A}\bar{C} + \bar{B}\bar{C} + \bar{B}D + \bar{A}B\bar{D}$.

(b) Hence, $F = (A + C)(B + C)(B + \bar{D})(A + \bar{B} + D)$. (c) $F = (AB + AC + BC + C)(AB + BD + A\bar{D} + \bar{B}\bar{D}) = (AB + C)(AB + BD + A\bar{D} + \bar{B}\bar{D}) = AB + ABD + AB\bar{D} + ABC + BCD + AC\bar{D} + \bar{B}C\bar{D} = AB + BCD + \bar{B}C\bar{D} + AC\bar{D} = AB + BCD + \bar{B}C\bar{D}$.

3.13

	A	B	C	D
$\bar{D}\bar{C}\bar{B}\bar{A}$ 1	$\bar{D}\bar{C}\bar{B}\bar{A}$	0000 10	1/2 00-0	10/14 -0-0
$\bar{D}\bar{C}\bar{B}A$ 2	$\bar{D}\bar{C}\bar{B}A$	0010 11	1/3 -000	17/18 -1-1
$\bar{D}\bar{C}B\bar{A}$ 3	$D\bar{C}\bar{B}\bar{A}$	1000 12	2/4 001-	
$\bar{D}C\bar{B}\bar{A}$ 4	$\bar{D}\bar{C}BA$	0011 13	2/6 -01-	
$D\bar{C}\bar{B}A$ 5	$\bar{D}C\bar{B}A$	0101 14	3/6 10-0	
$\bar{D}C\bar{B}\bar{A}$ 6	$D\bar{C}B\bar{A}$	1010 15	4/7 0-11	
$\bar{D}CBA$ 7	$\bar{D}CBA$	0111 16	5/7 01-1	
$D\bar{C}\bar{B}\bar{A}$ 8	$D\bar{C}BA$	1101 17	5/8 -101	
$D\bar{C}\bar{B}A$ 9	$DCBA$	1111 18	7/9 -111	
		19	8/9 11-1	

The prime implicants are: $\bar{A}\bar{B}\bar{C} + B\bar{C}\bar{D} + B\bar{C} + AB\bar{D} + AC\bar{D} + ACD + \bar{A}\bar{C} + AC$.

$\bar{A}\bar{B}\bar{C}-$	$-B\bar{C}\bar{D}$	$-B\bar{C}-$	$AB-\bar{D}$	$A-C\bar{D}$	$A-CD$	$\bar{A}-\bar{C}-$	$A-C-$
0000 0	0100 2	0100 2	1100 3	1010 5	1011 13	0000 0	1010 5
0001 8	1100 3	1100 3	1110 7	1110 7	1111 15	0100 2	1110 7
		0101 10				0001 8	1011 13
		1101 13				0101 10	1111 15

Prime implicant table:

	0	1	2	3	4	5	6	7	8	9	10	11	12	13	14	15
$\bar{A}\bar{B}\bar{C}$	✓								✓							
$B\bar{C}\bar{D}$			✓	✓												
$B\bar{C}$			✓	✓							✓			✓		
$AB\bar{D}$				✓				✓								
$AC\bar{D}$						✓		✓								
ACD														✓		✓
$\bar{A}\bar{C}$	✓		✓						✓		✓					
AC						✓		✓						✓		✓

The essential prime implicants give $F = B\bar{C} + \bar{A}\bar{C} + AC$.

3.14 The Karnaugh map is

CD\AB	00	01	11	10
00	0	0	0	0
01	0	1	1	1
11	1	1	1	1
10	0	0	0	0

From the map, SOP: $F = CD + BD + AD$.
POS: $F' = \bar{D} + \bar{A}\bar{B}\bar{C}$, so $F = D(A + B + C)$.

3.15 The truth table is

A	0	1	0	1	0	1	0	1	0	1	0	1	0	1	0	1
B	0	0	1	1	0	0	1	1	0	0	1	1	0	0	1	1
C	0	0	0	0	1	1	1	1	0	0	0	0	1	1	1	1
D	0	0	0	0	0	0	0	0	1	1	1	1	1	1	1	1
F	0	0	0	0	1	0	0	0	1	1	0	0	1	1	1	0
G	0	1	1	1	0	0	1	1	0	0	0	1	0	0	0	0
H	1	0	0	0	0	1	0	0	0	0	1	0	0	0	0	1

The Karnaugh maps for the three outputs are

F map:

CD \ AB	00	01	11	10
00	0	0	0	0
01	1	0	0	1
11	1	1	0	1
10	1	0	0	0

G map:

CD \ AB	00	01	11	10
00	0	1	1	1
01	0	0	1	0
11	0	0	0	0
10	0	1	1	0

H map:

CD \ AB	00	01	11	10
00	1	0	0	0
01	0	1	0	0
11	0	0	1	0
10	0	0	0	1

From the maps, $F = \bar{B}D + \bar{A}CD + \bar{A}\bar{B}C$, $G = BC + AB\bar{D} + A\bar{C}\bar{D}$, $H = \bar{A}\bar{B}\bar{C}\bar{D} + \bar{A}B\bar{C}D + AB\bar{C}\bar{D}$.

3.16 The truth table is

A	0	1	0	1	0	1	0	1	0	1	0	1	0	1	0	1
B	0	0	1	1	0	0	1	1	0	0	1	1	0	0	1	1
C	0	0	0	0	1	1	1	1	0	0	0	0	1	1	1	1
D	0	0	0	0	0	0	0	0	1	1	1	1	1	1	1	1
F'	C	1	D	0	C	1	D	0	C	1	D	0	C	1	D	0
F	0	1	0	0	1	1	0	0	0	1	1	0	1	1	1	0

The Karnaugh map is

$\begin{array}{c}AB\\CD\end{array}$	00	01	11	10
00	1	1	1	1
01	0	1	1	0
11	0	0	0	0
10	0	0	1	1

From the 1 cells, $F = \bar{C}\bar{D} + B\bar{C} + A\bar{D}$.

3.17 The truth table is

A	0	1	0	1	0	1	0	1	0	1	0	1	0	1	0	1
B	0	0	1	1	0	0	1	1	0	0	1	1	0	0	1	1
C	0	0	0	0	1	1	1	1	0	0	0	0	1	1	1	1
D	0	0	0	0	0	0	0	0	1	1	1	1	1	1	1	1
F	0	0	0	0	0	1	0	1	0	0	0	0	0	1	0	1
G	0	0	0	0	0	0	1	1	0	1	0	1	0	1	1	0
H	0	0	0	0	0	0	0	0	0	0	1	1	0	0	1	0
I	0	0	0	0	0	0	0	0	0	0	0	0	0	0	0	1

The Karnaugh maps are

F

$\begin{array}{c}AB\\CD\end{array}$	00	01	11	10
00	0	0	0	0
01	0	0	0	0
11	0	0	1	1
10	0	0	1	1

G

$\begin{array}{c}AB\\CD\end{array}$	00	01	11	10
00	0	0	0	0
01	0	0	1	1
11	0	1	0	1
10	0	1	1	0

H

$\begin{array}{c}AB\\CD\end{array}$	00	01	11	10
00	0	0	0	0
01	0	1	1	0
11	0	1	0	0
10	0	0	0	0

From the maps, $F = AC$, $G = \bar{A}BC + BC\bar{D} + A\bar{C}D + A\bar{B}D$, $H = B\bar{C}D + \bar{A}BD$, $I = ABCD$.

3.18 For each sum term write the binary value, thus: $F = (0101)(1010)(1110)(0011)$ $(0001)(1111)(0111)(1011)$. The mapping is

CD\AB	00	01	11	10
00	1	1	1	1
01	0	0	1	1
11	0	0	0	0
10	1	1	0	0

From the map, (a) $F = \bar{C}\bar{D} + A\bar{C} + \bar{A}D$. (b) $F' = \bar{A}D + CD + AC$ and $F = (\bar{A} + D)(C + D)(A + C)$

3.19 (a) $F = \Pi(0, 2, 3, 5) = (A + \bar{B} + C)(\bar{A} + \bar{B} + C)(\bar{A} + B + \bar{C})(A + B + C)$.
(b) The mapping is

C\AB	00	01	11	10
0	0	0	0	1
1	1	1	1	0

From the 0 cells, $\bar{F} = \bar{A}\bar{C} + B\bar{C} + A\bar{B}C$.

3.20 Figure 3.8 shows the output waveform when C changes from 0 to 1 and then from 1 to 0.

3.21 $\bar{A}B$ is necessary to cover 6. It also covers 2, 10 and 14. BD covers 10, 11 14 and 15. $A\bar{B}\bar{C}$ covers 1 and 9. $A\bar{B}D$ covers 5, $A\bar{C}D$ covers 8. This leaves 0 and this can be covered by either $\bar{A}\bar{C}\bar{D}$ or $\bar{B}\bar{C}\bar{D}$.

3.22 The output waveform is shown in Fig. 3.9.

Fig. 3.8

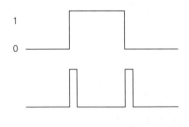

Fig. 3.9

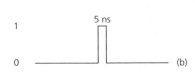

3.23 The truth table is

A	0	1	0	1	0	1	0	1
B	0	0	1	1	0	0	1	1
C	0	0	0	0	1	1	1	1
X	0	0	0	1	0	1	1	1
Y	0	1	0	1	0	1	1	1
Z	0	1	1	1	0	0	1	1 .

$X = AB\bar{C} + A\bar{B}C + \bar{A}BC + ABC = AC + AB + BC$
$Y = A\bar{B}\bar{C} + AB\bar{C} + A\bar{B}C + \bar{A}BC + ABC = A + BC$
$Z = A\bar{B}\bar{C} + \bar{A}B\bar{C} + AB\bar{C} + \bar{A}BC + ABC = B + AC.$

3.24 The truth table is

P		Q			Product			
B	A	E	D	C	Z	Y	X	W
0	0	0	0	0	0	0	0	0
0	0	0	0	1	0	0	0	0
0	0	0	1	0	0	0	0	0
0	0	0	1	1	0	0	0	0
0	0	1	0	0	0	0	0	0
0	0	1	0	1	0	0	0	0
0	1	0	0	0	0	0	0	0
0	1	0	0	1	0	0	0	1
0	1	0	1	0	0	0	1	0
0	1	0	1	1	0	0	1	1
0	1	1	0	0	0	1	0	0
0	1	1	0	1	0	1	0	1
1	0	0	0	0	0	0	0	0
1	0	0	0	1	0	0	1	0
1	0	0	1	0	0	1	0	0
1	0	0	1	1	0	1	1	0
1	0	1	0	0	1	0	0	0
1	0	1	0	1	1	0	1	0
1	1	0	0	0	0	0	0	0
1	1	0	0	1	0	0	1	1
1	1	0	1	0	0	1	1	0
1	1	0	1	1	1	0	0	1
1	1	1	0	0	1	1	0	0
1	1	1	0	1	1	1	1	1

From the truth table,
$W = A\bar{B}C\bar{D}\bar{E} + A\bar{B}CD\bar{E} + A\bar{B}CD\bar{E} + ABCD\bar{E} + ABCD\bar{E} + ABC\bar{D}\bar{E} = AC\bar{E} + AC\bar{D}$
$X = A\bar{B}\bar{C}D\bar{E} + A\bar{B}CD\bar{E} + \bar{A}BCD\bar{E} + \bar{A}BCD\bar{E} + \bar{A}BC\bar{D}E + ABC\bar{D}\bar{E} + AB\bar{C}D\bar{E} + ABCD\bar{E}$
$\quad = BC\bar{D} + A\bar{B}D\bar{E} + \bar{A}BC\bar{E} + AC\bar{D}\bar{E}$
$Y = A\bar{B}\bar{C}DE + A\bar{B}CDE + \bar{A}BC\bar{D}\bar{E} + \bar{A}BCD\bar{E} + AB\bar{C}D\bar{E} + ABC\bar{D}\bar{E} + ABCD\bar{E}$
$\quad = A\bar{D}E + \bar{A}BD\bar{E} + B\bar{C}D\bar{E}$
$Z = \bar{A}B\bar{C}DE + \bar{A}BC\bar{D}E + ABCD\bar{E} + AB\bar{C}DE + ABC\bar{D}E = B\bar{D}E + ABCD\bar{E}.$

3.25 The truth table is

	0	1	2	3	4	5	6	7	8	9
A	0	0	0	0	0	1	1	1	1	1
B	0	1	1	1	0	0	1	1	1	0
C	0	0	1	1	1	1	1	1	0	0
D	1	1	1	0	0	0	0	1	1	1
W	0	1	0	1	0	1	0	1	0	1
X	0	0	1	1	0	0	1	1	0	0
Y	0	0	0	0	1	1	1	1	0	0
Z	0	0	0	0	0	0	0	0	1	1

$$W = \bar{A}B\bar{C}D + \bar{A}BC\bar{D} + A\bar{B}C\bar{D} + ABCD + A\bar{B}\bar{C}D$$
$$X = \bar{A}BCD + \bar{A}BC\bar{D} + ABC\bar{D} + ABCD = BC$$
$$Y = \bar{A}\bar{B}C\bar{D} + A\bar{B}C\bar{D} + ABC\bar{D} + ABCD = ABC + \bar{B}C\bar{D}$$
$$Z = AB\bar{C}D + A\bar{B}\bar{C}D + A\bar{C}D.$$

3.26 The truth table is

D	C	B	A		Z	Y	X	W
0	0	0	0		0	0	1	1
0	0	0	1		0	1	0	0
0	0	1	0		0	1	0	1
0	0	1	1		0	1	1	0
0	1	0	0		0	1	1	1
0	1	0	1		1	0	0	0
0	1	1	0		1	0	0	1
0	1	1	1		1	0	1	0
1	0	0	0		1	0	1	1
1	0	0	1		1	1	0	0

The remaining numbers are 'don't cares', i.e. $10 = \bar{A}B\bar{C}D$, $11 = AB\bar{C}D$, $12 = \bar{A}\bar{B}CD$, $13 = A\bar{B}CD$, $14 = \bar{A}BCD$ and $15 = ABCD$. From the table and a Karnaugh mapping, $W = \bar{A}$, $X = AB + \bar{A}\bar{B}$, $Y = A\bar{C} + B\bar{C} + \bar{A}\bar{B}C$ and $Z = D + AC + BC$.

Chapter 4

4.1 (a) (i) $F' = C + (A + B)(\bar{A} + \bar{B}) = C + A\bar{B} + \bar{A}B$. $\bar{F} = \bar{C} + \bar{A}B + A\bar{B}$.

(ii) The map is

AB / C	00	01	11	10
0	0	0	0	0
1	1	0	1	0

From the 0 cells, $\bar{F} = \bar{C} + \bar{A}B + A\bar{B}$.

(b) $F' = A + C + AB + \bar{A}\bar{B} = A + C + \bar{A}\bar{B} = A + \bar{B} + C.$ $\bar{F} = \bar{A} + B + \bar{C}.$
The map is

AB C	00	01	11	10
0	0	0	0	0
1	0	0	0	1

From the 0 cells, $\bar{F} = \bar{C} + \bar{A} + B.$

4.2 The map is

AB CD	00	01	11	10
00	1	1	1	1
01	0	0	0	1
11	0	0	1	1
10	1	1	0	0

(a) From the map, $F = \bar{C}\bar{D} + ACD + A\bar{B}\bar{C}.$
(b) Add term $A\bar{B}D$ to give $F = \bar{C}\bar{D} + ACD + A\bar{B}D + A\bar{B}\bar{C}.$

4.3 The map is

AB CD	00	01	11	10
00	0	0	0	0
01	0	0	0	0
11	1	1	1	1
10	0	1	1	1

(a) From the 1 cells, $F = CD + BC + AC.$
(b) From the 0 cells $F' = \bar{C} + \bar{A}\bar{B}\bar{D}$, hence, $F = C(A + B + D).$

4.4 The map is

AB CD	00	01	11	10
00	0	0	0	0
01	1	1	0	0
11	1	1	1	1
10	1	0	1	1

(a) From the 1 cells, $F = \bar{B}C + AC + \bar{A}D$.

This requires five 2-input NAND gates and one 3-input gate but only 74HC00 2-input NAND gate ICs are available. The required circuit is shown in Fig. 4.24. It requires two 74HC00 ICs.

Fig. 4.24

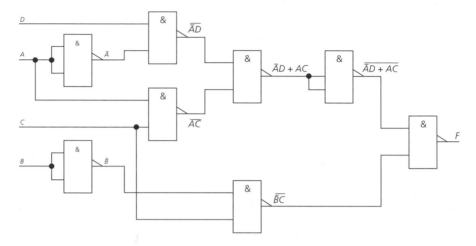

(b) From the 0 cells, $\bar{F} = \bar{C}\bar{D} + A\bar{C} + \bar{A}B\bar{D}$ and $F = \overline{\bar{C}\bar{D} + A\bar{C} + \bar{A}B\bar{D}} = (C + D)(\bar{A} + C)(A + \bar{B} + D)$. Implementation requires two 2-input NOR gates and two 3-input NOR gates, i.e. one 74HC02 and 74HC27.

4.5 (a) $F = \bar{A}\bar{B}\bar{C} + ABC$. (b) $F' = A\bar{B} + B\bar{C} + \bar{A}C$, hence, $F = (\bar{A} + B)(\bar{B} + C)(A + \bar{C})$.

4.6 $F_1 = \bar{A}\bar{B}D + \bar{B}C\bar{D}$ and $F_2 = \bar{A}BD + \bar{A}BC + A\bar{B}C\bar{D}$. F_1 can be implemented with one triple 3-input NAND gate, e.g. the 74HC10. F_2 requires one 74HC10 and one dual 4-input NAND gate IC, e.g. the 74HC20.

4.7 The map is

	AB			
CD	00	01	11	10
00	0	×	1	1
01	×	1	1	1
11	0	0	0	0
10	0	0	1	1

(a) From the map, NAND: $F = A\bar{C} + \bar{C}D + A\bar{D}$. NOR: $F' = \bar{A}\bar{B} + CD + \bar{A}C$, hence, $F = (A + B)(\bar{C} + \bar{D})(A + \bar{C})$.

(b) $F = \bar{A}\bar{B}\bar{C} + \bar{A}\bar{B}C + \bar{A}B\bar{C} + ABC$. The map is

	AB			
C	00	01	11	10
0	1	1	0	0
1	1	0	1	0

(i) NAND: $F = \bar{A}\bar{B} + \bar{A}\bar{C} + ABC$. (ii) NOR: $F' = A\bar{C} + A\bar{B} + \bar{A}BC$, hence, $F = (\bar{A} + C)(\bar{A} + B)(A + \bar{B} + \bar{C})$.

4.8 (a) $ACD + \bar{A}BCD + \bar{B}CD + \bar{A}B\bar{C}\bar{D} + \bar{A}\bar{B}CD + A\bar{B}C\bar{D} = ABCD + A\bar{B}CD + \bar{A}BCD + A\bar{B}\bar{C}D + \bar{A}\bar{B}\bar{C}D + \bar{A}\bar{B}CD + \bar{A}BCD + A\bar{B}C\bar{D} = CD(AB + A\bar{B} + \bar{A}B + \bar{A}\bar{B}) + \bar{C}D(A\bar{B} + \bar{A}\bar{B}) + C\bar{D}(A\bar{B})$. $\bar{C}\bar{D}$: no term, so connect D_0 to 0. $C\bar{D}$: $A\bar{B}$ connected to D_1. $\bar{C}D$: $(A\bar{B} + \bar{A}\bar{B})$ connected to D_2. CD: $(AB + A\bar{B} + \bar{A}B + \bar{A}\bar{B} = A(B + \bar{B}) + \bar{A}(B + \bar{B}) = \bar{A} + A = 1)$, connect D_3 to 1.

4.9 $F = A\bar{B}C + \bar{A}B\bar{C}$. The map is

$$
\begin{array}{c|c|c|c|c}
 & \multicolumn{4}{c}{AB} \\
C & 00 & 01 & 11 & 10 \\
\hline
0 & 0 & 1 & 0 & 0 \\
\hline
1 & 0 & 0 & 0 & 1 \\
\end{array}
$$

From the 0 cells, $F' = \bar{A}\bar{B} + A\bar{C} + BC$ and hence $F = (A + B)(\bar{A} + C)(\bar{B} + \bar{C})$.

4.10 $A\bar{B}C + \bar{A}B\bar{C} + A\bar{B}\bar{C} + \bar{A}BC = A\bar{B}(C) + \bar{A}B(\bar{C}) + A\bar{B}(\bar{C}) + (\bar{A}B)C$. $\bar{A}\bar{B}$: connect D_0 to C. $A\bar{B}$: connect D_1 to $C + \bar{C} = 1$. $\bar{A}B$: connect D_2 to $\bar{C}$. AB: connect D_3 to 0.

4.11 (a) 4-to-1 multiplexer. (b) Multiplexes four lines to one or performs parallel-to-serial conversion. (c) (i) 34.15 V. (ii) 4.4 V. (d) (i) 1.35 V. (ii) 0.1 V. (e) (i) 38 ns. (ii) 35 ns. (iii) 24 ns.

4.12 The truth table is given in Table 4.8. (a) From the table, $F = A\bar{B}\bar{C} + \bar{A}B\bar{C} + \bar{A}\bar{B}C + ABC$ NAND. (b) $F = \bar{C}(A\bar{B} + \bar{A}B) + C(\bar{A}\bar{B} + AB)$ NOR.

4.13 (a) Table 4.9 gives the truth table. From the truth table, connect: D_0 to D, D_1 to $\bar{D}$, D_2 to 1, D_3 to 0, D_4 to D, D_5 to 1, D_6 to $\bar{D}$ and D_7 to 1. (b) Simplifying gives $F = AC +$

Table 4.8

A	0	1	0	1	0	1	0	1
B	0	0	1	1	0	0	1	1
C	0	0	0	0	1	1	1	1
F	0	1	1	0	1	0	0	1

Table 4.9

A	0	1	0	1	0	1	0	1	0	1	0	1	0	1	0	1
B	0	0	1	1	0	0	1	1	0	0	1	1	0	0	1	1
C	0	0	0	0	1	1	1	1	0	0	0	0	1	1	1	1
D	0	0	0	0	0	0	0	0	1	1	1	1	1	1	1	1
F	0	1	1	0	0	1	1	1	1	0	1	0	1	1	0	1
	D_0	D_1	D_2	D_3	D_4	D_5	D_6	D_7	D_0	D_1	D_2	D_3	D_4	D_5	D_6	D_7

$A\bar{B}\bar{D} + \bar{A}B\bar{C} + \bar{A}BD + BC\bar{D}$. NAND gates required are: one 5-input, four 3-input and one 2-input.

4.14 $F = ABC + A\bar{B}C + AB\bar{C} + \bar{A}B\bar{C} + \bar{A}\bar{B}C$. $D_0 = D_1 = D_6 = 0$. Other data inputs = 1.

4.15 (a) 8-to-1 line multiplexer, quad 2-to-1 line multiplexer. (b) LOW for both. (c) G1 indicates AND dependency between the S input and the inputs I_0 and I_1 of each multiplexer. G denotes AND dependency between the S_0, S_1, S_2 and S_3 inputs and the seven data inputs. (d) 8-to-1 line has one HIGH and one LOW. The 2-to-1 line has active-HIGH output. (e) The circuit is enabled by $E_1.E_2E_3$.

4.16 Comparing with equation (4.2), D_0, D_2, D_4 and $D_7 = 0$, other data inputs = 1.

4.17 The map is

CD\AB	00	01	11	10
00	1	1	1	1
01	1	1	0	1
11	0	0	1	1
10	0	0	1	0

(a) From the 1 cells, $F = \bar{C}\bar{D} + \bar{A}\bar{C} + ABC + A\bar{B}D$. (b) From the 0 cells, $F' = \bar{A}C + \bar{B}C\bar{D} + AB\bar{C}D$, hence, $F = (A + \bar{C})(B + \bar{C} + D)(\bar{A} + \bar{B} + C + \bar{D})$.

Chapter 5

5.1 (a) $P = Q$ output is HIGH and $P > Q$ output is LOW. (b) $P = Q$ output is HIGH and so is the $P > Q$ output. (c) $P = Q$ output is LOW and $P > Q$ output is HIGH.

5.2 (a) Connect C_{in} of the least-significant adder (LSA) to earth, C_{out} of the LSA to C_{in} of the MSA. C_{out} of MSA gives any carry into a ninth bit.

5.3 Connect C_{in} of the least-significant adder to earth. For the other three adders connect the C_{in} terminal to the C_{out} terminal of the preceding adder. The C_{out} terminal of the most-significant adder is the carry-out of the 16-bit adder. Apply the 16-bit digital words to the A_1, A_2, A_3 and A_4/B_1, B_2 B_3 and B_4 terminals. The 16-bit sum is taken from the Σ terminals.

5.4 (a) Noise margin = $V_{IL(max)} - V_{OL(max)} = 1.35 - 0.33 = 1.02$ V, or $V_{OH(min)} - V_{IH(min)} = 3.84 - 3.15 = 0.69$ V. (b) (i) Y_5. (ii) Y_7. (c) All HIGH.

5.5 $F = \Sigma(0, 7)$ so connect outputs 0 and 7 to the inputs of a 2-input NAND gate. $G = \bar{A}\bar{B}C + A\bar{B}C + \bar{A}B\bar{C} + \bar{A}BC = \Sigma(2, 4, 5, 6)$. So connect outputs 2, 4, 5 and 6 to the inputs of a 4-input NAND gate.

5.6 Possible devices are: 75ALS157, 74AS157, 75F157, 74LS157, 74AHC157, 74AHCT157, 74HC157, 74HCT157 and 74LVC157. If 3.3 V operation is to be used then only the 74HC, 74AHC and 74LVC devices remain in contention. For operation at 30 MHz probably the best choice is the 74AHC device but it is a bit dearer and perhaps harder to obtain.

5.7 Connect the 32 inputs to four 74HC151 with their select inputs ABC connected in parallel. Connect the outputs of the four 151s to the inputs of the 153. Connect select inputs D and E to the 151 and take the output of the circuit from the output of the 151.

5.8 (a) G0/7: inputs select any one of the eight outputs. &: G_1 must be HIGH AND both $\overline{G2A}$ and $\overline{G2B}$ must be LOW for the decoder to be enabled. The outputs are active-LOW. (b) When the IC is used as a decoder the enable inputs function to enable/disable the circuit. When the device is used as a demultiplexer the enable inputs act as the data inputs. Any one of the enables can be used as the data input, then the circuit acts as a demultiplexer.

5.9 Parallel the A, B and C inputs. Connect input D to both $\overline{G2A}$ on IC A and to G_1 on the other IC B. Connect the data input to $\overline{G2B}$ on both IC A and IC B. Connect G_1 on IC A to 5 V and connect $\overline{G2A}$ on IC B to earth. The output is then all eight outputs on IC A plus the four lowest outputs on IC B

5.10 $A = B$: $(\overline{A_2 \oplus B_2})(\overline{A_1 \oplus B_1})(\overline{A_0 \oplus B_0})$.
$A \neq B$: first test the MSB. If not equal the number with MSB = 1 is the larger. If the MSBs are equal the next bit is tested and whichever number has 1 in this position is the larger. If both bits are equal test the LSB. Thus,
$A > B$: $A_2\overline{B_2} + (\overline{A_2 \oplus B_2})A_1\overline{B_1} + (\overline{A_2 \oplus B_2})(\overline{A_1 \oplus B_1})A_0\overline{B_0}$.
$B > A$: $\overline{A_2}B_2 + (\overline{A_2 \oplus B_2})\overline{A_1}B_1 + (\overline{A_2 \oplus B_2})(\overline{A_1 \oplus B_1})\overline{A_0}B_0$.

5.11 The truth table is

A	0	1	0	1	0	1	0	1
B	0	0	1	1	0	0	1	1
C	0	0	0	0	1	1	1	1
F	0	0	0	1	0	1	1	1

From the table, $F = AB\overline{C} + A\overline{B}C + \overline{A}BC + ABC = AB + AC + BC = (A + B)(A + C)(B + C)$.

5.12 The truth table is

A	0	1	0	1	0	1	0	1
B	0	0	1	1	0	0	1	1
C	0	0	0	0	1	1	1	1
X	0	0	1	0	1	0	1	0
Y	0	1	0	1	0	1	0	1

From the table, $X = \overline{A}B\overline{C} + \overline{A}\overline{B}C + \overline{A}BC = \overline{A}B + \overline{A}C$. $Y = A\overline{B}\overline{C} + AB\overline{C} + A\overline{B}C + ABC = A$.

Fig. 6.32

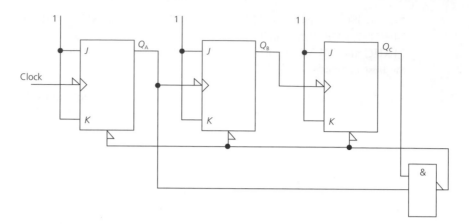

Fig. 6.33

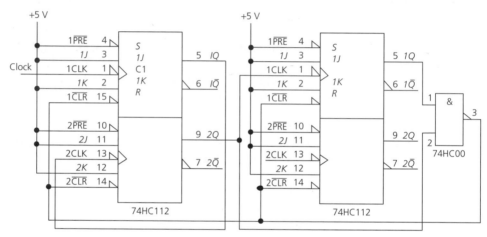

Chapter 6

6.1 (a) Maximum clock frequency $= 1/[14 \times 20 \times 10^{-9}] = 3.57$ MHz. (b) (i) Dual 4-bit non-synchronous counter. (ii) Trailing-edge triggered. (iii) No programming is possible. (iv) Non-synchronous.

6.2 (a) It is an octal D latch. (b) When $\overline{OE}$ is HIGH. (c) When LE is LOW. (d) High output capability.

6.3 The divide-by-5 ripple counter is shown in Fig. 6.32. Disadvantages are (i) that it is relatively slow and (ii) glitches occurs at short count 5.

6.4 The counter is shown in Fig. 6.33.

6.5 The circuit should clear when the count reaches 13. $13_{10} = 1101$ so connect outputs Q_A, Q_B and Q_D to the inputs of a 3-input NAND gate. The output of the NAND gate is then connected to the $\overline{CLR}$ input of each flip-flop.

6.6 The count sequence is shown in Table 6.6.

Table 6.6

Clock pulse	Q_D	Q_C	Q_B	Q_A
0	0	0	0	0
1	1	1	1	1
2	1	1	1	0
3	1	1	0	1
4	1	0	1	0

Initially, all stages are cleared so the outputs of all gates are 1. Hence, all stages will toggle at the leading edge of the first clock pulse. After one clock pulse all stages are set, hence all gate outputs are 0. Only flip-flop A toggles at clock 2. When clock pulse 3 arrives only Q_A is at 0 and hence only the first AND gate has an output at 1. This means that flip-flops A and B both toggle. For clock pulse 4, the outputs of the first two AND gates are at 1 and the output of the last AND gate is at 0. Stages A, B and C all toggle.

Carrying on in this way gives the count sequence as 0000, 1111, 1110, 1101, 1010, 0101, 1010, 0101 indefinitely.

6.7 Counts 1000 and 0000 do not occur.

6.8 Following the same procedure as in exercise 6.6 the count sequence is 100, 011, 010, 001, 000, 100, etc.

6.9 The counting sequence is 0, 1, 2, 3, 4, 5, 6, 7, 8, 9, 0, etc.

6.10 Use the 74HC191 4-bit synchronous up-down counter. Connect the D/$\overline{U}$ terminal at pin 6 to the Q output of the J K flip-flop 74HC112. Connect the clock generator to the CLK terminal at pin 14 and the RCO output to the CLK input of the J-K flip-flop. Connect both J and K terminals to 1 so that the flip-flop toggles. When the count reaches 15 (or 0) the RCO output will go LOW (or HIGH), and set or reset the flip-flop to alter the voltage level of the D/$\overline{U}$ pin.

6.11 $P = 5 \times I_{DC}$ per flip-flop $= 5 \times 80\ \mu A = 400\ \mu A$ static dissipation.

6.12 (a) Modulus $= 2^6 = 64$. (b) $f_{max} = 1/(nt_{pd}) = 1/(6 \times 40 \times 10^{-9}) = 4.17$ MHz.

6.13 (a) (i) $16^1 = 16 < 200 < 16^2 = 256$. Two ICs are needed. (ii) 3 ICs. (b)(i) $200_{10} = $ C4H $= 1100\ 0100$. Connect Q_D and Q_C of the higher-order IC to inputs of a NAND gate. Connect Q_C of the lower-order gate to the NAND gate. Connect the output of the gate to the $\overline{CLR}$ terminals. (ii) $4002_{10} = $ FA2H $= 1111\ 1010\ 0010$. Gating is required to obtain $F = \overline{Q_{D3}Q_{C3}Q_{B3}Q_{A3}}$, connect F to $\overline{CLR}$ terminals.

6.14 (a) $f_{max} = 1/(61 \times 10^{-9}) = 16.4$ MHz. (b) $I = 80\ \mu A$. $P = 5 \times 6 \times 80\ \mu A = 2.4$ mW. (c) $f_{max} = 1/[(20 + 50 + 50) \times 10^{-9}] = 8.33$ MHz.

6.15 Connect four J-K, or D, flip-flops in cascade, e.g. one 74HC109/112 IC or one 74HC174 IC. Decode Q_C and Q_D by applying to the inputs of a 2-input NAND gate, e.g. 74HC00. Connect the gate output to the $\overline{CLR}$ terminal of each flip-flop if 109/112 is used or to common $\overline{CLR}$ if 174 is used.

6.16 The state table is shown in Table 6.7.

Table 6.7

Clock pulse	Q_C	Q_B	Q_A
0	0	0	0
1	0	0	1
2	0	1	0
3	0	1	1
4	1	0	0
5	1	0	1
6	1	1	0
7	0	0	0

J_A/K_A: Q_A is to toggle so $J_A = K_A = 1$.

J_B/K_B: Q_B is to change when $Q_A = 1$ and $Q_C = 0$ or when $Q_A = Q_C = 1$. So $J_B = K_B = Q_A\bar{Q}_C + Q_A Q_C = Q_A$.

J_C/K_C: Q_C is to change when $Q_A = Q_B = 1$ or $Q_A Q_B = 1$.

6.17 (a) From the data sheet, $t_{w(min)} = 16$ ns but if the maximum frequency of 30 MHz is used $t_w = 0.5 \times (1/30 \text{ MHz}) = 16.16$ ns. (b) 20 ns. (c) 5 ns.

6.18 (a) 15 ns. (b) 4 ns. (c) 20.

6.19 (b) Brief data: 373 octal transparent D latch. OE (output enable) input allows eight outputs to be at a normal logical state or high impedance. 374 octal edge-triggered D flip-flop with three-state outputs and OE terminal. 533 similar to 373. 374 and 574 three-state true data with clear. 273 flip-flop, true data with clear. 377 flip-flop, true data with clock enable. 534 flip-flop, inverting data with clock enable. 373 transparent latch, true data. 573 transparent latch, true data with three-state outputs. 563 transparent latch, inverted data, three-state outputs. Is true or inverted data the requirement? Three-state outputs or not? Latch or flip-flop? Availability?

6.20 Three ICs are needed since $16^3 = 256$. $365 = 16DH = 0001\ 0110\ 1101$.

6.21 (a) $16^4 = 65\ 536$. (b) $77A1H = 7 \times 4096 + 7 \times 256 + 10 \times 16 + 1 = 30\ 625_{10}$. $65\ 536 - 30\ 625 = 34\ 911$. (c) $f_{out} = (15 \times 10^6)/(34\ 911) \approx 430$ Hz.

6.22 The state table is given in Table 6.8.

Table 6.8

	Present state		Present inputs		Next state		
Clock pulse	Q_B	Q_A	$D_B = Q_A + Q_B$	$D_A = \bar{Q}_A$	Q_B	Q_A	Count
0	0	0	0	1	0	1	0
1	0	1	1	0	1	0	1
2	1	0	1	1	1	1	2
3	1	1	0	0			0

Hence, count $= 3$.

Chapter 7

7.1 0, 1, 3, 7, 15, 31, 63, 127, 0, etc.

7.2 (a) No. (b) No. (c) (i) The data is entered. (ii) Data entry is inhibited. (d) When the clock is either HIGH or LOW. (e) 6 V. (f) 21 ns, 5 ns.

7.3 The division ratios are: binary counter; $2^8 = 256$: ring counter; 8: Johnson counter $2 \times 8 = 16$. Hence, the overall division ratio is $256 \times 8 \times 16 = 32\,768$. Therefore, output frequency $= (10 \times 10^6)/32\,768 \approx 305$ Hz.

7.4 (a) Add one more stage to Fig. 7.15. (b) 10.

7.6 (a) Connect the serial output terminal to the serial input terminal. (b) Connect the serial output to the serial input terminal via an inverter.

7.7 Starting from $Q_A = 1$ and all other stages cleared gives the sequence shown in Table 7.12.

Table 7.12					
Q_D	Q_C	Q_B	Q_A	$Q_A^+ = Q_A \oplus Q_D$	Count
0	0	0	1	1	1
0	0	1	1	1	3
0	1	1	1	1	7
1	1	1	1	0	15
1	1	1	0	1	14
1	1	0	1	0	13
1	0	1	0	1	10
0	1	0	1	1	5
1	0	1	1	0	11
0	1	1	0	0	6
1	1	0	0	1	12
1	0	0	1	0	9
0	0	1	0	0	2
0	1	0	0	0	4
1	0	0	0	1	8
0	0	0	1	1	1

Hence, the count is 1, 3, 7, 15, 14, 13, 10, 5, 11, 6, 12, 9, 2, 4, 8, 1, etc.

7.8 0, $Q_A Q_E$, $Q_B Q_C$; 4, $Q_D Q_E$; 6, $Q_A Q_B$; 8, $Q_C Q_D$.

7.9 The circuit is shown in Fig. 7.24.

Fig. 7.24

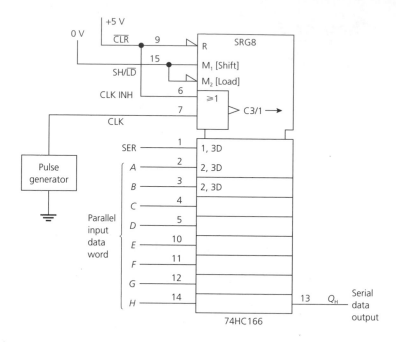

74HC166

Fig. 7.25

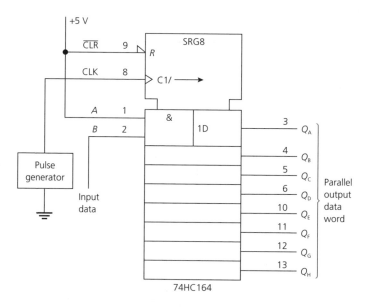

74HC164

7.10 (a) (i) $2n = 6$, or $n = 3$. (ii) $2n = 10$, $n = 5$. (iii) $2n = 20$, $n = 10$. (iv) $2n = 30$, $n = 15$. (b) (i) Count $= 2^3 = 8$. (ii) Count $= 2^5 = 32$. (iii) Count $= 2^{10} = 1024$. (iv) Count $= 2^{15} = 32\ 768$.

7.11 Clock the shift register at 1 MHz. Take outputs from Q_E, Q_G and Q_H.

7.12 (a) See Fig. 7.25.

(b) MSB. So that the MSB is in position Q_H when parallel read-out takes place.
(c) Connect input A to 1 and serial input data to input B. Connect both the clock and the strobe (enable) signal lines to the inputs of a 2-input AND gate. Connect the output of the gate to the CLK terminal. Clear the register by taking $\overline{\text{CLR}}$ LOW. After eight clock pulses take the strobe line LOW. The data is then held at the eight Q outputs and can be read out when required.

7.13 $180_{10} = 10110100$. See Table 7.13.

Table 7.13

Clock pulse	Q_H	Q_G	Q_F	Q_E	Q_D	Q_C	Q_B	Q_A
0	1	0	1	1	0	1	1	0
1	0	1	1	0	1	1	0	1
2	1	1	0	1	1	0	1	0
3	1	0	1	1	0	1	0	1
4	0	1	1	0	1	0	1	1
5	1	1	0	1	0	1	1	0
6	1	0	1	0	1	1	0	1
7	0	1	0	1	1	0	1	0
8	1	0	1	1	0	1	0	0

7.14 Entry of serial word takes 8 clock pulses. Transfer of parallel word takes 1 clock pulse. Total time = 9 clock periods. Each clock pulse occupies $1/(10 \times 10^{-6}) = 0.1$ μs. Hence, time per data word = 0.9 μs. Time taken to fill memory = $65\,536 \times 0.9 \times 10^{-6}$ = 59 μs.

Chapter 8

8.1 The state diagram is shown in Fig. 8.20. From the diagram the state table is given in Table 8.17.

Fig. 8.20

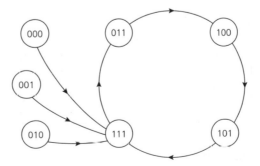

Table 8.17

	Present state			Next state			Required inputs		
	Q_C	Q_B	Q_A	Q_C^+	Q_B^+	Q_A^+	D_C	D_B	D_A
0	0	0	0	1	1	1	1	1	1
1	0	0	1	1	1	1	1	1	1
2	0	1	0	1	1	1	1	1	1
3	0	1	1	1	0	0	1	0	0
4	1	0	0	1	0	1	1	0	1
5	1	0	1	1	1	1	1	1	1
6	1	1	0	1	1	1	1	1	1
7	1	1	1	0	1	1	0	1	1

Mapping gives

$$D_A = \bar{Q}_A + \bar{Q}_B$$

$$D_B = \bar{Q}_A Q_B + \bar{Q}_A \bar{Q}_C + Q_A Q_C + Q_A \bar{Q}_B$$
$$= Q_A \oplus Q_B + \overline{Q_A \oplus Q_C}$$

$$D_C = \bar{Q}_C + \bar{Q}_A + \bar{Q}_B$$

8.2 The state diagram is shown in Fig. 8.20 above. The state table is given in Table 8.18.

Table 8.18

	Present state			Next state			Required inputs					
	Q_C	Q_B	Q_A	Q_C^+	Q_B^+	Q_A^+	J_C	K_C	J_B	K_B	J_A	K_A
0	0	0	0	1	1	1	1	×	1	×	1	×
1	0	0	1	1	1	1	1	×	1	×	×	0
2	0	1	0	1	1	1	1	×	×	0	1	×
3	0	1	1	1	0	0	1	×	×	1	×	1
4	1	0	0	1	0	1	×	0	0	×	1	×
5	1	0	1	1	1	1	×	0	1	×	×	0
6	1	1	0	1	1	1	×	0	×	0	1	×
7	1	1	1	0	1	1	×	1	×	0	×	0

The state maps are

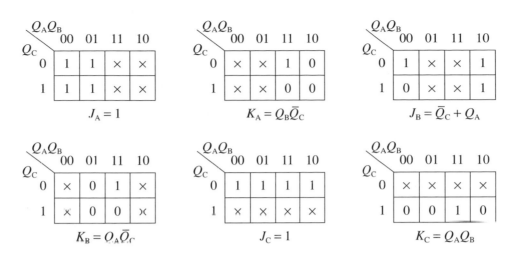

$J_A = 1$

$K_A = Q_B\bar{Q}_C$

$J_B = \bar{Q}_C + Q_A$

$K_B = Q_A\bar{Q}_C$

$J_C = 1$

$K_C = Q_AQ_B$

8.3 The state diagram is shown in Fig. 8.21. The state table is given in Table 8.19.

Fig. 8.21

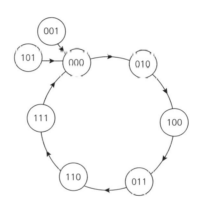

Table 8.19								
	Present state			Next state			Required inputs		
	Q_C	Q_B	Q_A	Q_C^+	Q_B^+	Q_A^+	D_C	D_B	D_A
0	0	0	0	0	1	0	0	1	0
1	0	0	1	0	0	0	0	0	0
2	0	1	0	1	0	0	1	0	0
3	0	1	1	1	1	0	1	1	0
4	1	0	0	0	1	1	0	1	1
5	1	0	1	0	0	0	0	0	0
6	1	1	0	1	1	1	1	1	1
7	1	1	1	0	0	0	0	0	0

Mapping gives

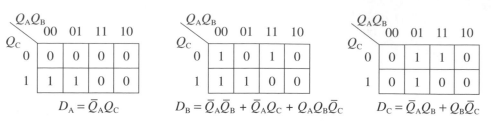

$$D_A = \bar{Q}_A Q_C \qquad D_B = \bar{Q}_A \bar{Q}_B + \bar{Q}_A Q_C + Q_A Q_B \bar{Q}_C \qquad D_C = \bar{Q}_A Q_B + Q_B \bar{Q}_C$$

8.4 The state diagram is shown in Fig. 8.22. From this diagram the state table (Table 8.20) is obtained.

Fig. 8.22

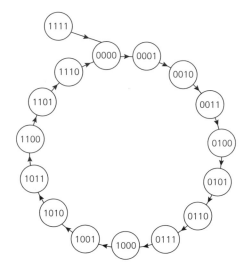

Table 8.20

	Present state				Next state				Required inputs							
	Q_D	Q_C	Q_B	Q_A	Q_D^+	Q_C^+	Q_B^+	Q_A^+	J_D	K_D	J_C	K_C	J_B	K_B	J_A	K_A
0	0	0	0	0	0	0	0	1	0	×	0	×	0	×	1	×
1	0	0	0	1	0	0	1	0	0	×	0	×	1	×	×	1
2	0	0	1	0	0	0	1	1	0	×	0	×	×	0	1	×
3	0	0	1	1	0	1	0	0	0	×	1	×	×	1	×	1
4	0	1	0	0	0	1	0	1	0	×	×	0	0	×	1	×
5	0	1	0	1	0	1	1	0	0	×	×	0	1	×	×	1
6	0	1	1	0	0	1	1	1	0	×	×	0	×	0	1	×
7	0	1	1	1	1	0	0	0	1	×	×	1	×	1	×	1
8	1	0	0	0	1	0	0	1	×	0	0	×	0	×	1	×
9	1	0	0	1	1	0	1	0	×	0	0	×	1	×	×	1
10	1	0	1	0	1	0	1	1	×	0	0	×	×	0	1	×
11	1	0	1	1	1	1	0	0	×	0	1	×	×	1	×	1
12	1	1	0	0	1	1	0	1	×	0	×	0	0	×	1	×
13	1	1	0	1	1	1	1	0	×	0	×	0	1	×	×	1
14	1	1	1	0	0	0	0	0	×	1	×	1	×	1	0	×
15	1	1	1	1	0	0	0	0	×	1	×	1	×	1	×	1

The state maps are

K_A Q_AQ_B / Q_CQ_D	00	01	11	10
00	×	×	1	1
01	×	×	1	1
11	×	×	1	1
10	×	×	1	1

$$K_A = 1$$

J_A Q_AQ_B / Q_CQ_D	00	01	11	10
00	1	1	×	×
01	1	1	×	×
11	1	0	×	×
10	1	1	×	×

$$J_A = \bar{Q}_B + \bar{Q}_C + \bar{Q}_D$$

K_B Q_AQ_B / Q_CQ_D	00	01	11	10
00	×	0	1	×
01	×	0	1	×
11	×	1	1	×
10	×	0	1	×

$$K_B = Q_A + Q_CQ_D$$

J_B Q_AQ_B / Q_CQ_D	00	01	11	10
00	0	×	×	1
01	0	×	×	1
11	0	×	×	1
10	0	×	×	1

$$J_B = Q_A$$

K_C Q_AQ_B / Q_CQ_D	00	01	11	10
00	×	×	×	×
01	×	×	×	×
11	0	1	1	0
10	0	0	1	0

$$K_C = Q_BQ_D + Q_AQ_B$$

J_C Q_AQ_B / Q_CQ_D	00	01	11	10
00	0	0	1	0
01	0	0	1	0
11	×	×	×	×
10	×	×	×	×

$$J_C = Q_AQ_B$$

K_D Q_AQ_B / Q_CQ_D	00	01	11	10
00	×	×	×	×
01	0	0	0	0
11	0	1	1	0
10	×	×	×	×

$$K_D = Q_CQ_B$$

J_D Q_AQ_B / Q_CQ_D	00	01	11	10
00	0	0	0	0
01	×	×	×	×
11	×	×	×	×
10	0	0	1	0

$$J_D = Q_AQ_BQ_C$$

8.5 The state diagram is shown in Fig. 8.23, and the state table is given in Table 8.21.

Fig. 8.23

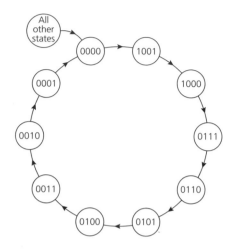

Table 8.21

	Present state				Next state				Required inputs			
	Q_D	Q_C	Q_B	Q_A	Q_D^+	Q_C^+	Q_B^+	Q_A^+	D_D	D_C	D_B	D_A
0	0	0	0	0	1	0	0	1	1	0	0	1
1	0	0	0	1	0	0	0	0	0	0	0	0
2	0	0	1	0	0	0	0	1	0	0	0	1
3	0	0	1	1	0	0	1	0	0	0	1	0
4	0	1	0	0	0	0	1	1	0	0	1	1
5	0	1	0	1	0	1	0	0	0	1	0	0
6	0	1	1	0	0	1	0	1	0	1	0	1
7	0	1	1	1	0	1	1	0	0	1	1	0
8	1	0	0	0	0	1	1	1	0	1	1	1
9	1	0	0	1	1	0	0	0	1	0	0	0
10	1	0	1	0	1	0	0	1	1	0	0	1
11	1	0	1	1	1	0	0	1	1	0	0	1
12	1	1	0	0	1	0	0	1	1	0	0	1
13	1	1	0	1	1	0	0	1	1	0	0	1
14	1	1	1	0	1	0	0	1	1	0	0	1
15	1	1	1	1	1	0	0	1	1	0	0	1

Mapping gives

Q_AQ_B

Q_CQ_D	00	01	11	10
00	1	1	0	0
01	1	1	1	0
11	1	1	1	1
10	1	0	1	0

D_A

$$D_A = \bar{Q}_A\bar{Q}_B + \bar{Q}_A\bar{Q}_C + Q_CQ_D + Q_AQ_BQ_C + Q_BQ_D$$

Q_AQ_B

Q_CQ_D	00	01	11	10
00	0	0	1	0
01	1	0	0	0
11	0	0	0	0
10	1	0	1	0

D_B

$$D_B = \bar{Q}_A\bar{Q}_B\bar{Q}_CQ_D + \bar{Q}_A\bar{Q}_BQ_C\bar{Q}_D + Q_AQ_B\bar{Q}_D$$

Q_AQ_B

Q_CQ_D	00	01	11	10
00	0	0	0	0
01	1	0	0	0
11	0	0	0	0
10	0	1	1	1

D_C

$$D_C = \bar{Q}_A\bar{Q}_B\bar{Q}_CQ_D + Q_BQ_C\bar{Q}_D + Q_AQ_C\bar{Q}_D$$

Q_AQ_B

Q_CQ_D	00	01	11	10
00	1	0	0	0
01	0	1	1	1
11	1	1	1	1
10	0	0	0	0

D_D

$$D_D = \bar{Q}_A\bar{Q}_B\bar{Q}_C\bar{Q}_D + Q_CQ_D + Q_BQ_D + Q_AQ_D$$

8.6 The state diagram is shown in Fig. 8.24. The state table obtained from the diagram is given in Table 8.22.

Fig. 8.24

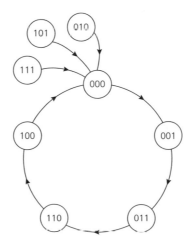

Table 8.22

	Present state			Next state			Required inputs		
	Q_C	Q_B	Q_A	Q_C^+	Q_B^+	Q_A^+	D_C	D_B	D_A
0	0	0	0	0	0	1	0	0	1
1	0	0	1	0	1	1	0	1	1
2	0	1	0	0	0	0	0	0	0
3	0	1	1	1	1	0	1	1	0
4	1	0	0	0	0	0	0	0	0
5	1	0	1	0	0	0	0	0	0
6	1	1	0	1	0	0	1	0	0
7	1	1	1	0	0	0	0	0	0

The state maps are

Q_AQ_B

D_A Q_C	00	01	11	10
0	1	0	0	1
1	0	0	0	0

$$D_A = \bar{Q}_B\bar{Q}_C$$

Q_AQ_B

D_B Q_C	00	01	11	10
0	0	0	1	1
1	0	0	0	0

$$D_B = Q_A\bar{Q}_C$$

Q_AQ_B

D_C Q_C	00	01	11	10
0	0	0	1	0
1	0	1	0	0

$$D_C = Q_AQ_B\bar{Q}_C + \bar{Q}_AQ_BQ_C$$

8.7 The state diagram is shown in Fig. 8.25. The state table is given in Table 8.23.

Fig. 8.25

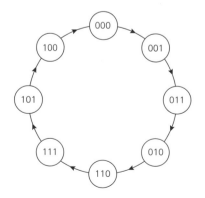

Table 8.23

	Present state			Next state			Required inputs		
	Q_C	Q_B	Q_A	Q_C^+	Q_B^+	Q_A^+	D_C	D_B	D_A
0	0	0	0	0	0	1	0	0	1
1	0	0	1	0	1	1	0	1	1
2	0	1	1	0	1	0	0	1	0
3	0	1	0	1	1	0	1	1	0
4	1	1	0	1	1	1	1	1	1
5	1	1	1	1	0	1	1	0	1
6	1	0	1	1	0	0	1	0	0
7	1	0	0	0	0	0	0	0	0

The state maps are

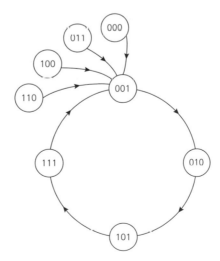

$$D_A = Q_B Q_C + \bar{Q}_B \bar{Q}_C$$

$$D_B = \bar{Q}_A Q_B + Q_A \bar{Q}_C$$

$$D_C = \bar{Q}_A Q_B + Q_A Q_C$$

8.8 The state diagram is shown in Fig. 8.26. The state table obtained from the state diagram is given in Table 8.24.

Fig 8.26

Table 8.24

	Present state			Next state			Required inputs					
	Q_C	Q_B	Q_A	Q_C^+	Q_B^+	Q_A^+	J_C	K_C	J_B	K_B	J_A	K_A
0	0	0	0	0	0	1	0	×	0	×	1	×
1	0	0	1	0	1	0	0	×	1	×	×	1
2	0	1	0	1	0	1	1	×	×	1	1	×
3	0	1	1	0	0	1	0	×	×	1	×	0
4	1	0	0	0	0	1	×	1	0	×	1	×
5	1	0	1	1	1	1	×	0	1	×	×	0
6	1	1	0	0	0	1	×	1	×	1	1	×
7	1	1	1	0	0	1	×	1	×	1	×	0

The state maps are

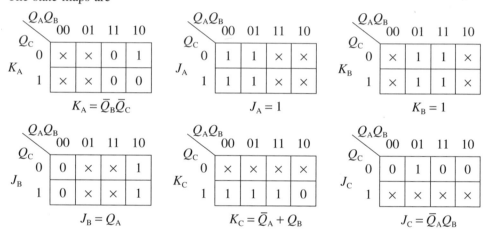

$$K_A = \bar{Q}_B\bar{Q}_C \qquad J_A = 1 \qquad K_B = 1$$

$$J_B = Q_A \qquad K_C = \bar{Q}_A + Q_B \qquad J_C = \bar{Q}_A Q_B$$

Alternatively, use the unwanted states as don't cares. This gives Table 8.25.

Table 8.25

	Present state			Next state			Required inputs					
	Q_C	Q_B	Q_A	Q_C^+	Q_B^+	Q_A^+	J_C	K_C	J_B	K_B	J_A	K_A
0	0	0	0	×	×	×	×	×	×	×	×	×
1	0	0	1	0	1	0	0	×	1	×	×	1
2	0	1	0	1	0	1	1	×	×	1	1	×
3	0	1	1	×	×	×	×	×	×	×	×	×
4	1	0	0	×	×	×	×	×	×	×	×	×
5	1	0	1	1	1	1	×	0	1	×	×	0
6	1	1	0	×	×	×	×	×	×	×	×	×
7	1	1	1	0	0	1	×	1	×	1	×	0

Mapping gives

$Q_A Q_B$
Q_C	00	01	11	10
0	×	×	×	1
1	×	×	0	0

K_A

$$K_A = \bar{Q}_C$$

$Q_A Q_B$
Q_C	00	01	11	10
0	×	1	×	×
1	×	×	×	×

J_A

$$J_A = 1$$

$Q_A Q_B$
Q_C	00	01	11	10
0	×	1	×	×
1	×	×	×	×

K_B

$$K_B = 1$$

$Q_A Q_B$
Q_C	00	01	11	10
0	×	×	×	1
1	×	×	×	1

J_B

$$J_B = 1$$

$Q_A Q_B$
Q_C	00	01	11	10
0	×	×	×	×
1	×	×	1	0

K_C

$$K_C = Q_B$$

$Q_A Q_B$
Q_C	00	01	11	10
0	×	1	×	0
1	×	×	×	×

J_C

$$J_C = Q_B$$

8.9 The state diagram is shown in Fig. 8.27 and the state table is given in Table 8.26.

Fig. 8.27

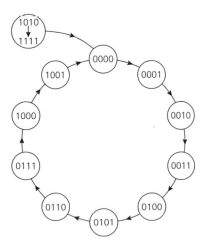

Table 8.26

	Present state				Next state				Required inputs			
	Q_D	Q_C	Q_B	Q_A	Q_D^+	Q_C^+	Q_B^+	Q_A^+	T_D	T_C	T_B	T_A
0	0	0	0	0	0	0	0	1	0	0	0	1
1	0	0	0	1	0	0	1	0	0	0	1	1
2	0	0	1	0	0	0	1	1	0	0	0	1
3	0	0	1	1	0	1	0	0	0	1	1	1
4	0	1	0	0	0	1	0	1	0	0	0	1
5	0	1	0	1	0	1	1	0	0	0	1	1
6	0	1	1	0	0	1	1	1	0	0	0	1
7	0	1	1	1	1	0	0	0	1	1	1	1
8	1	0	0	0	1	0	0	1	0	0	0	1
9	1	0	0	1	0	0	0	0	1	0	0	1
10	1	0	1	0	0	0	0	0	1	0	1	0
11	1	0	1	1	0	0	0	0	1	0	1	1
12	1	1	0	0	0	0	0	0	1	1	0	0
13	1	1	0	1	0	0	0	0	1	1	0	1
14	1	1	1	0	0	0	0	0	1	1	1	0
15	1	1	1	1	0	0	0	0	1	1	1	1

Mapping gives

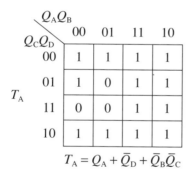

T_A

$Q_C Q_D$ \\ $Q_A Q_B$	00	01	11	10
00	1	1	1	1
01	1	0	1	1
11	0	0	1	1
10	1	1	1	1

$$T_A = Q_A + \bar{Q}_D + \bar{Q}_B \bar{Q}_C$$

T_B

$Q_C Q_D$ \\ $Q_A Q_B$	00	01	11	10
00	0	0	1	1
01	0	1	1	0
11	0	1	1	0
10	0	0	1	1

$$T_B = Q_B Q_D + Q_A Q_B + Q_A \bar{Q}_D$$

T_C

$Q_C Q_D$ \\ $Q_A Q_B$	00	01	11	10
00	0	0	1	0
01	0	0	0	0
11	1	1	1	1
10	0	0	1	0

$$T_C = Q_C Q_D + Q_A Q_B \bar{Q}_D$$

T_D

$Q_C Q_D$ \\ $Q_A Q_B$	00	01	11	10
00	0	0	0	0
01	0	1	1	1
11	1	1	1	1
10	0	0	1	0

$$T_D = Q_C Q_D + Q_B Q_D + Q_A Q_B Q_C + Q_A Q_D$$

8.10 The state diagram is shown in Fig. 8.28. The state table is given in Table 8.27.

Fig. 8.28

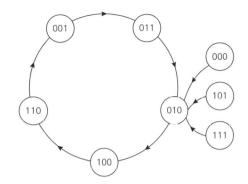

Table 8.27

	Present state			Next state			Required inputs					
	Q_C	Q_B	Q_A	Q_C^+	Q_B^+	Q_A^+	J_C	K_C	J_B	K_B	J_A	K_A
0	0	0	0	0	1	0	0	×	1	×	0	×
1	0	0	1	0	1	1	0	×	1	×	×	0
2	0	1	0	1	0	0	1	×	×	1	0	×
3	0	1	1	0	1	0	0	×	×	0	×	1
4	1	0	0	1	1	0	×	0	1	×	0	×
5	1	0	1	0	1	0	×	1	1	×	×	1
6	1	1	0	0	0	1	×	1	×	1	1	×
7	1	1	1	0	1	0	×	1	×	0	×	1

Mapping and simplifying gives $J_A = Q_B Q_C$, $K_A = Q_B + Q_C$, $J_B = 1$, $K_B = Q_A$, $J_C = Q_A Q_B$, $K_C = Q_A + Q_B$.

8.11 Figure 8.29 shows how a counter can be started.

8.12 Figure 8.30 shows the state diagram.

Fig. 8.29

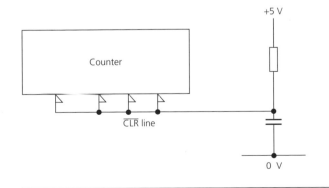

Fig. 8.30

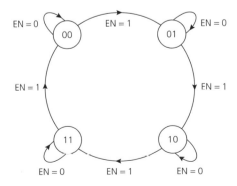

Chapter 9

9.1 The state table is shown in Table 9.22 (P = pulse).

Table 9.22

Present state				Next state			Required inputs		
Q_C	Q_B	Q_A	P	Q_C^+	Q_B^+	Q_A^+	D_C	D_B	D_A
0	0	0	0	0	0	0	0	0	0
0	0	1	0	0	0	1	0	0	1
0	1	0	0	0	1	0	0	1	0
0	1	1	0	0	1	1	0	1	1
1	0	0	0	1	0	0	1	0	0
1	0	1	0	1	0	1	1	0	1
1	1	0	0	1	1	0	1	1	0
1	1	1	0	1	1	1	1	1	1
0	0	0	1	0	0	1	0	0	1
0	0	1	1	0	1	0	0	1	0
0	1	0	1	0	1	1	0	1	1
0	1	1	1	1	0	0	1	0	0
1	0	0	1	1	0	1	1	0	1
1	0	1	1	1	1	0	1	1	0
1	1	0	1	1	1	1	1	1	1
1	1	1	1	0	0	0	0	0	0

The state maps are

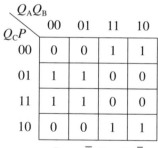

$Q_A Q_B$

$Q_C P$	00	01	11	10
00	0	0	1	1
01	1	1	0	0
11	1	1	0	0
10	0	0	1	1

$$D_A = \bar{Q}_A P + Q_A \bar{P}$$

$Q_A Q_B$

$Q_C P$	00	01	11	10
00	0	1	1	0
01	0	1	0	1
11	0	1	0	1
10	0	1	1	0

$$D_B = \bar{Q}_A Q_B + Q_B \bar{P} + Q_A \bar{Q}_B P$$

$Q_A Q_B$

$Q_C P$	00	01	11	10
00	0	0	0	0
01	0	0	1	0
11	1	1	0	1
10	1	1	1	1

$$D_C = Q_C \bar{P} + \bar{Q}_B Q_C + Q_A Q_B \bar{Q}_C P + \bar{Q}_A Q_C$$

9.2 The state diagram is shown in Fig. 9.25. There are six states, so three flip-flops are required. The state table is given in Table 9.23, assuming that J-K flip-flops are employed.

Fig. 9.25

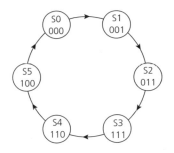

Table 9.23

	Present state			Next state			Required inputs						Output
	Q_C	Q_B	Q_A	Q_C^+	Q_B^+	Q_A^+	J_C	K_C	J_B	K_B	J_A	K_A	F
S0	0	0	0	0	0	1	0	×	0	×	1	×	0
S1	0	0	1	0	1	1	0	×	1	×	×	0	1
S2	0	1	1	1	1	1	1	×	×	0	×	0	1
S3	1	1	1	1	1	0	×	0	×	0	×	1	1
S4	1	1	0	1	0	0	×	0	×	1	0	×	1
S5	1	0	0	0	0	0	×	1	0	×	0	×	1

The state maps are

J_A

Q_C \ Q_AQ_B	00	01	11	10
0	1	×	×	×
1	0	0	×	×

$$J_A = Q_C$$

K_A

Q_C \ Q_AQ_B	00	01	11	10
0	×	×	0	0
1	×	×	1	×

$$K_A = Q_C$$

J_B

Q_C \ Q_AQ_B	00	01	11	10
0	0	×	×	1
1	0	×	×	×

$$J_B = Q_A$$

K_B

Q_C \ Q_AQ_B	00	01	11	10
0	×	×	0	×
1	×	1	0	×

$$K_B = \bar{Q}_A$$

J_C

Q_C \ Q_AQ_B	00	01	11	10
0	0	×	1	0
1	×	×	×	×

$$J_C = Q_B$$

K_C

Q_C \ Q_AQ_B	00	01	11	10
0	×	×	×	×
1	1	0	0	×

$$K_C = \bar{Q}_B$$

F

Q_C \ Q_AQ_B	00	01	11	10
0	0	×	1	1
1	1	1	1	×

$$F = Q_A + Q_C$$

9.3 The state diagram is shown in Fig. 9.26. The state table is given in Table 9.24.

Fig. 9.26

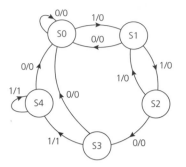

Table 9.24

	Present state			Next state $x = 0$						$x = 1$						Output
	Q_C	Q_B	Q_A	Q_C	Q_B	Q_A	D_C	D_B	D_A	Q_C	Q_B	Q_A	D_C	D_B	D_A	F
S0	0	0	0	0	0	0	0	0	0	0	0	1	0	0	1	0
S1	0	0	1	0	0	0	0	0	0	0	1	1	0	1	1	0
S2	0	1	1	0	1	0	0	1	0	0	0	1	0	0	1	0
S3	0	1	0	0	0	0	0	0	0	1	1	0	1	1	0	0
S4	1	1	0	0	0	0	0	0	0	1	1	0	1	1	0	1

The state maps are

D_A

$Q_C x$ \ $Q_A Q_B$	00	01	11	10
00	0	0	0	0
01	1	0	1	1
11	×	0	×	×
10	×	0	×	×

$$D_A = \bar{Q}_B x + Q_A x$$

D_B

$Q_C x$ \ $Q_A Q_B$	00	01	11	10
00	0	0	0	0
01	1	1	0	1
11	×	1	×	×
10	×	0	×	×

$$D_B = \bar{Q}_A x + \bar{Q}_B x$$

D_C

$Q_C x$ \ $Q_A Q_B$	00	01	11	10
00	0	0	0	0
01	0	1	0	0
11	×	1	×	×
10	×	0	×	×

$$D_C = \bar{Q}_A Q_B x$$

F

$Q_C x$ \ $Q_A Q_B$	00	01	11	10
00	0	0	0	0
01	0	0	0	0
11	×	1	×	×
10	×	1	×	×

$$F = Q_C$$

9.4 The state diagram is given in Fig. 9.27.

Fig. 9.27

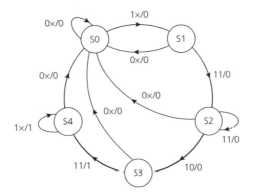

9.5 The state table is given in Table 9.25.

Table 9.25

Present state			Next state								
Q_C	Q_B	Q_A	Q_C	Q_B	Q_A	J_C	K_C	J_B	K_B	J_A	K_A
0	0	0	1	1	0	1	×	1	×	0	×
0	0	1	1	0	0	1	×	0	×	×	1
0	1	0	0	0	0	0	×	×	1	0	×
0	1	1	1	0	1	1	×	×	1	×	0
1	0	0	0	0	0	×	1	0	×	0	×
1	0	1	1	1	1	×	0	1	×	×	0
1	1	0	0	1	1	×	1	×	0	1	×
1	1	1	0	0	1	×	1	×	1	×	0

The state maps are

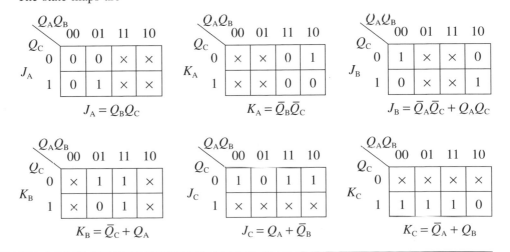

$$J_A = Q_B Q_C \qquad K_A = \bar{Q}_B \bar{Q}_C \qquad J_B = \bar{Q}_A \bar{Q}_C + Q_A Q_C$$

$$K_B = \bar{Q}_C + Q_A \qquad J_C = Q_A + \bar{Q}_B \qquad K_C = \bar{Q}_A + Q_B$$

9.6 The state table is given in Table 9.26.

Table 9.26

Present state			Next state					
Q_C	Q_B	Q_A	Q_C	Q_B	Q_A	D_C	D_B	D_A
0	0	0	1	1	0	1	1	0
0	0	1	1	0	0	1	0	0
0	1	0	0	0	0	0	0	0
0	1	1	1	0	1	1	0	1
1	0	0	0	0	0	0	0	0
1	0	1	1	1	1	1	1	1
1	1	0	0	1	1	0	1	1
1	1	1	0	0	1	0	0	1

Mapping gives

D_A

Q_C \ Q_AQ_B	00	01	11	10
0	0	0	1	0
1	0	1	1	1

$D_A = Q_BQ_C + Q_AQ_B + Q_AQ_C$

D_B

Q_C \ Q_AQ_B	00	01	11	10
0	1	0	0	0
1	0	1	0	1

$D_B = \bar{Q}_A\bar{Q}_B\bar{Q}_C + \bar{Q}_AQ_BQ_C + Q_A\bar{Q}_BQ_C$

D_C

Q_C \ Q_AQ_B	00	01	11	10
0	1	0	1	1
1	0	0	0	1

$D_C = \bar{Q}_B\bar{Q}_C + Q_A\bar{Q}_C + Q_A\bar{Q}_B$

9.7 The state diagram is shown in Fig. 9.28.

Fig. 9.28

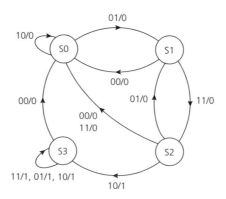

There are four states so that two flip-flops are required.
The state table is given in Table 9.27.

		Next state														Output
		A = 1								A = 0						
		B = 0				B = 1				B = 0				B = 1		
Present state																
Q_B	Q_A	Q_B	Q_A	D_B	D_A	Q_B	Q_A	D_B	D_A	Q_B	Q_A	D_B	D_A	Q_B	Q_A	D_B	D_A	F
S0	0	0	0	0	0	0	0	0	0	0	0	0	0	0	0	1	0	1	0
S1	0	1	0	0	0	0	1	0	1	0	0	0	0	0	0	0	0	0	0
S2	1	0	1	1	1	1	0	0	0	0	0	0	0	0	0	1	0	1	0
S3	1	1	1	1	1	1	1	1	1	1	0	0	0	0	1	1	1	1	1

The state maps are

$$D_A = \bar{Q}_A\bar{A}B + Q_AQ_BB + Q_BA\bar{B}$$

$$D_B = Q_BA\bar{B} + Q_AAB + Q_AQ_BB$$

$$F = Q_AQ_B$$

9.8 There are four states: S0; input 0, output 0: S1; input 1 (first time), output 0: S2; input 0 (after a 1), output 0: S3; input 1 (second time), output 1. State S0 follows state S3. Let S0 be 00, S1 be 01, S2 be 11 and S3 be 10, then the flow matrix is

xy A	00	01	11	10
0	(S0)	S2	(S2)	S0
1	S1	(S1)	S3	(S3)

The excitation map is

A \ xy	00	01	11	10
0	00		11	
1		01		10

and adding the unstable states gives

A \ xy	00	01	11	10
0	00	11	11	00
1	01	01	10	10

Writing separate maps for X and Y:

X, A \ xy	00	01	11	10
0	0	1	1	0
1	0	0	1	1

Y, A \ xy	00	01	11	10
0	0	1	1	0
1	1	1	0	0

Hence, $X = Ax + xy$ and $Y = \bar{x}y + \bar{A}y$. Output $F = S3 = Ax\bar{y}$.

9.9 The state diagram is shown in Fig. 9.29. Since there are six states three flip-flops are needed. The state table is given in Table 9.28.

Fig. 9.29

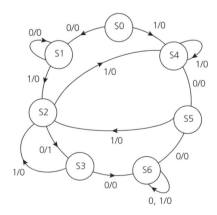

Table 9.28

Present state				Next state Input = 0				= 1			Output
Q_C	Q_B	Q_A		Q_C^+	Q_B^+	Q_A^+		Q_C^+	Q_B^+	Q_A^+	
S0	0	0	0	S1	0	0	1	S4	1	0	0	0
S1	0	0	1	S1	0	0	1	S2	0	1	0	0
S2	0	1	0	S3	0	1	1	S4	1	0	0	1
S3	0	1	1	S6	1	1	0	S2	0	1	0	0
S4	1	0	0	S5	1	0	1	S4	1	0	0	0
S5	1	0	1	S6	1	1	0	S2	0	1	0	0
S6	1	1	0	S6	1	1	0	S6	1	1	0	0

The required D inputs are the same as the next state Q outputs. Mapping gives

D_A

$Q_C x$ \ $Q_A Q_B$	00	01	11	10
00	1	1	0	1
01	0	0	0	0
11	0	0	×	0
10	1	0	×	0

$$D_A = \bar{Q}_B\bar{Q}_C\bar{x} + Q_A\bar{Q}_C\bar{x} + \bar{Q}_A\bar{Q}_B\bar{x}$$

D_B

$Q_C x$ \ $Q_A Q_B$	00	01	11	10
00	0	1	1	0
01	0	0	1	1
11	0	1	×	1
10	0	1	×	1

$$D_B = Q_B\bar{x} + Q_B Q_C + Q_A x + Q_A Q_C$$

D_C

$Q_C x$ \ $Q_A Q_B$	00	01	11	10
00	0	0	1	0
01	1	1	0	0
11	1	1	×	0
10	1	1	×	1

$$D_C = Q_C\bar{x} + \bar{Q}_A Q_C + \bar{Q}_A x + Q_A Q_B\bar{x}$$

$$F = Q_A Q_B \bar{Q}_C.$$

9.10 The state table is shown in Table 9.29.

Table 9.29

Present state		Next state Input x = 0						= 1						Output
Q_A	Q_B	Q_B	Q_A	J_B	K_B	J_A	K_A	Q_B	Q_A	J_B	K_B	J_A	K_A	
S0	0	0	0	0	0	×	0	×	0	1	0	×	1	×	0
S1	0	1	1	1	1	×	×	0	0	1	0	×	×	0	0
S2	1	1	0	0	×	1	×	1	1	0	×	0	×	1	0
S3	1	0	1	1	×	0	1	×	0	0	×	1	0	×	0 x = 0, 1 x = 1

J_A

x \ Q_AQ_B	00	01	11	10
0	0	1	×	×
1	1	0	×	×

K_A

x \ Q_AQ_B	00	01	11	10
0	×	×	1	0
1	×	×	1	0

J_B

x \ Q_AQ_B	00	01	11	10
0	0	×	×	1
1	0	×	×	0

K_B

x \ Q_AQ_B	00	01	11	10
0	×	0	1	×
1	×	1	0	×

From the mappings, $J_A = Q_B\bar{x} + \bar{Q}_Bx$, $K_A = Q_B$, $J_B = Q_A\bar{x}$, $K_B = Q_A\bar{x} + \bar{Q}_Ax$.

9.11 The possible coin combinations are: 20p + 20p, 20p + 10p + 10p, 10p + 10p + 20p, 10p + 20p + 10p and 10p + 10p + 10p + 10p. The state diagram is shown in Fig. 9.30. Since there are five states three flip-flops are needed. The state table is given in Table 9.30.

Fig. 9.30

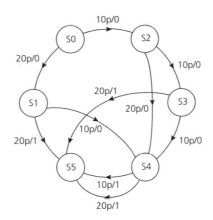

Table 9.30

	Present state			Inputs			Next state			Required inputs			Output
	Q_C	Q_D	Q_A	20p	10p		Q_C^+	Q_B^+	Q_A^+	D_C	D_B	D_A	
S0	0	0	0	0	0	S0	0	0	0	0	0	0	0
	0	0	0	1	0	S1	0	0	1	0	0	1	0
	0	0	0	0	1	S2	0	1	0	0	1	0	0
S1	0	0	1	1	0	S5	1	0	1	1	0	1	1
	0	0	1	0	1	S4	1	0	0	1	0	0	0
S2	0	1	0	1	0	S4	1	0	0	1	0	0	0
	0	1	0	0	1	S3	0	1	1	0	1	1	0
S3	0	1	1	1	0	S1	0	0	1	0	0	1	0
	0	1	1	0	1	S4	1	0	0	1	0	0	0
S4	1	0	0	1	0	S5	1	0	1	1	0	1	1
	1	0	0	0	1	S5	1	0	1	1	0	1	1

The mapping is

Q_C20p \ Q_AQ_B	00	01	11	10
00	0	×	×	×
01	1	0	1	1
11	1	×	×	×
10	×	×	×	×

D_A

$\overline{10}p$

Q_C20p \ Q_AQ_B	00	01	11	10
00	0	1	0	0
01	×	×	×	×
11	×	×	×	×
10	1	×	×	×

10p

D_B — $\overline{10}p$

Q_C20p \ Q_AQ_B	00	01	11	10
00	0	×	×	×
01	0	0	0	0
11	0	×	×	×
10	×	×	×	×

$10p$

Q_C20p \ Q_AQ_B	00	01	11	10
00	1	1	0	0
01	×	×	×	×
11	×	×	×	×
10	0	×	×	×

D_C — $\overline{10}p$

Q_C20p \ Q_AQ_B	00	01	11	10
00	0	×	×	×
01	0	1	0	1
11	1	×	×	×
10	×	×	×	×

$10p$

Q_C20p \ Q_AQ_B	00	01	11	10
00	0	0	1	1
01	×	×	×	×
11	×	×	×	×
10	1	×	×	×

From the maps, $D_A = Q_C + Q_A\overline{10}p + \bar{Q}_B20p + \bar{Q}_AQ_B10p$, $D_B = \bar{Q}_A\bar{Q}_C10p$, $D_C = Q_C + Q_A\bar{Q}_B + Q_A10p + \bar{Q}_AQ_B10p$.

Chapter 10

10.1 $F = AB\overline{C} + A\overline{B}C + ABC + \overline{A}BC$.

10.2 (a) (i) 12, 6. (ii) 16, 2. (iii) 20, 8. (b) (i) 12, 10. (ii) 20, 8. (iii) 16, 8.

10.3 $F = \overline{A}B\overline{C} + A\overline{B}C + A\overline{B}\overline{C}$.

10.6 MODULE Logic function
Device 'P16V8'
Title logic circuit
"Input pins
A, B, C, D, PIN 2, 3, 4, 5;
"Output pins
F, G, H, I, J, PIN 18, 19, 20, 21;
"Boolean equations
F = A#B, G = C&D, H = !A&!B, I = !C#!D, J = !(A$C);

10.7 MODULE Logic
Device 'P16V8'
Title Combinational logic
A, B, C, D, F PIN 2, 3, 4, 5, 18;
F = A&B&!C&!D#A&!B&C&!D#A&B&C&D#!A&B&C&D#A&!B&C&D#
!A&!B&C&D#A&B&!C&D#!A&!B&!C&D#A&!B&!C&D#!A&B&C&!D

10.8 This equation cannot be implemented directly since there are more than eight terms. Hence, re-write in the form $Y = \overline{ABCDEFGHI}$. Then,
MODULE
Title
Device 'P16V8'
A, B, C, D, E, F, G, H, I, PIN 2, 3, 4, 5, 6, 7, 8, 9, 10,
Y, PIN 17;
!F = A&B&C&!D&!E&F&G&!H&!I
END

10.9 (a) MODULE Ex-OR
Description Exclusive-OR gate
Device 'P22V10'
TRUTH_TABLE [A, B] → [F]
[0, 0] → [0];
[0, 1] → [1];
[1, 0] → [1];
[1, 1] → [0]:
A, B, F, PIN 2, 3, 18;
END,

(b) TRUTH_TABLE [A, B] → [S, C_{out}]
[0, 0] → [0, 0];
[1, 0] → [1, 0];
[0, 1] → [1, 0];
[1, 1] → [0,1];

10.11 MODULE GATES
Title 'AND and Exclusive-NOR gates'
device 'P16V8'
"inputs
A, B, PIN 1, 2;
"output
F, G, PIN 18, 19;
Equations
F = A&B;
G = !(A&B);
END

10.12 MODULE MUX
Title 1-to-4 multiplexer
Multiplexer device 'P16V8'

D0, D1, D2, D3, PIN 2, 3, 4, 5;
Y, PIN 16;
S0, S1, PIN 17, 18;
TRUTH_TABLE [S0, S1] → [D0, D1, D2, D3]
[0, 0] → [D0];
[0, 1] → [D1];
[1, 0] → [D2];
[1, 1] → [D3];
END

Chapter 11

11.1 Within ± LSB.

11.2 (c) 2^{16} = 65 536. (f) (i) $5/(500 \times 2^{12})$ = 2.44 μV. (ii) $5/(500 \times 2^{16})$ = 152.6 nV.

11.3 The minimum sampling frequency is 20 kHz, so assume it to be 20 per cent higher at 24 kHz. There is then one sample taken every $1/(24 \times 10^3) \approx 42$ μs. Hence, time taken $= (512 \times 8/13) \times 42 \times 10^{-6} = 13.23$ ms.

11.6 FS = $(2^{12} - 1)/2^{12} \times 12 = 11.997$ V. Digital word = 62H = 98_{10}. V_{out} = (98/4095) × 11.997 = 0.287 V.

11.7 There are 2^{10} = 1024 different codewords. (a) Resolution = $16/(2^{10} - 1)$ = 15.64 mV. (b) Percentage resolution = $(15.64 \times 10^{-3})/16$ = 0.098 per cent. (c) Maximum conversion time = 1024 clock periods = 1024 ms.

11.8 (a) Resolution = 5/15 = 0.33 V. (b) Conversion time = 5 clock periods = 5 μs.

11.9 Resolution = $10 \times 10^{-3} = 10/(2^n - 1)$. 2^n = 1001 and n = 10.

11.10 Required conversion time = 0.01 s = 1024 × T. Clock frequency = 1024/0.01 = 102.4 kHz.

11.11 (a) 0110 0100 = 100_{10} = 2.0 V. 1011 0110 = 176_{10}. Hence, V = 2.0 × 176/100 = 3.52 V. Alternatively, LSB = 2/100 = 20 mV. Step 176 = $176 \times 20 \times 10^3$ = 3.52 V. (b) 1000 1110 = 142_{10}. V = 2 × 142/100 = 2.84 V.

11.12 The voltage at the − terminal must also be 0.4 V. Hence $(5 - 0.4)/R_3 = 0.4/666.7$, or R_3 = 7668 Ω.

11.13 (a) R_{in} = 10 kΩ, so $I_{IN} = 5/(10 \times 10^3)$ = 500 μA. (b) (i) $31.25 \times 10^{-6} \times 30 \times 10^3$ = 0.625 V. (ii) $125 \times 10^{-6} \times 20 \times 10^3$ = 2.5 V. (iii) $20 \times 10^3 (31.25 + 62.5 + 125 + 250) \times 10^{-6}$ = 9.375 V.

11.14 Output voltages are: (data/256) × 10. Calculate each and plot.

11.15 (a) Resolution = $(1/2^8) \times 100$ = 0.39 per cent. (b) Resolution = $(1/2^{12}) \times 100$ = 0.024 per cent. (c) Resolution = $(1/2^{16}) \times 100 = 1.53 \times 10^{-3}$ per cent.

11.16 Step size = 6/256 = 23.44 mV. 1010 1010 = 170_{10} so the output voltage ought to be $170 \times 23.44 \times 10^{-3} = 3.9848$ V. The actual output voltage may be $3.9848 \pm 0.75/100 \times 3.9848 = 3.9848 \pm 0.03 = 3.9548$ to 4.0148 V.

11.17 (a) Step size = $10/2^8 = 39.06$ mV. Steps for 3.6 V = $3.6/(39.06 \times 10^{-3}) \approx 92$. Hence, digital word = 0101 1100. (b) 92 clock pulses so conversion time = $92 \times 1/(1.5 \times 10^6) = 61.3$ μs. (c) Resolution = 39 mV.

11.18 (a) $V_{out} = -[10/160 \times 0 + 10/80 \times 5 + 10/40 \times 0 + 10/20 \times 5] = -5[1/8 + 1/2] = -3.125$ V. (b) The nominally 80 kΩ resistor is actually 92 kΩ. Hence, $V_{out} = -5[10/92 + 10/20] = -3.044$ V.

11.19 (a) The offset error. (b) The gain error. (c) The non-monotonic error. (d) The non-linearity error. (e) The differential non-linearity error.

Chapter 12

12.2 At the load $\rho_L = 1$. $V_L = 2 + (1 \times 2) = 4$ V.

12.3 (a) $C = 1/(6 \times 80 \times 12 \times 10^6) = 174$ pF. (b) $P = 25 \times 12 \times 10^6 \times 174 \times 10^{-12} = 52.2$ mW.

12.4 $Z_o - [87/(\sqrt{4.9} + 1.41)]\{\log_e [5.98 \times 0.75)/(0.8 \times 0.5 + 0.1)]\} - 52.65$ Ω.

12.7 $B = 1/(100 \times 10^{-6}) = 10$ kHz.

12.8 $Z_o = \sqrt{(L/C)} = 70$ Ω, hence, $L = 4900C$. $t_D = 5 \times 10^{-9} = \sqrt{(LC)} - \sqrt{(4900C^2)} = 70C$. Hence, $C = 71.43$ pF/m and $L = 350$ nH/m. Added capacitance = $(25 \times 6)/6 = 25$ pF/m giving a total capacitance of 96.43 pF/m. $t_D - \sqrt{(96.43 \times 10^{-12} \times 350 \times 10^{-9})} - 5.81$ ns/m. $Z_o = \sqrt{(350 \times 10^{-9})/(96.43 \times 10^{-12})} = 60.3$ Ω.

12.9 (a) $1 = (2 \times 10^{-9})/(2 \times 5 \times 10^{-9}) = 0.2$ m. (b) $1 = (2 \times 10^{-9})/(2 \times 20 \times 10^{-9}) = 0.05$ m.

12.10 Draw the Bergeron diagram from which (a) $V_s = 2.8$ V, $I_s = 14$ mA. (b) $V_i = 2.5$ V, $I_i = 10$ mA.

12.11

Fig. 12.33

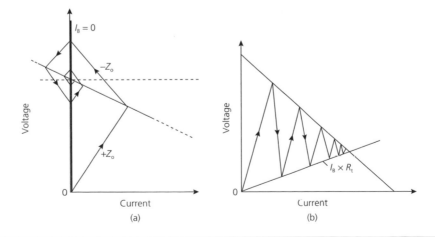

(a)   (b)

12.12 (b) Steps last for twice the propagation delay of the line. (c) (i) $\rho_L = (3R_o - R_o)/(3R_o + R_o) = 0.5$. $V_s = 6$ V. (ii) $\rho_L = (R_o/3 - R_o)/(R_o/3 + R_o) = -0.5$. $V_s = 2$ V.

12.13 $v = 1/(4.92 \times 10^{-9}) = 2.032 \times 10^8$ m/s. Hence, $l_{max} = 2.023 \times 10^8 \times (1.6 \times 10^{-9})/2 = 0.162$ m.

12.14 $V_D/R_D = 5/100 = 50$ mA. Steady-state voltage $= (5 \times 220)/(100 + 220) = 3.44$ V. The Bergeron plot is shown in Fig. 12.34(a) and from this the waveforms at each end of the line have been obtained (see Fig. 12.34(b)).

Fig. 12.34

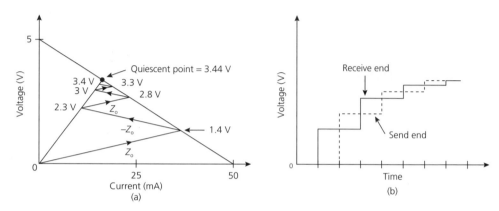

(a)

(b)

Index